HERBAL BIOACTIVES AND FOOD FORTIFICATION

EXTRACTION AND FORMULATION

Nutraceuticals: Basic Research/Clinical Applications
Series Editor: Yashwant Pathak, PhD

Herbal Bioactives and Food Fortification: Extraction and Formulation
D. Suresh Kumar

Handbook of Metallonutraceuticals
Yashwant V. Pathak and Jayant N. Lokhande

Nutraceuticals and Health: Review of Human Evidence
Somdat Mahabir and Yashwant V. Pathak

Marine Nutraceuticals: Prospects and Perspectives
Se-Kwon Kim

NUTRACEUTICALS Basic Research/Clinical Applications

HERBAL BIOACTIVES AND FOOD FORTIFICATION

EXTRACTION AND FORMULATION

D. Suresh Kumar

CRC Press
Taylor & Francis Group
Boca Raton London New York

CRC Press is an imprint of the
Taylor & Francis Group, an **informa** business

CRC Press
Taylor & Francis Group
6000 Broken Sound Parkway NW, Suite 300
Boca Raton, FL 33487-2742

Printed on acid-free paper
Version Date: 20150717

International Standard Book Number-13: 978-1-4822-5363-4 (Hardback)

Visit the Taylor & Francis Web site at
http://www.taylorandfrancis.com

and the CRC Press Web site at
http://www.crcpress.com

Printed and bound by CPI Group (UK) Ltd, Croydon, CR0 4YY

Ayurveda, Unani, Siddha medicine and allied streams were kept vibrant over the centuries by the efforts of their practitioners. The prestigious status enjoyed by these systems in contemporary India is a fruit of their service. To those ordinary and extraordinary physicians, I dedicate this humble work!

Contents

Series Preface

Nutraceuticals: Basic Research/Clinical Applications

The nutraceuticals and functional food industries have grown significantly in the last two decades. The acceptance of these products by larger populations, especially in the West, is ever increasing. The scope of the CRC series on Nutraceuticals: Basic Research/Clinical Applications aims at bringing out a range of books edited by distinguished scientists and researchers from both academia and industry who have significant knowledge and experience in scientific pursuit and critical analysis.

This series will address various aspects of nutraceutical products, including the historical perspective, traditional knowledge base, analytical evaluations, green food to processing, and clinical applications. The series will not only be very useful to researchers and academicians, but it will also provide valuable reference books for personnel in the nutraceuticals and food industries.

Several books have already been published in this series, including *Nutraceuticals and Health: Review of Human Evidence*, *Handbook of Metallonutraceuticals*, *Marine Nutraceuticals*, and so on. This series has been well received by the academia as well as the scientific and industrial community in the field of nutraceuticals.

The newest addition to the series is the present book *Herbal Bioactives and Food Fortification: Extraction and Formulation*, which is edited by Dr. Suresh Kumar. The book explains the fundamental steps for formulating food supplements and fortified food using extracts of herbs. The book addresses the way to choose potential herbs, optimize the extraction process, scale-up the process, analyze the extract, and use the extract for fortifying food and beverages. The common man nowadays consumes fortified beverages, fortified spreads, and tablets or capsules containing herbal extracts. Many aspects of this technology are scattered throughout the literature, but we felt there was a need to bring the information together in an authentic manner for the consumer as well as the layperson. The book presents different aspects of herbal drugs together so that it benefits those interested in this area of food technology. This book will be very useful to pharmaceutists, food technologists, cosmetologists, nutritionists, researchers of Ayurveda, Unani, and Siddha medicine, ethnopharmacologists, phytotherapists, and biologists.

Dr. Suresh Kumar received his PhD from Banaras Hindu University, one of the oldest universities in India. Subsequently, he worked as a postdoc at the University of Aston in the United Kingdom. Later, a chance encounter with some religious persons introduced him to the study of Ayurveda, the traditional medical system of India. He undertook a survey of the state of Ayurveda in the province and published his findings in provincial and national weeklies. Since 2003, he has worked in industry.

I think this book will be a great addition to the series and it will be a very useful resource for readers and scientists. We encourage scientists to communicate with us if they are interested in sharing their knowledge and contributing to the series.

Yashwant Pathak, MPharm, EMBA,
MS (Conflict Management), PhD
Series Editor
College of Pharmacy, University of South Florida Health
Tampa, Florida

Preface

The association between humans and medicinal plants dates back to prehistoric times. Paleodietary studies demonstrate that prehistoric populations that lived 10,000 years ago in the northern Chihuahuan desert of Mexico consumed many plants rich in inulin-type fructans. Archaeological oncology research has demonstrated that remnants of alcoholic beverages from ancient Egypt and China contained many plant-derived compounds with lung and colon cancer-fighting activity. Using sophisticated analytical tools, the residue from an ancient Egyptian wine obtained from an urn dating to 3150 BC that was found in the tomb of Pharaoh Scorpion I of Dynasty 0 was identified as fortified grape wine. Medical traditions prevalent in many parts of the world continue to cater to the health needs of the people of those countries.

The World Health Organization estimates that 80% of the population of Asian and African countries use herbal medicine for some aspect of primary health care. Synthetic medicines are prohibitively expensive for most of the world's population, half of which lives on less than $2 per day. In comparison, herbal medicines can be prepared from herbs grown inexpensively or gathered from nature at no cost.

In recent times, a major shift in global health care management policy has been instrumental in renewing interest in herbal medicine. To encourage national and international action to develop and implement primary health care throughout the world, the World Health Organization convened the International Conference on Primary Health Care (September 6–12, 1978) at Alma Ata, in the former Soviet Republic of Kazakhstan. The conference adopted the famous Alma Ata Declaration, which called on member nations to formulate national policies, strategies, and plans to launch and sustain primary health care. Members were especially encouraged to mobilize their own national resources. The Alma Ata Declaration ignited worldwide interest in traditional medicine.

Consequently, by 1980, the therapeutic use of natural products was widespread in Europe and the United States. This necessitated the recognition of these products as a separate segment. At a biomedical management course held in Villa Olmo, Como, Italy, Stephen DeFelice gave the name *nutraceutical* to a nutritional product that has clinically proven medical benefits but which its manufacturer cannot claim to the public or the medical practitioner on account of existing regulations. The term *nutraceutical* includes isolated nutrients, dietary supplements, specific diets, genetically engineered designer foods, plant products, and processed foods such as soups, cereals, and beverages. The development and manufacture of nutraceuticals is a vibrant industry all over the world.

The technical aspects of the development of products from herbs are scattered in the literature. Several scholarly works are available. However, they deal with isolated areas. There is a clear need for a book that provides information on all aspects of the extraction of biological actives from plants and the development of dietary supplements and fortified food. This book is intended to meet that need. It is organized into six chapters.

Chapter 1 presents a brief survey of the use of herbs in different civilizations and traces the evolution of herbal medicine. It portrays how the discipline of ethnopharmacology and the Alma Ata Declaration of 1978 caused the emergence of nutraceuticals. It provides some examples of nutraceuticals described in Ayurveda, Siddha, and *Unani* medicine. It concludes with a future outlook.

Plants are amazing chemical laboratories. This aspect is highlighted in Chapter 2. The functions of secondary metabolites, the factors that cause their variation, the extraction of essential oils, the various methods for the solvent extraction of bioactives, and the equipment used are described.

Chapter 3 addresses the topic of the extraction of bioactives from plants. It describes various aspects including the selection of the plant species, the quality control of raw materials, the comminution of herbs, the selection of solvents, and so on. The optimization of the extraction of herbs with reference to response surface methodology is described in detail. This is followed by large-scale extraction, an analysis of the extract, stability studies, and the toxicology of the products. The control of critical points in industrial extraction is illustrated with four examples.

There is growing concern all over the world regarding the quality, safety, and toxicity of foods and medicines accompanied by a greater preference for "natural" substances. Supercritical fluid extraction technology is a major technology that has emerged for the production of solvent-free substances. Chapter 4 discusses the supercritical fluid extraction process, its applications, and its advantages and disadvantages.

Herbal extracts can be used in the manufacture of food supplements, according to the principles of modern pharmaceutics, and using various dosage forms to deliver the medicinal substances. Chapter 5 provides an overview of the formulation of tablets, effervescent tablets, rapidly dissolving tablets, capsules, and syrups. Special attention is paid to the masking of tastes and the stability of food supplements.

Nowadays, elderly people worldwide are more concerned about the quality of the food and beverages that they consume. Many of them periodically monitor their biomarkers in an effort to reduce the risk of diseases. Functional foods have therefore emerged as an effective means for the prevention of diseases. Various beverages and foods such as soups, yogurt, sauces, mayonnaise, pickles, chutneys, cheese, sausage, bread, and biscuits can be fortified using herbal extracts. Chapter 6 describes the benefits of these fortified foods and reviews the scientific studies on fortified food and beverages. The various methods of their analysis are also suggested.

Acknowledgments

Several people have helped in the making of this book. Foremost among them is Dr. Muhammed Majeed, founder of Sabinsa Corporation, New Jersey, and Sami Labs Ltd., Bangalore. A technocrat and an imaginative marketing strategist, Dr. Muhammed Majeed is largely responsible for popularizing herbal dietary supplements in North America, long before the word *nutraceuticals* was coined. It was in his R&D laboratory that I was introduced to the fascinating world of the solvent extraction of herbs. I have gained from that association and am indebted to him.

Dr. Y.S. Prabhakar of the Division of Medicinal Chemistry, Central Drug Research Institute (CDRI), Lucknow, and Dr. Mohammad Mansoor Ansari, of Sami Labs Ltd., Bangalore, constantly encouraged me during the writing of this book. Dr. Mrs. M. Angela A. Meireles, professor of food engineering, University of Campinas, Brazil, graciously read Chapter 4 critically and offered constructive suggestions. I am grateful to her for this kind gesture. The preparation of the final manuscript files and graphics was an important part of this work, for which I am beholden to Alex Thomas, toxicologist, and C.M. Muhamed Ibnas, clinical research associate with CARe Keralam Ltd. Finally, I am grateful to Stephen Zollo and Amber Donley of Taylor & Francis Group for patiently guiding me to make this project a success.

This book explains the fundamental steps for formulating food supplements and fortified food using extracts of herbs. It shows the reader the way to choose potential herbs, optimize the extraction process, scale-up the process, analyze the extract, and use the extract for fortifying food and beverages. This book is presented before pharmaceutists, food technologists, nutritionists, researchers of Ayurveda, *Unani*, and Siddha medicine, ethnopharmacologists, phytotherapists, and biologists with the hope that it will advance the progress of food fortification.

D. Suresh Kumar

Author

D. Suresh Kumar was born on September 21, 1949, in the southern Indian province of Kerala, where he received his early education. He obtained a BSc degree in zoology from the University of Kerala (1969) and secured MSc (1972) and PhD degrees (1977) from Banaras Hindu University, Varanasi. His doctoral thesis was on the hormonal control of oxidative metabolism in reptiles. Thereafter, he spent two years as a postdoctoral fellow in the Department of Biological Sciences, University of Aston in Birmingham, England, investigating the pancreatic physiology of the rainbow trout. He returned to India in 1980 and joined the Department of Zoology, University of Calicut, Kerala, as pool officer in the scientist pool of the Council of Scientific and Industrial Research, New Delhi. During his stay there, a chance encounter with some religious persons introduced him to the study of Ayurveda, the traditional medical system of India. He undertook a survey of the state of Ayurveda in the province and published his findings in provincial and national weeklies. In 1986, he joined the International Institute of Ayurveda, Coimbatore, as a research officer in the Department of Physiology. From 1986 to 2003, he carried out research on various aspects of Ayurveda. In collaboration with Dr. Y.S. Prabhakar, presently at CDRI Lucknow, he proposed the first mathematical model for the ayurvedic concept of *tridōṣa* in the disease state. He also offered a novel definition for the ayurvedic class of medicine *arka*, based on his study of the Sanskrit text *Arkaprakāśa*. In 2003, he joined Sami Labs Ltd., Bangalore, as senior scientist in the R&D laboratory. He spent several years in the company working on various aspects of new product development. Since 2012, he has been employed at the Ayurveda consortium, Confederation for Ayurveda Renaissance Keralam Ltd., Koratty, Kerala. As head of its R&D laboratory, he directs research projects beneficial to the Ayurveda industry of Kerala.

1

Medical Herbalism through the Ages

Vis medicatrix naturae curavit!
(The healing power of nature cures!)

Roman proverb

1.1 Introduction

The association between humans and medicinal plants dates back to prehistoric times. Ancient people eagerly collected herbs because plants were the main source of their medicines. According to the *Book of Jubilees*, well before the great deluge, the angel Raphael communicated to Noah the remedies obtainable from trees, plants, and roots and their use. Noah recorded them in a book, which he later passed on to his son Shem. This was the beginning of the medical knowledge of the Israelites (Charles, 1917).

1.2 The "Iceman" used natural laxatives

Study of the pollen grains of several plants found in a middle Paleolithic cave in Iraq revealed that the prehistoric inhabitants were familiar with the medicinal use of herbs including *Centaurea solstitialis* and *Ephedra altissima* (Lietava, 1992). Long ago, archaeologists unearthed some human bones in China, dating back to 2000 BC. These "oracle bones" had the names of medicinal plants etched on them (Weiss and Weiss, 1992). Samples of a medicinal mushroom *Piptoporus betulinus* were found with the Austrian/Italian "iceman" of the Alps of Oetztal (3000 BC). Capasso (1998) suggests that the "iceman" might have employed this mushroom to treat gastrointestinal problems caused by the whipworm, *Trichuris trichuria*.

Paleodietary studies reveal that prehistoric populations who lived 10,000 years ago in the semi-arid northern Chihuahua desert in Mexico, consumed a variety of plants including agave (*Agave lechuguilla*), sotol (*Dasylirian* sp.), and onion (*Allium drummondii*), which are rich in inulin-type fructans (Sobolik, 1994; Leach and Sobolik, 2010). An analysis of 359 desiccated human feces (coprolites) from archaeological sites confirm the consumption of these prebiotic-rich herbs, as the coprolites contain undigested fragments and DNA remnants of these plant materials (Sobolik, 1996). Isotopic analysis of human skeletons and the coprolites and macrobotanical remnants indicate that the prehistoric Chihuahuans consumed about 135 g of prebiotic inulin-like fructans per day. This was far higher than the present-day average dietary intake of <20 g in the United States (Leach and Sobolik, 2010).

1.3 Herbs in different civilizations

Like the medical traditions of China and India, the Greek and Egyptian systems also considered nature as the greatest healer. All these systems believed that the human body is a replica of the universe. This belief is reflected in the Eastern mystic's view of "as above so below." Egyptian medicine benefited from trade that by 1500 BC linked Egypt to many other lands. Coptic travelers to Africa brought home many plant medicines such as myrrh gum, olibanum, and sandalwood. Traders from the southwestern tip of the Arabian desert introduced Egypt to their own native frankincense and many herbs of India. Cinnamon, ginger, the pomegranate tree, and calamus thus entered Egyptian medicine. Through contacts with the Minoan culture of Crete, Egyptian medicine was enriched with saffron, sage, and henna (Nunn, 2002).

1.3.1 Archaeological evidence

A study combining archaeology with analytical chemistry was recently reported by Salih et al. (2009). A cache of black cumin seeds (*Nigella sativa*), obtained from a pilgrim flask of the Old Hittite Period excavated at Boyali Höyük in the north-central region of Anatolia in Turkey, was extracted with ethanol and subjected to a gas chromatography–mass spectrometry (GC-MS) study. A sample of an ethanol extract of recently collected black cumin seeds was studied for comparison. Both the chromatograms were surprisingly similar, indicating that even though the seeds had been partially burnt in a fire in the temple about 3600 years ago, they had retained enough natural compounds to facilitate their identification by GC-MS analysis (**Figure 1.1**). The Hittites had obviously used black cumin seeds in combination with bee propolis to treat their ailments.

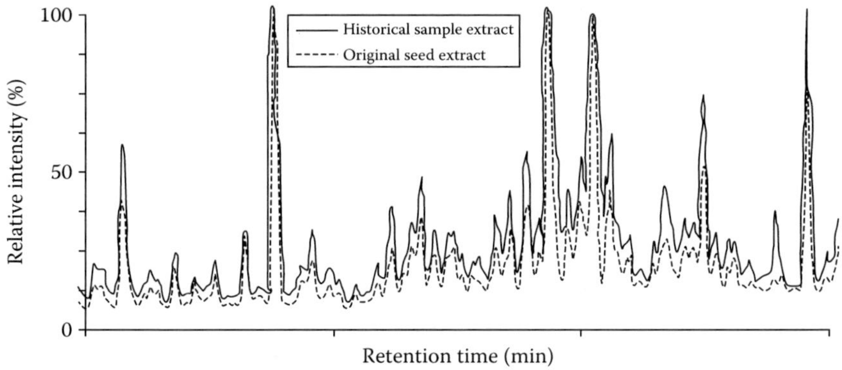

Figure 1.1 *A part of a GC-MS chromatogram of ancient* Nigella sativa *extract in ethanol. (Reproduced from* Journal of Ethnopharmacology, *124, Salih, B., Sipahi, T., Dönmez, E. O. Ancient nigella seeds from Boyali Höyük in north-central Turkey, 416–420, Copyright (2009), with permission from Elsevier.)*

Archaeological and chemical evidence is available for the use of cocoa fruits or beans in ancient Mesoamerica (the region extending south and east from central Mexico to include parts of Guatemala, Belize, Honduras, and Nicaragua). Using liquid chromatography–mass spectrometry (LC-MS) and GC-MS, Henderson et al. (2007) demonstrated the presence of theobromine, the alkaloid that is predominant in cocoa beans, in small and elegant serving vessels from excavation sites in Puerto Escondido of present-day Honduras. This suggests that cocoa beverages were being made in South America before 1000 BC. Hurst et al. (2002) have also reported similar findings. Using high-performance liquid chromatography (HPLC) coupled with atmospheric-pressure chemical-ionization mass spectrometry, they demonstrated the presence of theobromine in spouted ceramic vessels from 600 BC, excavated from northern Belize in Central America. Frothed chocolate drinks made from cocoa seeds were central to the social and ritual life throughout South America. Interestingly, the word *chocolate* is derived from the Aztec *chocolatl* (Henderson et al., 2007).

1.3.2 Greek medicine

The origins of Western medicine can be traced back to the teachings of the Greek physician Hippocrates (460–377 BC), who believed that diseases resulted from an imbalance in the four body humors and that they could be corrected by the judicious use of medicinal plants. Hippocrates' medical system was based on rational principles and he advocated the use of around 300 plants to heal his patients (Sumner, 2000).

Theophrastus (371–287 BC) was a student of Plato and Aristole. He wrote extensively on the medicinal uses of plants and he is considered to be the first botanist. Theophrastus' book *Historia Plantarum* covered the collection and preparation of herbs, medicines, spices, and fragrances. This work was used as a reliable reference for 2000 years (Levey, 1973a).

The Greek tradition was carried forward by Pedanius Dioscorides (40–90 AD). He composed *De Materia Medica*, which is a comprehensive treatise on the cultivation and qualities of nearly 600 herbs (Osbaldeston and Wood, 2000). Dioscorides observed that the time of day and the flowering time could influence the potency of medicines. An example he provided was the common opium poppy, whose alkaloid content has been demonstrated by modern methods to follow a chronobiological rhythm. Dioscorides recommended the decoction of willow bark for painful gout. Centuries later, it was discovered that salicylic acid is the active ingredient in willow bark and, in 1899, Friedrich Bayer & Co. of Germany synthesized acetylsalicylic acid (aspirin), thus ushering in a new era in therapeutics (Sneader, 2000; Rainsford, 2004). *De Materia Medica* was used as a resource work for the next 15 centuries.

Galen (129–199 AD) employed about 304 medicinal plants and complex mixtures, which were specially designed for each kind of therapy that he advocated. Thus, he became the founder of "galenics" (Wink, 1998).

1.3.3 Greco-Arabic medicine

From the decline of the Roman Empire to the European Renaissance in the fifteenth century, the Islamic world was an important center of medical knowledge. Arabian medicine was significantly influenced by pre-Islamic traditional Arabian medicine, Greek medicine, Ayurveda, and the medical system of ancient Iran. Many of the works of Greek and Roman physicians such as Hippocrates, Dioscorides, Soranus, Celsus, and Galen were translated into Arabic and these had a lasting influence on Arabic medicine (Prioreschi, 2001).

In *Al-Quanoon* (The Canon), Abu-ibn-Sina (Avicenna to Europeans), the doyen of Arabian medicine, recommended the leaves of the European yew (*Taxus baccata*) for treating cardiac ailments. This was many centuries before the calcium channel blocking property of *T. baccata* was reported (Tekol, 2007).

Although all Arab scholar-physicians included medicinal plants in their formulations, special mention must be made of the Andalusian–Arab physician Abu al-Abbas al-Nabati. He introduced scientific techniques in the description, testing, and identification of numerous medicinal herbs. His student Ibn-Baitar (1197–1248 AD) published *Kitab-al-Jami-fi-al-Adwiya-al-Mufrada*, which is considered one of the greatest botanical compilations. Among many other subjects, it contains medical information on more than 500 herbs. This work was translated into Latin in 1758 (Afnan, 1958).

It is generally believed that Arab knowledge of single herb remedies came from Greek sources. Nevertheless, using the novel method of botanonymy, or combining botanical and philological knowledge, Levey (1973b) inferred that early Arabic knowledge of simples came mostly from Mesopotamia and Asia. In Al-Kindi's book on medical formulary, 31% of the Arabic names for herbs are of Mesopotamian origin. This flow of knowledge was facilitated through Syriac, Aramaic, Hebrew, Persian, and in some cases, Greek.

1.3.4 Chinese medicine

Traditional Chinese medicine originated as a mystical medical system and later evolved into a system of herbal medicine. The Chinese materia medica was periodically enriched in different dynasties by the reevaluation and addition of uses for existing herbs. *Xin Xiu Ben Cao* (695 AD) and *Ben Cao Gang Mu* (1596 AD) are two such prominent works. This tradition is still followed in modern times. An example is the publication of the monumental work *Zhang Yao Da Ci Dian* (1977) (Huang, 1999). The number of herbs mentioned in the Chinese materia medica varies considerably. For example, *Shen Nang Ben Cao Jing* (220 AD) describes 365 herbs, whereas *Xin Xiu Ben Cao* (695 AD) mentions 659 herbs. Currently, about 400–600 herbs are used, 150 of which are high priority (Zhou, 1998).

1.3.5 Ayurveda

Ayurveda is based on the basic axiom that diseases spring from a disturbance of *vāta*, *pitta*, and *kapha*, collectively known as *tridōṣa*. Medicinal preparations derived from herbs are generally employed to correct the destabilized *tridōṣa*. More than 500 herbs are mentioned in Sanskrit medical texts. Nevertheless, contemporary Ayurveda makes use of nearly 127 herbs (Sharma, 2000).

Various ayurvedic lexicons such as *Abhidānamañjari*, *Bhāvaprakāśa Nighaṇṭu*, *Dhanvantari Nighaṇṭu*, and *Rāja Nighaṇṭu* provide information on the botanical characteristics and medicinal properties of herbs. This information is usually given in a string of Sanskrit synonyms, composed in a metrical style, much akin to that of a religious hymn. These synonyms denote the morphological peculiarities, ayurvedic properties, and therapeutic value of the herb in question. For example, *Abhidānamañjari* calls turmeric *Gauri* (of white complexion), *Haridra* (the yellow one), *Rajani* (night), *Pīta* (yellow), *Piṇḍa* (lump), *Kāñcani* (gold), *Strīvallabha* (liked by ladies), and *Varṇavati* (improver of complexion). Because of their linguistic and metrical characteristics, it is easy to memorize these quatrains. In ancient times, when there were no printed books, students of Ayurveda were encouraged to memorize entire texts. The essential ayurvedic information on all herbs used in the system is now available in a five-volume work (Warrier et al., 1994, 1996a, 1996b, 1997a, 1997b).

1.3.6 Tamil medicine

Another important system of Indian medicine is the Tamil tradition, popularly known as *Siddha medicine*. Although in its basics it is indistinguishable from

Ayurveda, through the ages the Tamil tradition has incorporated many elements of the alchemical practices of the Jains, Buddhists, and Saivites (Pillai, 1931).

According to Venkatraman (1990), alchemy and esoteric practices (tantrism), which are hallmarks of the Tamil Siddha system, owe their influences to northern India. Around 1200 AD, the Buddhist monasteries of Udantapura and Vikramasila (in present-day Bihar) were destroyed by invaders and the monks fled southward, carrying the esoteric knowledge. Nathism or the cult of Nathasiddhas, which was already in existence in the north, rose to prominence following the downfall of the Buddhist institutions. These Nathasiddhas are held to be directly responsible for the development of the Tamil Siddha system (Majeed and Kumar, 2008). Exponents of Siddha medicine consider the system to have developed independently from Ayurveda. However, Majeed and Kumar (2008) recently surveyed the medieval text *Vaittiyacintāmaṇi*-800 of Yukimuni and provided evidence for the first time that Siddha medicine is nothing but Ayurveda in a different form.

The Tamil physicians made attempts to study the therapeutic value of many plants predominant in the South and not utilized by Ayurveda. Examples are *Acalypha indica*, *Alangium salvifolium*, *Azima tetracantha*, *Borassus flabellifer*, *Cassia auriculata*, *Delonix alata*, *Dichrostachys cineria*, *Ervatamia coronaria*, *Lippia nodiflora*, *Marsilea quadrifolia*, *Mukia maderasapatana*, *Solanum trilobatum*, and *Wedelia calendulacea* (Kumar, 1995; Tirunanam, 1997). History teaches us that many Hindu kingdoms flourished in Cambodia, Laos, Malaya, Siam, and Vietnam (Lamb, 1975). Most of them were ruled by Tamil kings or their successors. It is possible that the Tamil physicians who visited these Southeast Asian countries made themselves familiar with many oriental medical practices, which finally made their way into Tamil Siddha medicine texts.

1.3.7 Unani

Although the Muslim presence in India is said to have begun with the military campaign of Mahmud of Ghazna (1014 AD), there is evidence to suggest that the interaction between Hindus and Muslims began much earlier. During the caliphate of Umar (636 AD), the governor of Bahrain attacked Thana (Bombay) and later Bharuch and Daybu on the Gujarat coast (Ahmad, 1988). Greco-Arabic medicine reached India with these visitors. It is said that the new system of medicine was not easily accepted because of the temperament of the people and the relatively superior nature of Ayurveda. Therefore, a hybrid of Greco-Arabic medicine and Ayurveda was slowly produced. This new medical system later became known as *Unani Tibb* or *Tibbi* medicine (Verma and Kewsani, 1974).

Under the patronage of Muslim rulers, scholars translated many Sanskrit texts into Arabic or composed *Unani* treatises, borrowing profusely from Ayurveda.

Zia Muhammed Mubarak, a courtier of Muhammed Tughlaq (1325–1351 AD), composed the recently discovered *Majma-e-Ziayi* (*Collections of Zia*), which had a separate chapter on medicine as prescribed by Nagarjuna and other sages of India (Verma and Keswani, 1974).

Aurangazeb's reign (1658–1707 AD) created an atmosphere that was conducive to the popularization of *Unani* (Rizvi, 1975). A famous physician of his court, Muhammed Akbar Arzani produced about eight Persian medical compilations. One of them, *Tibb-e-Hindi* (*Medicine of the Hindus*), deals with drugs of the ayurvedic formulary (Verma and Keswani, 1974). The Moghul period was marked by the translation into Persian of most of the medical texts written in Arabic, as Persian was the court language of the time. By the time Aurangazeb ascended the throne, all Arabic texts used in the *Unani* system were available in Persian. During the nineteenth century, many of these works were translated into Urdu, the popular language of the Muslims of northern India (Verma and Keswani, 1974).

As a result of the interaction between Ayurveda and Greco-Arabic medicine, two new dosage forms were incorporated into *Unani*. *Khamiras* or fermented products were developed on the lines of *āsava* and *ariṣṭa* of Ayurveda. The Moghul nobility had an aversion to drinking bitter decoctions of drugs and the Persian physicians circumvented this problem by developing *khamiras*, which are more palatable (Said, 1978; Kumar, 1992).

Ma'jun are electuaries developed by *Unani* physicians. They are very similar to the *lehyams* of Ayurveda. Examples are *Ma'jun Jograj*, *Gujul Ma'jun*, *Ma'jun-e-Hamal Alawi Khani*, and *Ma'jun Rah-al-Muminin*. Hakim Azad Khan, who composed the text *Muhit-i-Azam*, is credited with designing many electuaries. Although the *Unani* physicians had taken their cue from Ayurveda, they ingeniously formulated many novel *ma'jun*, which have few parallels in ayurvedic pharmacy. An example is *Ma'jun Murawwah-al-Arwah*, which has more than 100 ingredients including exotic items such as camel milk cheese (*mayashutr A'rabi*), dried turtle eggs (*baiza sang pusht khushk kia hua*), mongoose flesh (*ibn irs*), and sparrow brain (*maghz sar kunjashk*). A cursory look at the list of ingredients reveals the acceptance into *Unani* of drugs from several countries (Said, 1978; Kumar, 1992).

The *Unani* formulary was enriched by the inclusion of many of the plants used in Ayurveda. Ali (1990) has identified 210 such plants. In the majority of cases, the *Unani* names are persianized Sanskrit words. Examples are *Bish* (*Aconitum ferox*), *Wuz* (*Acorus calamus*), and *Mothoo* (*Cyperus rotundus*).

The industrial manufacture of *Unani* medicines was first started by Hakim Hafiz Abdul Majeed (1883–1922) from Pilbhit in present-day Uttar Pradesh. He started his career as an apprentice to Hakim Ajmal Khan. In 1906, in one of the alleys of old Delhi, he started a clinic and a pharmacy, which he called Hamdard Dawakhana. The Urdu word *hamdard* means "we shoulder thy pain" and it symbolizes his vision. This humble business concern, founded by Hakim Abdul Majeed, later grew to be the largest manufacturer of *Unani* medicines in the country. After his passing, the business was carried forward by his elder son Hakim Abdul Hameed who was instrumental in establishing Jamia Hamdard or Hamdard University in Delhi. After the partition of India in 1947, Hakim Hameed's younger brother, Hakim Mohammed Said, migrated to Pakistan and established Hamdard Laboratories Pakistan. Madinat-al-Hikmat University and Hamdard Foundation Pakistan, established in 1964, have been active in the promotion of the medical system. The prestigious research quarterly *Hamdard Medicus* is published in Pakistan.

1.4 Herbal medicine in medieval Europe

The tradition of herbal medicine was kept alive in the monasteries of medieval Europe. A common feature of all monasteries was a herb garden, where herbs were grown to treat patients and to teach the novices about medicinal plants. Charles the Great (Charlemagne, 747–814 AD) enacted the royal decree *Capitulare de Villis*, which ordered that medicinal plants should be grown in the king's garden and monasteries. It listed 24 species that were to be grown in such gardens. Walahfried Strabo (809–849 AD) is remembered for his *Liber de Cultura Hortum* (Book on the growing of plants), which is considered the first textbook of medical botany (Heinrich et al., 2004).

The Arabian influence on Spain and Italy had a beneficial effect on herbalism as well. Arabs are believed to have contributed to the founding of the great medical school at Salerno, Italy. The illustrious physicians of the school composed *Antidotarium Nicolai* (1130 AD), an early European work on medicinal herbs. *Antidotarium Nicolai* was one of the most influential books of the school of Salerno and by 1270 AD it was part of the curriculum at the University of Paris. The success of this book is indicated by the fact that it was translated into Arabic at a time when medical books were being translated from Arabic to Latin. The *Antidotarium* is a collection of pre-Constantinian herbal recipes from ancient to medieval authors. It describes many medicinal formulas, including a curious one called *Spongia somnifera* (soporific sponge), whose application to the patient's nostrils was alleged to induce anesthesia quickly (Prioreschi, 2003).

Up to the beginning of the sixteenth century, the botanical literature in general depended solely on the work of Dioscorides. The tradition of publishing herbals or illustrated guides to medicinal plants began with that of Apuleius Platonicus. It was followed by the Latin *Herbarius* (1484), the German *Herbarius* (Gart de Gesundheit) (1485), and the *Hortus Sanitatis* (1491). Although these herbals revolutionized the dissemination of knowledge of medicinal plants, very often their illustrations were inaccurate and their descriptions were imaginary. The credit for publishing the first authentic, illustrated herbal goes to the German physician Leonhart Fuchs (1501–1566). He published *De Historia Stirpium* (The history of plants) in 1542. He employed three engravers to prepare woodcuts of the 497 medicinal plants that illustrated his Latin text (Swan, 2006) (**Figures 1.2 and 1.3**). *De Historia Stirpium* was followed by several illustrated herbals, prominent among which are the three-volume *Herbarium Vivae Eicones* of Otto Brunfeld and *Das Neue Kraeuterbuch* of Hieronymous Bock (Sumner, 2000).

***Figure 1.2** Frontispiece of* De Historia Stirpium. *(Reproduced with permission from Antiquariaat Sanderus, Belgium; www.sanderusmaps.com.)*

A fillip to the publication of herbals was provided by the invention of the printing press by Johannes Gutenberg. Prominent among them are *The Dispensatorium* (1546) by Valerius Cordus, *Herbal* (1525) by Richard Banke, *The Grete Herball* (1526) by Peter Treveris, *New Herball* (1551, 1562, 1568) by William Turner, *Theatrum Botanicum* (1640) by John Parkinson, and *The English Physitian* (1652) by Nicolas Culpeper.

Carolus Linnaeus, who is credited with the system of binomial nomenclature of plants published *Materia Medica* (1749), a reference work for physicians. This invaluable work included descriptions of illnesses and specific medicines, medicinal effects, dosages, and the country of origin of these medicinal plants (Sumner, 2000).

1.5 Herbal medicine in America

1.5.1 Thomson's botanic medicine

Early America's foremost herbalist was Samuel Thomson (1760–1843). Born in New Hampshire, he studied with a midwife and some Red Indian healers. In 1800, he happened to cure his daughter's illness, whose condition was pronounced to be incurable by the local physician. Thereafter, he developed a healing system based on herbs and hot baths inspired by European and Red Indian herbalism.

Samuel Thomson was an ingenious medical marketing strategist. In 1813, he was granted an American patent for "Thomson's Inspired System of Botanic Practice of Medicine." This allowed him to sell his new medical system across the United States. He sold "family rights" to about 3 million women who happily helped to spread the new system. Initially, he made his medicines from 65 herbs bought from the Shakers. Later, he packaged his own herbal formulas, which were part of his patented system. His formulas were called *patented medicines*. After Thomson's death in 1843, his medical system became less fashionable. However, it was revived and preserved by his follower Dr. John Kellogg of Michigan, who is also famous as the inventor of America's first health food, the Kellogg cornflakes (Castleman, 2001).

Figure 1.3 *Engraving of red chilli from* De Historia Stirpium. *(Reproduced with permission from Pangloss Books, London; www.pangloss-books.com.)*

1.5.2 Eclectic medicine

The mainstream medicine of America in the nineteenth century was called *heroic medicine* on account of the "heroism" shown by patients to tolerate several therapeutic measures such as bloodletting and cathartics and toxic substances such as mercurials. Despite the predominance of heroic medicine, herbal medicine was very much in vogue. In 1830, a group of anti-heroic practitioners, including Thomsonians, Red Indian herbalists, and disillusioned practitioners, met in New York and founded the Reformed Medical School. The new system was called *eclectic medicine* and it was an admixture of European, Asian, Red Indian, and African (slave) herbalism. Their school in Cincinnati was called The Eclectic Medical Institute (Castleman, 2001).

The Eclectics were scientific herbalists who experimented with herbs, extracting and analyzing their active constituents. Their findings were published in the research journals of the day. Two of the most important Eclectics were John King (1813–1893) and John Uri Lloyd (1849–1936). John King is the author of *King's American Dispensatory*, modern America's most comprehensive herbal (Felter and Lloyd, 1984). John Lloyd was a pharmacist who specialized in the chemistry of herbs. He was an avid reader and together with his brothers, he established the Lloyd Library. For many years, the Lloyd Library published the journal *Lloydia*, now known as the *Journal of Natural Products* (Castleman, 2001).

Eclectic medicine remained popular up to 1900, when it was overshadowed by synthetic drugs. The institute's last batch of students graduated in 1939 and eclectic medicine's scientific herbalism breathed its last (Castleman, 2001).

1.5.3 Contribution of the Shakers

Founded in the early 1770s by an English Quaker, Ann Lee, the Shakers sect arrived in New York and set up their first community near Albany in 1774. The Shakers practiced strict pacifism, gender equality, and segregation. They launched the American medicinal herb industry in 1799, initially collecting herbs from the wild and later growing them. The herb business was profitable and, in 1824, the Sabbathday Lake community built a large building for drying and processing the herbs. The building is a standing testimony to the philosophy of the Shakers.

The Shakers sold their herbs through a widely distributed catalogue. They offered 142 herbs, roots, barks, and seeds. The Shakers soon developed a reputation for the honesty and purity of their herbs. During the American Civil War, they sold medicinal herbs to the Union army. After the war, they sold medicinal herbs directly to hospitals and America's nascent pharmaceutical industry.

The Shakers also invented the modern pill. They drilled holes in wooden frames, put measured amounts of herb powder in them, and then hammered pegs into the holes to compress the pills. The Shakers' herb business waned after World War II. However, the Sabbathday Lake community revived it in the 1960s (Castleman, 2001).

1.6 Decline of herbal medicine

In 1828, the German organic chemist Friedrich Woehler accidentally made urea in his laboratory and this opened up a new path to therapeutics through synthetic chemistry. By 1850, more organic compounds were being synthesized in chemical laboratories.

After the civil war, Americans were more interested in medical training in Germany, as the teaching there was more rigorous, it took longer, and the clinical training was intensive. Soon there was an influx of German-trained American physicians. A German-style medical school and hospital were established in Baltimore—Johns Hopkins Medical School. Thereafter, Harvard University devised a graduate program in medical training. Other medical schools started emulating the Harvard–John Hopkins model and invested heavily in well-equipped laboratories and teaching hospitals (Harvey, 2000).

The herbal medicine schools could not compete with the medical schools in infrastructure and with the federal licensing policy of favoring the Harvard–Johns Hopkins model, the other medical schools started closing down. The influential Carnegie Foundation's Flexner Report praised the Harvard–Johns Hopkins model and condemned all others (Flexner, 1910). The Carnegie Foundation and other wealthy foundations favored the Harvard-Johns Hopkins-type of medical schools. By 1940, American medical schools offered training in the Harvard–Johns Hopkins graduate program model. No medical school offered training in herbal medicine (Castleman, 2001).

Despite these setbacks, herbal medicine continued to be practiced by a few diehards. Dr. Bernard Lust (1869–1945), founder of modern naturopathy, opened the first health-food store in America and his nephew John Lust wrote *The Herb Book*. In England, C.F. Hilda Leyel single-handedly revived British herbalism by opening her Culpeper shops in 1927 (Castleman, 2001).

The first edition of the *United States Pharmacopoeia* of 1820 listed 650 drugs, of which 70% (455) were from plants. The 11th edition of the *United States Pharmacopoeia* of 1936 listed 570 drugs; however, the percentage of plant-derived medicines had dropped to only 40% of the total (260). Between 1920 and 1936, many plant-derived compounds were no longer recognized as effective medicines by the American government. On the contrary, they were being replaced by synthetic organic compounds (Sumner, 2000).

1.7 Renewed interest in herbal medicine

The trend toward synthetic medicines was partially reversed in the years before World War II, when Western medicine discovered antibiotics such as penicillin. With renewed interest in medicines from natural products, more plant-derived drugs have appeared on the scene.

1.7.1 Emergence of standardized extracts

The use of standardized extracts for improving human health and nutrition started with the pioneering work of the French chemist Jack Masquelier. In 1948, he published his doctoral thesis on the successful isolation and chemical description of the phytonutrients known today as oligomeric proanthocyanidins (OPCs). Masquelier's research demonstrated that these compounds strengthen the blood vessels, thereby improving vascular function. In 1951, he was granted a French patent to manufacture a standardized OPC fraction and employ it in therapeutics under the trade name Flavay® (French Patent 1965, 1966). Masquelier was granted an American patent in 1969 and this heralded the widespread use of OPCs (Masquelier, 1969). Following in the footsteps of Masquelier, a series of natural products having specific applications in therapeutics and nutrition were introduced to the market.

The veracity of Masquelier's claim has now been established. An international group of genomic scientists recently reported that daily supplements of OPC from grape seeds may change the gene expression associated with cardiovascular disease pathways. OPCs decrease the adhesion of immune cells to the vascular endothelium and potentially lower the infiltration of these immune cells into the vascular wall, which is the primary step in the development of atherosclerosis (Milenkovic et al., 2014).

1.7.2 The discipline of ethnopharmacology

With the increasing interest in herbs from indigenous medical systems, a new discipline known as ethnopharmacology also made its appearance. The term *ethnopharmacology* was introduced at an international symposium held in San Francisco in 1967 (Efron et al., 1967). This symposium was related to traditional psychoactive drugs such as kava and coca. Nevertheless, the area has now expanded. When introduced initially, ethnopharmacology meant a multidisciplinary area of research concerned with the observation, description, and experimental investigation of indigenous drugs and their biological activities (Rivier and Bruhn, 1979). Ethnopharmacology was redefined in 1981 as "the interdisciplinary scientific exploration of biologically active agents traditionally employed or observed by man" (Bruhn and Holmstedt, 1981).

1.7.3 Alma Ata Declaration

A major shift in global health care management policy was also instrumental in renewing interest in herbal medicine in general and herbs in particular. To encourage national and international action to develop and implement primary health care throughout the world, the World Health Organization (WHO) convened the International Conference on Primary Health Care (September 6–12, 1978) at Alma Ata, in the former Soviet Republic of Kazakhstan (Hall and Taylor, 2003). The conference adopted the famous Alma Ata Declaration, which called on member nations to formulate national policies, strategies, and plans to launch and sustain primary health care. The members were especially encouraged to mobilize their own national resources (Anonymous, 1978). The Western world was thus encouraged to study in depth the various traditional medical systems of the world.

In June 1974, Dr. N.R. Farnsworth led a 12-member herbal pharmacology study group to China. The aim of the delegation was to learn something about Chinese approaches to the use of herbal medicines (Anonymous, 1975). The *Journal of Ethnopharmacology*, founded shortly thereafter, continues to contribute greatly to the dissemination of validated knowledge on medicinal herbs and their formulations from diverse cultures.

Over the years, mainstream medicine has become very expensive in the West. To avoid becoming too dependent on physicians and to save money, many Westerners have started looking toward alternative systems including herbal medicine. This enthusiasm is also partly generated by several spiritual organizations such as the International Society for Krishna Consciousness (ISKCON), Osho Foundation, Maharishi Foundation, International Sivananda Yoga Vedanta Center, and The Art of Living International Center. These organizations have helped to popularize Ayurveda in the West.

Another major reason for the revival of interest in medicinal herbs is the growing concern about the undesirable side effects associated with synthetic medicines. Aspirin, nonsteroidal anti-inflammatory drugs, statins, cholesterol-lowering drugs, antidepressants, and antihypertensives cause serious side effects (Anonymous, 2015). One study has found that 2,216,000 Americans suffer a serious condition and are permanently disabled from a prescribed drug every year (Lazarou et al., 1998). However, these figures are only a fraction of the reality. Prominent medical researchers state that only 1 in 20 adverse reactions are reported for fear of a lawsuit (Bates, 1998). A French study observed that only 1 out of 24,433 adverse drug reactions is reported (Moride et al., 1997).

1.8 Emergence of nutraceuticals

1.8.1 Emergence of new terminology

By 1980, the therapeutic use of natural products was widespread in Europe and the United States. This necessitated the recognition of these products as a separate segment. Stephen DeFelice was the first person to coin the term *nutraceutical* to denote these substances. At a biomedical management course held in Villa Olmo, Como, Italy, DeFelice gave the name *nutraceutical* to a nutritional product (a single entity or a combination that includes special diets) that has clinically proven medical benefits but which its manufacturer cannot claim to the public or the medical practitioner on account of existing regulations (DeFelice, 1992). In the contemporary food and nutrition industry, such products may range from isolated nutrients, dietary supplements, and specific diets to genetically engineered designer foods, plant products, and processed foods such as soups, cereals, and beverages. Some examples are flavonoid antioxidants, α-linolenic acid from flax seeds, curcumin from turmeric, β-carotene from marigold petals, anthocyanins from berries, and carotenoids from chillies. Such specific constituents of herbal, animal, or microbial origin are also called *cosmeceuticals*, *dermaceuticals*, *phytochemicals*, *phytonutrients*, *phytofoods*, *designer foods*, *f(ph)armafoods*, *medifoods*, *vitafoods*, *fortified foods*, and *functional foods* (Roberfroid, 2000).

Nutraceuticals come under the purview of the Dietary Supplement Health and Education Act of 1994 of the United States of America. According to this act, a dietary supplement is "a product that is intended to supplement the diet and contains any one of the following ingredients like a vitamin, a mineral, an amino acid, a dietary substance for use by people to supplement the diet by increasing the total dietary intake, or a concentrate, metabolite, constituent, extract, or combination of any of the above." However, the manufacturer cannot make any label claim to diagnose, mitigate, treat, cure, or prevent a specific disease or class of diseases (Anonymous, 1994).

1.8.2 Associations and periodicals

The emergence of nutraceuticals has necessitated the formation of professional bodies and periodicals. The American Nutraceutical Association (ANA) and the National Nutritional Foods Association (NNfA) cater to the needs of the nutraceutical industry. ANA was formed in 1997 with the mission of "providing quality education to health care professionals, consumers and manufacturers involved in the emerging nutraceutical industry" (Anonymous, 2009a). It also publishes the free online journal, *The Journal of the American Nutraceutical Association*.

The NNfA, founded in 1936, is America's largest and oldest nonprofit organization dedicated to the natural products industry. The NNfA represents more than 8000 retailers, manufacturers, wholesalers, and distributors of natural products, including foods, dietary supplements, and health/beauty aids (Anonymous, 2009b).

The nutraceutical revolution has also given birth to several journals devoted to the theme: *Nutraceuticals World*, *Nutrients*, *The Journal of the American Nutraceutical Association*, *Food Product Design*, *Functional Ingredients*, *Journal of Medicinal Food*, *Journal of Nutraceuticals*, *Functional & Medical Foods*, *Journal of Dietary Supplements*, and *Journal of Functional Foods* (Shahidi, 2009).

1.8.3 European Scientific Cooperation on Phytotherapy

The European Scientific Cooperation on Phytotherapy (ESCOP) was founded in 1989 by six national scientific associations; many more joined in later years. ESCOP provides scientifically based assistance for the harmonized assessment of herbal products. Currently, national associations of Austria, Belgium, Denmark, France, Germany, Ireland, Italy, the Netherlands, Norway, Portugal, Sweden, Switzerland, and Great Britain are members of ESCOP. According to the definition offered by ESCOP (2000), phytomedicines or herbal medicinal products are "medicinal products containing as active ingredients, only plants, parts of plants or plant materials, or combinations thereof, whether in the crude or processed state."

Members of ESCOP meet formally at annual general meetings. The scientific committee of ESCOP, consisting of delegates from each member country, undertakes the arduous task of compiling European monographs summarizing the medicinal uses of herbal medicines and their safety. The first ESCOP monographs were released at a symposium in 1990. The first six fascicules, each containing 10 ESCOP monographs, have been published (Vincieri and Riva, 2003). These monographs are now gaining statutory value.

1.9 Nutraceuticals in Indian medicine

1.9.1 Ayurveda and Siddha medicine

It may be recalled at this point that the concept of nutraceutical was already embodied in Ayurveda, Unani, and Siddha medicine and allied streams. These systems do not find much difference between food and medicine.

Therefore, it is not surprising to find references to medicinal food. For example, *Aṣṭāṅgahṛdayam* recommends the daily consumption of buttermilk to cure many diseases of the digestive tract (Vaidyan, 1990). There are numerous references to herbs boiled in milk and the decoctions indicated in specific diseases. The Sanskrit text *Vaidyamanōrama* contains many recipes of medicated milk, yogurt, buttermilk, and porridges (**Table 1.1**).

1.9.2 Unani medicine

Unani went a step further and devised four separate dosage forms (*halwa*, *rubub*, *sharbat*, and *murabba*) that are true nutraceuticals by modern definition. *Halwa* are prepared by bringing a mixture of flour, milk, clarified butter, sugar, and herbs to a thick consistency. Examples are *Halwa-i-Badam*, *Halwa-i-Murgh*, and *Halwa-i-Ghaikwar* (Said, 1997a). A *sharbat* is a sweet beverage containing sugar, citric acid, and extracts of herbs. Common examples are *Sharbat Alu Balu*, *Sharbat Afsantin*, and *Sharbat Sandal* (Said, 1997b). As both *halwa* and *sharbat* are sweet or less bitter in taste, they are easily accepted by patients.

1.9.2.1 Halwa

The word *halwa* is derived from the Arabic root *hulw*, which means "sweet." In seventh-century Arabia, the word meant a paste of dates kneaded with milk. Perhaps inspired by the ancient Iranian sweetmeat *afroshag*, the Arabians designed *halwa* in the ninth century. It is generally prepared from semolina or flour, sugar, and clarified butter, worked into a stiff paste, which is then cut into bars or molded into fanciful shapes such as fish (Davidson, 1999). This sweet delicacy came to India along with Arabian culture. *Unani* medicine has a rich collection of recipes for this delectable confection. The following is a brief description of the recipe for *Halwa-i-Baiza-i- Murgh* (Said, 1997c).

The seeds of *Elettaria cardamomum* (2.5 g), aril, and the seed of *Myristica fragrans* (5 g each) are finely powdered and set aside. Wheat flour (400 g) and condensed milk (800 g) are mixed together and kneaded well with clarified butter (100 g) and powdered gum resin of *Boswellia serrata* (6 g). This mixture is also set aside. The yolks of 40 hen's eggs are beaten well and mixed with sugar (400 g) and the essence of *keora* (essence of the flowers of *Pandanus tectorius*). Clarified butter (300 g) is poured into a cauldron and heated. The egg yolk–sugar–*keora* mixture is added to the hot clarified butter and stirred continuously with a strong ladle, till the mixture separates freely from the ladle. The wheat flour–condensed milk mixture is added and stirring is continued. The cauldron is removed from the fire when the mass becomes semisolid. The powdered seeds of *Elettaria cardamomum* and *Myristica fragrans* are added and mixed well. The mass of *halwa* is then put into a tray, the inside of which has been smeared with clarified butter. The *halwa* is spread evenly in the tray. Fried bits of kernels of *badam* (*Prunus amygdalus*)

Table 1.1 Some Illustrative Nutraceutical Formulas from *Vaidyamanōrama*

Sl No.	Description of Formula	Reference
1	Consumption of milk boiled with roots of *Asparagus racemosus* cures urinary diseases.	Moos (1978: 3)
2	Fruits of *Vitis vinifera* are ground into a paste and mixed with yogurt. Consumption of this preparation along with candied sugar cures dysuria.	Moos (1978: 6)
3	Consumption of 15 g of the seeds of *Strychnos potatorum* and a small quantity of honey suspended in buttermilk cures all chronic polyuric diseases.	Moos (1978: 15)
4	Consumption of a paste of the roots or leaves of *Salacia reticulata* mixed with buttermilk cures all kinds of polyuric diseases.	Moos (1978: 15)
5	A paste of the seeds of *Ficus religiosa* and deer horn is suspended in buttermilk. Consumption of this cures polyuric diseases quickly.	Moos (1978: 15)
6	Consumption of the powdered seeds of *Embelia ribes* mixed with buttermilk cures urinary stones.	Moos (1978: 16)
7	Leaves of *Tabernaemontana divaricata* and rice are ground together, rolled into small balls, and fried in sesame oil. Consumption of these cakes will cure 20 types of polyuric diseases.	Moos (1978: 17)
8	*Loranthus elasticus* growing on *Ficus racemosa* is ground into a paste and mixed with buttermilk. Consumption of this cures diabetic carbuncles and diabetes mellitus.	Moos (1978: 19)
9	The powder of the *Asparagus racemosus* roots (60 g) is suspended in milk and consumed. Regular use of this formula for a month rejuvenates the body.	Moos (1978: 25)
10	Consumption of ginger powder in milk cures advanced jaundice.	Moos (1978: 67)
11	Dry ginger, pericarp of *Terminalia chebula*, heartwood of *Cedrus deodara*, and the roots of *Boerhaavia diffusa* are ground into a paste and suspended in buttermilk. Consumption of this drink cures edema all over the body.	Moos (1978: 75)
12	Drinking buttermilk mixed with a paste of *Cedrus deodara* (heartwood) cures edema completely.	Moos (1978: 76)
13	Consumption of the paste of *Tamarindus indicus* roots mixed with milk cures amenorrhea.	Moos (1978: 146)
14	The powder of *Piper nigrum* fruits administered in yogurt cures fever.	Moos (1979: 31)
15	Mix buttermilk with the ground leaves of *Murraya koenigii, Allium sativum* (bulbs), *Zingiber officinale* (rhizome), *Piper longum* (fruit), *Piper nigrum* (fruit), and a decoction of *Dolichos biflorus* seeds. Consumption of this drink cures fever.	Moos (1979: 32)
16	Consumption of the paste of the tender leaves of *Ficus religiosa* suspended in milk cures nocturnal fever.	Moos (1979: 38)
17	*Loranthus elasticus* growing on the *Aegle marmelos* tree is powdered and drunk in buttermilk, early in the morning. This cures irregular fever.	Moos (1979: 39)
18	The powder of *Terminalia chebula* (pericarp), *Tinospora cordifolia* (stem), rock-salt, *Zingiber officinale* (rhizome), and *Trachyspermum ammi* (seeds) mixed with buttermilk improves stamina and digestive efficiency.	Moos (1979: 43)
19	A paste of *Sauvarcala* salt (sodium sulfate mixed with sodium chloride), *Cuminum cyminum* (seeds), *Glycyrrhiza glabra* (root), and *Kaempferia rotunda* (tuber) is mixed with yogurt and a small quantity of honey. Consumption of this cures dysfunctional bleeding.	Moos (1979: 63–64)

(Continued)

Table 1.1 (Continued) Some Illustrative Nutraceutical Formulas from *Vaidyamanōrama*

Sl No.	Description of Formula	Reference
20	Consumption of powdered *Zingiber officinale* (rhizome) mixed with sugar and yogurt cures respiratory distress.	Moos (1979: 72–73)
21	Consumption of powdered *Piper longum* (fruits) mixed with sugar and yogurt cures respiratory distress.	Moos (1979: 72–73)
22	Drinking buttermilk mixed with a paste of *Eclipta alba* (whole herb) cures hoarseness of the voice.	Moos (1979: 112)
23	Consumption of milk boiled with the roots of *Desmodium gangeticum* cures heart disease.	Moos (1979: 134)
24	The roots of *Plumbago zeylanica* are detoxified, ground into a paste, and mixed with milk. The consumption of yogurt prepared from this milk cures hemorrhoids.	Moos (1979: 148)
25	Consumption of powdered *Terminalia chebula* (pericarp), *Holarrhena antidysenterica* (seeds), *Embelia ribes* (seeds), and *Plumbago zeylanica* (root) suspended in buttermilk cures hemorrhoids.	Moos (1979: 148–149)
26	Consumption of milk boiled with 20 tubers of *Cyperus roundus* cures painful dysentery.	Moos(1979: 165)
27	200 tubers of *Cyperus rotundus* are boiled in 1 L of milk. The strained milk is fermented. Consumption of the yogurt prepared from this milk cures dysentery.	Moos (1979: 169)

Source: Adapted from Moos, N.S., *Vaidyamanōrama*, Kottayam: Vaidya Sarathi Press, 1978; Moos, N.S., *Vaidyamanōrama*, Kottayam: Vaidya Sarathi Press, 1979. With permission.

are stuck on the *halwa* and while still hot, the *halwa* is cut into small pieces of convenient size. Each morning, 25 g of this *halwa* is to be eaten, followed by a glass (150 mL) of milk. When used regularly, it improves stamina, vigor, and vitality (Said, 1997c).

1.9.2.2 Rubub

The Arabic word *rub* means extract (*rubub* = plural). The juice expressed from fresh fruits and vegetables is concentrated with or without the addition of sugar. This concentrated extract is known as *rub*. In this way, the medicinal principles of fruits and vegetables are made available to patients even in off-seasons (Said, 1997d). The mode of preparation of *Rub-e-Angur* is described next.

Fresh grapes are washed in water and their juice is extracted. Grape juice (10 L) is mixed with cane sugar (3 kg) and stirred well. The mixture is heated over a slow fire to prevent charring. Heating is stopped when the volume is reduced by half. *Nitrun bunjawi* (sodium benzoate) (35 g) and aqua *Onosma bracteatum* (100 mL) are added, mixed well, cooled, and stored in glass containers. A dose of 25 mL of this preparation dissolved in a glass (150 mL) of water is to be consumed twice a day. *Rub-e-Angur* is a cardiac tonic, an appetizer, and a refrigerant (Said, 1997d).

1.9.2.3 Sharbat

Sharbat owes its origin to Arabia, Turkey, and Persia, where sweetened and diluted fruit juices were popular. The word *sharbat* is derived from the Arabic root *sarba*, which means "drink." This pleasant drink reached Europe through the Ottoman Empire of Turkey. Medicated *sharbat*s are popular in *Unani* medicine and around 60 are in use today (Said, 1997b).

Sharbat Gazar is prepared in the following way. Cane sugar (4 kg) is added to fresh carrot juice (12 L). The mixture is heated over a slow fire and on attaining a thick consistency, 8 g of *sat limun* (citric acid) and 6 g of *nitrun bunjawi* are added. The *sharbat* is filtered through a cloth and on cooling, it is stored in containers. A dose of 50 mL of the *sharbat* diluted in cold water is administered to correct disorders of the heart (Said, 1997b).

1.9.2.4 Murabba

The fruit pickle *murabba* is believed to have originated in Gurjistan, the present-day Republic of Georgia. The Gurjaris, who were inhabitants of this area, fled their homeland to escape from invaders and settled in present-day Gujarat and adjoining areas. Thereafter, *Murabba* became a popular confection and was adopted as a functional food in *Unani*. *Murabbas* are usually made from the rhizomes of *Zingiber officinale* and the fresh fruits of *Emblica officinalis*, *Ananas sativus*, *Aegle marmelos*, *Benincasa cerifera*, and *Terminalia chebula* (Said, 1997e). *Murabba-i-Amla* is prepared in the following way.

Fresh amla fruits are washed. Using a knife, three or four incisions are made on each fruit. The fruits are then put in a vessel with water and cooked till tender. The water is drained and the cooked fruits are allowed to cool. A syrup of cane sugar is prepared and poured into a stainless steel container. The cooked amla fruits are dropped into the syrup and are allowed to remain submerged in it for a week, by which time they are ready for use. The consumption of one amla fruit every morning will enliven the heart and the liver. It is also believed to cure bleeding piles (Said, 1997e).

1.10 Proven benefits of nutraceuticals

1.10.1 Phytosterols

Numerous research reports and clinical studies indicate the beneficial effects of nutraceuticals. For example, a 30-day trial showed that 1.7 g of phytosterols containing 20% sitostanol and 80% other phytosterols (primarily sitosterol and campesterol) reduced low-density lipoprotein (LDL) cholesterol by 24.4% in hypercholesterolemic men compared with 8.9% on the control diet (Jones et al., 1999).

Interestingly, Sierksma et al. (1999) designed a 9-week, double-blind, crossover study to assess the cholesterol-lowering effect of two spreads fortified with free soybean oil sterols or with sheanut oil sterols. Plasma total cholesterol and LDL cholesterol concentrations were significantly reduced by 3.8% and 6%, respectively, for the spread enriched with free soybean oil sterols compared with the control spread.

Phytosterols play an important role in the regulation of cardiovascular disease and also exhibit anticancer properties (Day, 1998; Jones and AbuMweis, 2009). These and several other aspects of phytosterols in nutrition have been reviewed by Chadwick et al. (2003).

1.10.2 Curcumin

Similarly, clinical studies have shown that curcumin can be effective in the treatment of several diseases (Goel et al., 2008). Deodhar et al. (1980) compared its antirheumatic potential with that of phenylbutazone in a short-term, double-blind, crossover study involving 18 young subjects. A daily dose of 1200 mg curcumin exerted an antirheumatic effect comparable with that of phenylbutazone. A later study by Satoskar et al. (1986) confirmed the anti-inflammatory action of the molecule in 46 men who had undergone surgical repair of an inguinal hernia or a hydrocele.

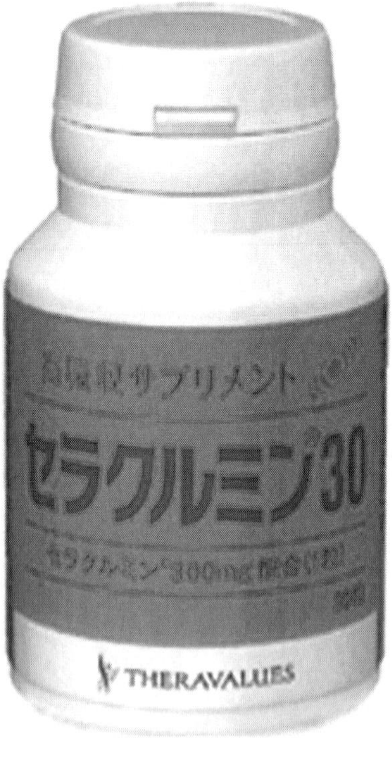

Figure 1.4 A successful nutraceutical, Theracurmin 30, highly bioavailable curcumin indicated in hypertension, fatigue, cancer, and arthritis. (Reproduced with permission from Theravalues Corporation, Tokyo, Japan.)

Curcumin has also been found to be useful in palliative therapy for cancerous skin lesions (Kuttan et al., 1987), in lowering cholesterol (Soni and Kuttan, 1992; Ramirez Bosca et al., 2000), in irritable bowel syndrome (Bundy et al., 2004), and in improving early renal graft function (Shoskes et al., 2005). A recent population-based study of 1010 elderly Asians with no dementia suggests that curcumin may improve cognitive function in the elderly, emphasizing its usefulness in the prevention of Alzheimer's disease (Ng et al., 2006) (Figure 1.4).

1.10.3 Lycopene

Lycopene is a carotenoid that is present in tomatoes, pink grapefruit, pink guava, and watermelon. It is one of the rare carotenoids absorbed through the intestine. It is also the most predominant carotenoid present in plasma. As humans cannot synthesize this nutrient, they must acquire it through their diet (Bramley, 2000). Lycopene is now reported to have a beneficial effect on coronary heart disease; this effect is related to its antioxidant property (Bramley, 2000, Stahl and Sies, 1999). Epidemiological evidence suggests that it can prevent cancer of the prostate, digestive tract, pancreas, bladder, cervix, skin, lung, and breast (Hasler, 2000; Larsson et al., 2010).

In a mouse model of allergic asthma, Hazlewood et al. (2010) assessed the effect of lycopene supplementation on inflammatory cell influx into the bronchoalveolar lavage fluid, lung tissue and blood, mucus-secreting cell numbers in the airways, lymph node drainage, ovalbumin-specific cytokine release, serum immunoglobulin G1 (IgG1) levels, and lung function. Lycopene supplementation significantly reversed all the pathological changes. Lycopene supplementation can offer protection from asthma.

In addition to coloring and stabilizing food products, lycopene is now known to have definite therapeutic properties. Clinical trials have shown that lycopene is useful in the management of carcinoma of the prostate gland (Kucuk et al., 2002; Ansari and Gupta, 2003). The results of a recent clinical study involving 50 patients suggest that lycopene has a potential therapeutic benefit in the adjunct management of high-grade gliomas (Puri et al., 2010).

1.10.4 Resveratrol

Chronic drinking of alcohol is known to cause oxidative damage to many organs of the body. This is due to the ability of alcohol to generate free radicals and enhance lipid peroxidation. Oxidative stress taking place at the subcellular level is the basis of many health problems such as cardiovascular diseases, aging, and neurodegenerative diseases. Despite its high-fat diet, the incidence of coronary heart disease in France is strikingly low. Interest in understanding this “French paradox” has identified resveratrol and other antioxidant polyphenols in grape skin as the agents responsible for causing this phenomenon (Sun et al., 2002). Though grapes are considered to be the major source of resveratrol, it is also present in many other fruits such as mulberry, bilberry, and jackfruit, several flowers, and nuts such as peanut (Appendino and Taglialatela-Scafeti, 2003; Lekli et al., 2010).

Recent studies have shown that resveratrol has estrogenic, antiplatelet, and anti-inflammatory properties (Lekli et al., 2010). It inhibits apoptotic cell death and thus offers protection from diseases such as myocardial ischemia, atherosclerosis, ventricular arrhythmias, and cerebral ischemia (Wang et al., 2002; Dudley et al., 2008; Kennedy et al., 2010).

1.11 Outlook for the future

The search for new bioactive molecules has a long history in the pharmaceutical industry of Europe and America. In their erudite and profusely illustrated work, Nicoloaou and Montagnon (2008) detail the adventures of chemists and biologists who discovered new molecules for therapy. Nevertheless, such efforts are infrequent in the world of nutraceuticals. The global herbs and nutraceuticals market is yet to realize its full potential. New herbal extracts and new benefits from already known extracts are being reported (Gruenwald, 2008). With the vast legacy of Ayurveda at its command, the herbal industry of India can become a phenomenal success in the nutraceutical world, if it subjects the traditional knowledge to dispassionate research. Three problems, however, need to be seriously addressed.

1.11.1 Strengthening of research

Many herbal companies are aware of the need for more research in this area, while others are totally oblivious to this aspect. Research and development efforts in many Ayurveda companies are limited to some rudimentary analysis of their finished products. Certificates of analysis are unheard of in such places! A few others borrow research information from other companies or institutes and try to promote their products in the market. Herbal medicine manufacturers should join forces with experts from the food and pharmaceutical industries.

1.11.2 Development of analytical methods

The nutraceutical industry is still in need of robust analytical methods. The official methods of analysis of the Association of Official Agricultural Chemists (AOAC) International, Gaithersburg, Maryland, are a reliable source (Horowitz, 2000). However, effective methods need to be developed to detect the intentional adulteration of products. The current tests for saw palmetto quality are based on total fatty acid content. However, this does not give an accurate assessment of the quality of the extract. Owing to its similarity to the fatty acid content of saw palmetto extract, palm oil can easily pass as genuine saw palmetto. Thus, unscrupulous manufacturers mix the extract with palm oil and escape prosecution. Other areas of attention include the adulteration of ginkgo, rutin, powders of berries, and natural dyes.

1.11.3 Clinical validation of claims

Currently, most nutraceuticals are sold on the strength of advertising campaigns. Many products that were launched with great enthusiasm soon turned out to be damp squibs. The only solution for this is controlled clinical trials. Although the extract of *Bacopa monnieri* has been in the Western market since 1980, proof of its memory-enhancing property has only recently become available (Stough et al., 2001; Roodenrys et al., 2002; Stough et al., 2008; Calabrese et al., 2008). The clinical efficacy of many ayurvedic herbs can be proved with similar studies.

1.12 Concluding remarks

Nutraceuticals represent the fastest-growing segment in the contemporary food industry. Its estimated market size is $30 billion, growing at 5% per annum. Thus, over the course of time, the interface between food and drugs will narrow. As Andlauer and Fuerst (2002) have opined, future consumers will be presented with a variety of products such as nutraceutical soups, nutraceutical processed meat, and fortified bread and sausages. As herbs remain the major source of nutraceuticals, there is greater scope to isolate their chemical constituents and study their biological effects for possible use in the prevention and treatment of diseases and the enhancement of beauty.

References

Afnan, S.M. 1958. *Avicenna—His Life and Works*. London: Allen & Unwin.

Ahmad, M. 1988. A critical study of Arab-o-Hind ke T'aalluqat. *Journal of Oriental Institute* 38: 343–354.

Ali, M. 1990. Ayurvedic drugs in Unani materia medica. *Ancient Science of Life* 9: 191–201.

Andlauer, W., Fuerst, P. 2002. Nutraceuticals: A piece of history, present status and outlook. *Food Research International* 35: 171–176.

Anonymous. 1975. *Herbal Pharmacology in the People's Republic of China. A Trip Report of the American Herbal Pharmacology Delegation submitted to the Committee on Scholarly Communication with the People's Republic of China*. Washington, DC: National Academy of Sciences.

Anonymous. 1978. Declaration of Alma-Ata International Conference on Primary Health Care, Alma-Ata, USSR, September, 6–12, 1978. http://www.who.int/hpr/NPH/docs/declaration_almaata.pdf. Accessed March 12, 2009.

Anonymous. 1994. Dietary Supplement Health and Education Act of 1994, Public Law 103–417, 103rd Congress. http://www.fda.gov/opacom/laws/dshea.html. Accessed March 25, 2009.

Anonymous. 2009a. http://www.ana-jana.org. The American Nutraceutical Association. Accessed March 7, 2009.

Anonymous. 2009b. http://www.naturalproductsassoc.org/site/PageServer?pagename=nf homepage. Natural Products Association. Accessed March 9, 2009.

Anonymous. 2015. Death and harm from prescription drugs. http://ibeatemphysemaand-copd.com/some-points-about-real-healing/. Accessed May 22, 2015.

Ansari, M.S., Gupta, N.P. 2003. A comparison of lycopene and orchidectomy vs. orchidectomy alone in the management of advanced prostate cancer. *BJU International* 92: 375–378.

Appendino, G., Taglialatela-Scafeti, O. 2003. Drug-like compounds from plants and spices. In *Dietary Supplements of Plant Origin*, ed. M. Maffei, 43–74. New York: Taylor & Francis.

Bates, D.W. 1998. Drugs and adverse drug reactions: How worried should we be? *Journal of the American Medical Association* 279: 1216–1217.

Bramley, P.M. 2000. Is lycopene beneficial to human health? *Phytochemistry* 54: 233–236.

Bruhn, J.G., Holmstedt, B. 1981. Ethnopharmacology, objectives, principles and perspectives. In *Natural Products as Medicinal Agents*, ed. J.L. Beal and E. Reinhard, 405–430. Stuttgart: Hippokrates Verlag.

Bundy, R., Walker, A.F., Middleton, R.W., Booth, J. 2004. Turmeric extract may improve irritable bowel syndrome symptomatology in otherwise healthy adults: A pilot study. *Journal of Alternative and Complementary Medicine* 10: 1015–1018.

Calabrese, C., Gregory, W.L., Leo, M., Kraemer, D., Bone, K., Oken, B. 2008. Effects of a standardized *Bacopa monnieri* extract on cognitive performance, anxiety and depression in the elderly: A randomized, double-blind, placebo-controlled trial. *Journal of Alternative and Complementary Medicine* 14: 707–713.

Capasso, L. 1998. 5300 years ago, the iceman used natural laxatives and antibiotics. *Lancet* 352: 1894.

Castleman, M. 2001. From magic to medicine: 5,000 years of herbal healing. In *The New Healing Herbs: The Classic Guide to Nature's Best*, 26–37. Emmaus, PA: Rodale Press.

Chadwick, R., Henson, S., Moseley, B., et al. 2003. Phytosterols enriched functional foods. In *Functional Foods*, ed. R. Chadwick et al., 183–205. Heidelberg: Springer.

Charles, R.H. 1917. Chapter 10. In *The Book of Jubilees*, 80–81. New York: The Macmillan Company.

Davidson, A. 1999. *Oxford Companion to Food*, 367. Oxford: Oxford University Press.

Day, C. 1998. Traditional plant treatments for diabetes mellitus: Pharmaceutical foods. *British Journal of Nutrition* 80: 5–6.

DeFelice, S.L. 1992. The nutraceutical initiative: A recommendation for U.S. economic and regulatory reforms. *Genetic Engineering News* 12: 13–15.

Deodhar, S.D., Sethi, R., Srimal, R.C. 1980. Preliminary study on antirheumatic activity of curcumin (diferuloyl methane). *Indian Journal of Medical Research* 71: 632–634.

Dudley, J., Das, S., Mukherjee, S., Das, D.K. 2008. Resveratrol, a unique phytoalexin present in red wine, delivers either survival signal or death signal to the ischemic myocardium depending on dose. *The Journal of Nutritional Biochemistry* 20: 443–452.

Efron, D.H., Holmstdt, B., Kline, N.S. 1967. *Ethnopharmacological Search for Psychoactive Drugs*. Publication No. 1645. Washington, DC: US Department of Health, Education and Welfare, Government Printing Office.

ESCOP (European Scientific Cooperation on Phytotherapy). 2000. http://www.escop.com. Accessed September 21, 2008.

Felter, H., Lloyd, J.U. 1984. *King's American Dispensatory*, Vols. 1 & 2. Sandy, OR: Eclectic Medical Publications.

Flexner, A. 1910. Medical education in the United States and Canada. A report to the Carnegie Foundation for the Advancement of Teaching. The Carnegie Foundation Bulletin No. 4. New York: Carnegie.

French Patent. 1965. Procede pour l'obtention d'hydroxyflavanne-3-4-diols. P.V. No. 998.508, No. 1.427.100.

French Patent. 1966. Noveau bioflavonide ou facteur vitaminique. P.V. No. 9.054, No. 4.482 M.

Goel, A., Kunnumakkara, A.B., Aggarwal, B.B. 2008. Curcumin as "Curecumin": From kitchen to clinic. *Biochemical Pharmacology* 75: 787–809.

Gruenwald, J. 2008. The global herbs & botanicals market. *Nutraceutical World* July/August.

Hall, J.J., Taylor, R. 2003. Health for all beyond 2000: The demise of the Alma-Ata Declaration and primary health care in developing countries. *Global Health* 178: 17–19.

Harvey, J.C. 2000. The six transformations of American health care. In *The Health Care Professional as Friend and Healer: Building on the Work of Edmund D. Pellegrino*, ed. D.C. Thomasma and J.L. Kissell, 113–132. Washington, DC: Georgetown University Press.

Hasler, C.M. 2000. Plants as medicine: The role of phytochemicals in optimal health. In *Phytochemicals and Phytopharmaceuticals*, ed. F. Shahidi and C.T. Ho, 1–12. Champaign, IL: AOAC Press.

Hazlewood, L.C., Wood, L.G., Hansbro, P.M., Foster, P.S. 2010. Dietary lycopene supplementation suppresses Th2 responses and lung eosinophilia in a mouse model of allergic asthma. *The Journal of Nutritional Biochemistry* 22: 95–100.

Heinrich, M., Barnes, J., Gibbons, S., Williamson, E.M. 2004. Pharmacognosy and its history: People, plants and natural products. In *Fundamentals of Pharmacognosy and Phytotherapy*, 12–14. Philadelphia, PA: Churchill Livingstone/Elsevier Science.

Henderson, J.S., Joyce, R.A., Hall, G.R., Hurst, W.J., McGovern, P.E. 2007. Chemical and archaeological evidence for the earliest cacao beverages. *Proceedings of the National Academy of Sciences* (U.S.A.) 104: 18937–18940.

Horowitz, W. 2000. *Official Methods of Analysis of AOAC International*, 17th edn, Vols. 2. Rockville, MD: AOAC International.

Huang, K.C. 1999. A brief history of Chinese medicine. In *The Pharmacology of Chinese Herbs*, 2nd edn, 7–14. Washington, DC: CRC Press.

Hurst, W.J., Tarka, S.M., Jr., Powlis, T.G., Valdez, F., Jr., Hester, T.R. 2002. Cacao usage by the earliest Maya civilization. *Nature* 418: 289–290.

Jones, P.J., AbuMweis, S.S. 2009. Phytosterols as functional food ingredients: Linkages to cardiovascular disease and cancer. *Current Opinion in Clinical Nutrition & Metabolic Care* 12: 147–151.

Jones, P.J.H., Ntanios, F.Y., Raeni-Sarjaz, M., Vanstone, C.A. 1999. Cholesterol-lowering efficacy of a sitostanol-containing phytosterol mixture with a prudent diet in hyperlipdemic men. *American Journal of Clinical Nutrition* 69: 1144–1150.

Kennedy, D.O., Wightman, E.L., Reay, J.L., et al. 2010. Effects of resveratrol on cerebral blood flow variables and cognitive performance in humans: A double-blind, placebo-controlled, crossover investigation. *American Journal of Clinical Nutrition* 91: 1590–1597.

Kucuk, O., Sarkar, F.H., Djuric, Z., et al. 2002. Effects of lycopene supplementation in patients with localized prostate cancer. *Experimental Biology and Medicine* (Maywood) 227: 881–885.

Kumar, D.S. 1992. Glimpses of interaction between Ayurveda and Unani. *Aryavaidyan* 6: 109–116.

Kumar, D.S. 1995. Some preliminary observations on the chronology of Tamil medical texts. In *Glimpses of Indian Ethnopharmacology*, ed. P. Pushpangadan, U. Nyman, and V. George, 77–84. Trivandrum: TBGRI.

Kuttan, R., Sudheeran, P.C., Joseph, C.D. 1987. Turmeric and curcumin as topical agents in cancer therapy. *Tumori* 73: 29–31.

Lamb, A. 1975. Indian influence in ancient South-East Asia. In *A Cultural History of India*, ed. A.L. Basham, 442–454. Oxford: Clarendon Press.

Larsson, S.C., Bergkvist, L., Wolk, A. 2010. Dietary carotenoids and risk of hormone receptor-defined breast cancer in a prospective cohort of Swedish women. *European Journal of Cancer* 46: 1079–1085.

Lazarou, J., Pomeranz, B.H., Corey, P.N. 1998. Incidence of adverse drug reactions in hospitalized patients—A meta-analysis of prospective studies. *Journal of the American Medical Association* 279: 1200–1205.

Leach, J.D., Sobolik, K.D. 2010. High dietary intake of prebiotic inulin-type fructans in the prehistoric Chihuahuan desert. *British Journal of Nutrition* 103: 1558–1561.

Lekli, I., Ray, D., Das, D.K. 2010. Longevity nutrients, resveratrol, wines and grapes. *Genes & Nutrition* 5: 55–60.

Levey, M. 1973a. Pre-Islamic pharmacology. In *Early Arabic Pharmacology*, 1–32. Leiden: E.J. Brill.

Levey, M. 1973b. Botanonymy. In *Early Arabic Pharmacology*, 54–65. Leiden: E.J. Brill.

Lietava, J. 1992. Medicinal plants in a middle Paleolithic grave Shanidar IV. *Journal of Ethnopharmacology* 35: 263–266.

Majeed, M., Kumar, D.S. 2008. Classification of diseases in the Tamil medical work *Vaittiyacintamani*-800 of Yukimuni I: Introduction. *Traditional South Asian Medicine* 8: 65–76.

Masquelier, J. 1969. Hydroxyflavin 3,4-diols, a method of producing them and medicament based thereon. US Patent No. 3,436,407.

Milenkovic, D., Berghe, W.V., Boby, C., et al. 2014. Dietary flavanols modulate the transcription of genes associated with cardiovascular pathology without changes in their DNA methylation state. *PLoS ONE* 9, e95527.

Moos, N.S. 1978. *Vaidyamanōrama*. Kottayam: Vaidya Sarathi Press.

Moos, N.S. 1979. *Vaidyamanōrama*. Kottayam: Vaidya Sarathi Press.
Moride, Y., Haramburu, F., Requejo, A.A., Bégaud, B. 1997. Under-reporting of adverse drug reactions in general practice. *British Journal of Clinical Pharmacology* 43: 177–181.
Ng, T.P., Chiam, P.C., Lee, T., Chua, H.C., Lim, L., Kua, E.H. 2006. Curry consumption and cognitive function in the elderly. *American Journal of Epidemiology* 164: 898–906.
Nicolaou, K.C., Montagnon, T.C. 2008. *Molecules that Changed the World*. Weinheim: Wiley-VCH Verlag.
Nunn, J.F. 2002. Drug therapy. In *Ancient Egyptian Medicine*, 136–158. Norman, OK: University of Oklahoma Press.
Osbaldeston, T., Wood, R.P.A. 2000. *De Materia Medica* [Modern English translation]. Johannesburg: IBIDIS.
Pillai, T.V.S. 1931. *Tamil-English Dictionary of Medicine, Chemistry, Botany and Allied Sciences*, Vol. 1, 1–114. Madras: Research Institute of Siddhar's Science.
Prioreschi, P. 2001. Medicine. In *A History of Medicine*, Vol. 4: Byzantine and Islamic Medicine, 199–392. Omaha, NE: Horatius Press.
Prioreschi, P. 2003. Medicine of the early Middle Ages (5th–12th centuries). In *A History of Medicine*, Vol. 5: Medieval Medicine, 230–235. Omaha, NE: Horatius Press.
Puri, T., Goyal, S., Julka, P.K., Nair, O., Sharma, D.N., Rath, G.K. 2010. Lycopene in treatment of high-grade gliomas: A pilot study. *Neurology India* 58: 20–23.
Rainsford, K.D. 2004. History of development of the salicylates. In *Aspirin and Related Drugs*, ed. K.D. Rainsford, 1–23. Boca Raton, FL: CRC Press.
Ramirez Bosca, A., Soler, A., Carrion-Gutierrez, M.A., Pamies Mira, D., Pardo Zapata, J., Diaz-Alperi, J. 2000. An hydroalcoholic extract of *Curcuma longa* lowers the abnormally high values of human-plasma fibrinogen. *Mechanisms of Ageing and Development* 114: 207–210.
Rivier, J., Bruhn, J.G. 1979. Editorial. *Journal of Ethnopharmacology* 1: 1.
Rizvi, S.A.A. 1975. The Muslim ruling dynasties. In *Cultural History of India*, ed. A.L. Basham, 245–265. Oxford: Clarendon Press.
Roberfroid, M.B. 2000. Defining functional foods. In *Functional Foods: Concept to Product*, ed. G.R. Gibson and C.M. Williams, 9–27. Boca Raton, FL: CRC Press.
Roodenrys, S., Booth, D., Bulzomi, S., Phipps, A., Micallef, C., Smoker, J. 2002. Chronic effects of Brahmi (*Bacopa monnieri*) on human memory. *Neuropsychopharmacology* 27: 279–281.
Said, M. 1978. Traditional medicine in the service of health. Proceedings of the 26th International Congress of History of Medicine, Plodiv, Bulgaria, August 20–25.
Said, M. 1997a. *Hamdard Pharmacopoeia of Eastern Medicine*, 115–119. Delhi: Sri Satguru Publications.
Said, M. 1997b. *Hamdard Pharmacopoeia of Eastern Medicine*, 169–189. Delhi: Sri Satguru Publications.
Said, M. 1997c. *Hamdard Pharmacopoeia of Eastern Medicine*, 115–116. Delhi: Sri Satguru Publications.
Said, M. 1997d. *Hamdard Pharmacopoeia of Eastern Medicine*, 136–138. Delhi: Sri Satguru Publications.
Said, M. 1997e. *Hamdard Pharmacopoeia of Eastern Medicine*, 251–253. Delhi: Sri Satguru Publications.
Salih, B., Sipahi, T., Dönmez, E.O. 2009. Ancient nigella seeds from Boyali Höyük in north-central Turkey. *Journal of Ethnopharmacology* 124: 416–420.
Satoskar, R.R., Shah, S.J., Shenoy, S.G. 1986. Evaluation of anti-inflammatory property of curcumin (diferuloyl methane) in patients with postoperative inflammation. *International Journal of Clinical Pharmacology, Therapy and Toxicology* 24: 651–654.
Shahidi, F. 2009. Launch of *Journal of Functional Foods*. *Journal of Functional Foods* 1: 1.
Sharma, P.V. 2000. Dravyaguṇa (pharmacology). In *Essentials of Ayurveda*, 17–33. Delhi: Motilal Banarsidass.

Shoskes, D., Lapierre, S., Cruz-Correa, M., Muruve, N., Rosario, R., Fromkin, B. 2005. Beneficial effects of the bioflavonoids curcumin and quercetin on early function in cadaveric renal transplantation: A randomized placebo-controlled trial. *Transplantation* 80: 1556–1559.

Sierksma, A., Westrstrate, J.A., Meijer, G.W. 1999. Spreads enriched with plant sterols, either esterified 4,4-dimethylsterols or free 4-desmethylsterols, and plasma total and LDL-cholesterol concentrations. *British Journal of Nutrition* 82: 273–282.

Sneader, W. 2000. The discovery of aspirin: A reappraisal. *British Medical Journal* 321: 1591–1594.

Sobolik, K.D. 1994. Paleonutrition of the lower Pecos region of the Chihuahuan desert. In *Paleonutrition: The Diet and Health of Prehistoric Americans*, ed. K.D. Soboilik, 247–264. Carbondale, IL: Southern Illinois University.

Sobolik, K.D. 1996. Nutritional constraints and mobility patterns of hunter-gatherers in the northern Chihuahuan desert. In *Case Studies in Environmental Archaeology*, ed. E.J. Reitz, L.A. Newsom, and L.J. Scudder, 195–214. New York: Plenum Press.

Soni, K.B., Kuttan, R. 1992. Effect of oral curcumin administration on serum peroxides and cholesterol levels in human volunteers. *Indian Journal of Physiology and Pharmacology* 36: 273–275.

Stahl, W., Sies, H. 1999. Carotenoids: Occurrence, biochemical activities and bioavailability. In *Antioxidant Food Supplements in Human Health*, ed. L. Packer, M. Hiramatsu, and T. Toshikawa, 183–202. San Diego, CA: Academic Press.

Stough, C., Lloyd, J., Clarke, J., et al. 2001. The chronic effects of an extract of *Bacopa monniera* (Brahmi) on cognitive function in healthy human subjects. *Psychopharmacology* (Berlin) 156: 481–484.

Stough, S., Downey, L.A., Lloyd, J., et al. 2008. Examining the nootropic effects of a special extract of *Bacopa monniera* on human cognitive functioning: 90 day double-blind placebo-controlled randomized trial. *Phytotherapy Research* 22: 1619–1634.

Sumner, J. 2000. A brief history of medicinal botany. In *The Natural History of Medicinal Plants*, 15–38. Portland, OR: Timber Press Inc.

Sun, A.Y., Simonyi, A., Sun, G.Y. 2002. The "French Paradox" and beyond: Neuroprotective effects of polyphenols. *Free Radical Biology and Medicine* 32, 314–318.

Swan, C. 2006. The uses of realism in early modern illustrated botany. In *Visualizing Medieval Medicine and Natural History, 1200–1550*, ed. J.A. Givens, K. Reeds, and A. Touwaide, 240–243. Aldershot: Ashgate Publishing.

Tekol, Y. 2007. The medieval physician Avicenna used an herbal calcium channel blocker, *Taxus baccata* L. *Phytotherapy Research* 21: 701–702.

Tirunanam, S. 1997. *Mūlikai Maruttuvam*, 1–237. Tiruchirapalli: Selvi Pathippakam.

Vaidyan, P.M.G. 1990. *Aṣṭāṅgahṛdayam*. Alleppey: Vidyarambham, *Sūtrasthāna*, Chapter 5, verses 38–39, p. 79.

Venkatraman, R. 1990. Features of the Tamil Siddha cult. In *A History of the Tamil Siddha Cult*, 76–165. Madurai: Ennes Publications.

Verma, R.L., Keswani, N.J. 1974. Unani medicine in medieval India—Its teachers and texts. In *The Science of Medicine and Physiological Concepts in Ancient and Medieval India*, ed. N.H. Keswani, 125–142. New Delhi: All India Institute of Medical Science.

Vincieri, F., Riva, A. 2003. ESCOP, the European Commission, consumer policy and health protection. In *Dietary Supplements of Plant Origin*, ed. M. Maffei, 25–30. London: Taylor & Francis.

Wang, Q., Xu, J., Rottinghaus, G.E., Simonyi, A., Lubahn, D., Sun, G.Y., Sun, A.Y. 2002. Resveratrol protects against global cerebral ischemic injury in gerbils. *Brain Research* 27: 439–447.

Warrier, P.K., Nambiar, V.P.K., Ramankutty, C. 1994. *Indian Medicinal Plants—A Compendium of 500 Species*, Vol. 1, 1–420. Hyderabad: Orient Longman.

Warrier, P.K., Nambiar, V.P.K., Ramankutty, C. 1996a. *Indian Medicinal Plants—A Compendium of 500 Species*, Vol. 3, 1–423. Hyderabad: Orient Longman.

Warrier, P.K., Nambiar, V.P.K., Ramankutty, C. 1996b. *Indian Medicinal Plants—A Compendium of 500 Species*, Vol. 5, 1–592. Hyderabad: Orient Longman.

Warrier, P.K., Nambiar, V.P.K., Ramankutty, C. 1997a. *Indian Medicinal Plants—A Compendium of 500 Species*, Vol. 2, 1–416. Hyderabad: Orient Longman.

Warrier, P.K., Nambiar, V.P.K., Ramankutty, C. 1997b. *Indian Medicinal Plants—A Compendium of 500 Species*, Vol. 4, 1–444. Hyderabad: Orient Longman.

Weiss, G., Weiss, S. 1992. Healing gardens and herbal medicine in history. In *Growing and Using the Healing Herbs*, 1–14. Avenel, NJ: Wing Books.

Wink, M. 1998. A short history of alkaloids. In *Alkaloids—Biochemistry, Ecology, and Medicinal Applications*, ed. M.F. Roberts and M. Wink, 11–44. New York: Plenum Press.

Zhou, Y.P. 1998. Introduction. In *Chinese Materia Medica—Chemistry, Pharmacology and Applications*, 1–9. Amsterdam: Harwood and Academic Publishers.

2

Bioactives from Plants

Except during the nine months before he draws his first breath, no man manages his affairs as well as a tree does.

George Bernard Shaw (1856–1950)
Maxims for Revolutionists

2.1 Amazing chemical laboratories

Plants are amazing laboratories of chemical synthesis. When supplied with the simplest and least expensive starting materials, such as carbon dioxide, sunlight, water, and minerals, they synthesize thousands of chemical compounds of differing molecular structures. A group of synthetic chemists provided with the same raw materials can never dream of achieving the same results in their lifetimes.

Some of the chemical compounds produced by plants for their basic metabolic needs are the same chemical compounds found in the animal and microbial worlds—amino acids, sugars, nucleic acids, lipids, and so on. They are products of the primary metabolism or chemical transformations fundamental to the existence of the plant. Nevertheless, plants differ from other life-forms in the great diversity of the additional chemical products that they produce. So far, about 50,000 distinct chemical compounds have been identified in the plant kingdom. These are products of the secondary metabolism (Ohlrogge and Chrispeels, 2002).

These secondary metabolites are only produced in specific types of cells and at specific times. They are broadly classified into various classes such as lipids, phenols, tannins, proteins, alkaloids, glycosides, volatile oils, resins, resin combinations, balsams, and mucilage (D'Amelio, 1999). Brielman (1999) provides a detailed classification of these phytochemicals.

2.2 Biogenesis of secondary metabolites

Plants utilize very specific enzymes to synthesize a multitude of compounds. These enzymes catalyze reactions in a well-defined sequence of steps. The biochemical sequence of the reactions involved in the synthesis of one compound is called *a metabolic pathway*. A few enzymes may be the components of a metabolic pathway, an example being the synthesis of starch from adenosine diphosphate. The three enzymes that take part in this bioconversion are adenosine diphosphate glucose pyrophosphorylase, starch synthase, and starch-branching enzyme (Martin and Smith, 1995). On the other hand, several enzymes can be present in a metabolic pathway, as exemplified by the synthesis of gibberellins from mevalonic acid (Yamaguchi, 2008). Enzymes are also found to catalyze reactions underlying the degradation and storage of the compound in question. The content of a given compound in a plant is regulated by the enzymes involved in the various metabolic pathways pertaining to that compound.

Plants, microorganisms, and animals have many unique metabolic pathways. However, the pathways common to these three forms of living organisms are the pentose phosphate pathway, glycolysis, and Krebs cycle. Plants synthesize all of the secondary metabolites using sugars, acetyl coenzyme A (CoA), and amino acids, which are the basic components of the primary metabolism. It is amazing that plants can convert carbon dioxide into such an array of complex compounds just by utilizing the energy of the sun, minerals, and water.

The biosynthetic pathway of monoterpene olefins and abietic acid explains how plants synthesize complex molecules from simple precursors. The starting material in this case is acetyl CoA, which is converted to geranyl pyrophosphate via mevalonic acid, isopentyl pyrophosphate, and dimethylallyl pyrophosphate. The monoterpene olefins α-pinene, β-pinene, 3-carene, β-phellandrene, and limonene are synthesized from geranyl pyrophosphate, catalyzed by monoterpene cyclases (Funk et al., 1994).

Alternatively, geranyl phosphate is transformed into farnesyl pyrophosphate and geranyl geranyl pyrophosphate, which is enzymatically converted to abietadiene, abietadienol, abietadienal, and abietic acid. These four enzymatic steps involve one cyclase, two hydroxylases, and a dehydrogenase (Funk et al., 1994).

2.3 Functions of secondary metabolites

2.3.1 As sources of metabolic energy

Plants devote much of their energy to the synthesis of pools of compounds that store the energy of the sun. Each of these stored metabolic products is broken down by specific enzymes when there is a need for energy, such as during the night, when sunlight is not available. Sugars, acetyl CoA, and amino acids, which are the basic units of these compounds, enter the primary metabolism of the plant and generate adenosine triphosphate (ATP).

2.3.2 Structural support

The bulk of plant biomass is made up of polysaccharides such as cellulose, hemicellulose, and pectins. These compounds impart strength to the plant cell wall, which is further strengthened by lignification and solidification, thereby turning the plant into a kind of raised framework on which are hung photosynthetic tissues such as leaves and stems in the most suitable orientation to trap sunlight and carbon dioxide. Terrestrial plants support the weight of their leaves with this structural support. Aquatic plants do not face this problem and they do not produce support compounds such as lignin (Cseke and Kaufman, 1999).

2.3.3 Pools of genetic information

The ability to transfer to its progenies the information for the production, localization, and function of all proteins in the body of an organism decides its survival in an environment. This genetic information mutates over time and this change in genetic information (evolution) is a never-ending process that produces new combinations. Some of these combinations may work in a given environment and others may not. Only those individuals that have a genetic combination that works in the new environment will survive. Others fade away from existence. Therefore, over time, plants have evolved several methods of sexual reproduction that facilitate the sharing of genetic information among individuals of a species. Thus, combinations of efficient enzymatic reactions are spread throughout a population (Cseke and Kaufman, 1999).

2.3.4 Keeping predators at bay

Plants arm themselves with chemical compounds to ward off insect predators, herbivores, and pathogenic microorganisms. One way of ensuring effective deterrence is through structural defense, making use of lignification, silicification, callose formation, and the deposition of wax.

The second approach is to employ species-specific chemicals that disrupt at least one critical biochemical pathway in the attacking organism. Examples of such compounds are alkaloids, proteins (lectins), saccharides and polysaccharides, flavonoids, terpenoids, cyanide-releasing compounds, organic acids, and long-chain carbon compounds (Cseke and Kaufman, 1999). Gums and mucilage are highly branched heteropolysaccharides. These substances are difficult for animals to digest. One advantage of producing such indigestible polymers is that the animal will not find the plant appealing. Like most pharmacologically active secondary metabolites of plants, the tetrahydrocannabinol in *Cannabis sativa* is assumed to be involved in self-defense, perhaps against herbivores (Pate, 1994).

Pine trees produce oleoresin, which is a mixture of turpentine and diterpene resin acids (rosin). This is secreted in response to physical wounding or an attack by fungi and insects. The wounded areas on the bark are filled with the oleoresin from which turpentine evaporates, leaving behind the solidified rosin that seals the wound (Johnson and Coroteau, 1987).

2.3.5 Attraction and deterrence of pollinators

The constituents of essential oils attract insects for pollination. For example, flowers containing linalool-rich essential oils attract moths during the night. Some species that may bear similarly colored flowers and grow in the same area do not produce linalool and do not attract moths. These flowers are pollinated by butterflies and bees during the daytime (Raguso et al., 1996). Limonene, geraniol, vanillin, and eugenol are also known to attract pollinators (Larcher, 1995; Theis and Lerdau, 2003; Dudareva and Negre, 2005; Cseke et al., 2007).

Pollinators are attracted to brightly colored flowers. The color of flowers can be due to carotenoids, phlobaphenes, flavonoids, betacyanins, and betaxanthines. Moths are attracted to lightly colored flowers, such as white and yellow flowers, as they are more visible at night. Birds are generally attracted to red, while bees do not visit red-colored flowers. Some anthocyanins are feeding deterrents and they are also believed to offer the plant protection from ultraviolet (UV) radiation (Holton and Cornish, 1995). When fruits ripen, chlorophyll pigments break down and reveal the carotenoid pigments. This change in color attracts animals.

Plants produce many storage forms of lipids, for example, those found in fruits such as avocado and olives and seeds such as sesame, sunflower, and peanuts. Some of them may serve as a reward for animals that help in the propagation of these plants. The carotenoid pigments seen on fruits attract animals, which help in seed dispersal. Many fruits also produce odoriferous monoterpenes that attract animals, the sugar stored in the fruit serving as a reward to them (Cseke and Kaufman, 1999).

2.3.6 Allelopathic action

Some plants disseminate into soils chemicals that are harmful to other plants or that prevent others from becoming established in their vicinity. Such chemicals are called *allelopathic chemicals*. A classic example of such a compound is naphthalene glucoside, which is produced by the leaves and roots of the walnut (*Juglans regia*). This glucoside itself is not allelopathic. It must undergo hydrolysis and oxidation by soil microbes resulting in juglone, which prevents germination of many plants. Other examples include essential oils, phenolic compounds, alkaloids, and steroids (Cseke and Kaufman, 1999).

2.3.7 Attraction of symbionts

Bacteria and fungi are often found to be very efficient in the absorption or production of many nutrients that are required by plants. There are many instances of a symbiotic relationship between plants and fungi or bacteria. Many species of the nitrogen-fixing bacterial genus *Brachyrhizobium* associate themselves with the roots of host plants and cause the development of root nodules where they reside, absorbing atmospheric nitrogen and storing it as reduced nitrogen in the form of NH_4. The reduced nitrogen is absorbed by the host plant and utilized for synthesizing amino acids. The roots of soybeans produce large amounts of the flavonoid daidzein, which is a signal for the bacteria to find the roots and establish a symbiotic relationship. Such symbiosis with bacteria is found in the deciduous Alder tree (*Alnus rubra*) and some species of grasses. An association with mycorrhizal fungi is seen in the root system of the pine tree. All these instances of plant–microbial symbiosis are triggered by specific chemical signals produced by the host plant (Cseke and Kaufman, 1999; Brundrett, 2002; Gianiazzi-Pearson, 2002).

2.4 Variation in chemical content of herbs

Biosynthetic pathways determine the production of secondary metabolites that are controlled by genetic and ontogenic factors.

2.4.1 Genetic factors

Genetic factors are responsible for the underlying variability of individual organisms. A good example is the opium poppy. The poppy exhibits a wide variation in its morphological characteristics such as flower shape, color of petals, shape and serration of leaves, and size and color of capsules and seeds. A significant variation is also observed in characteristics such as the height of the plant and the quantity of latex. The content of morphine in poppies can vary from 3% to 12% (Vanhaelen et al., 1991; Anonymous, 2004).

The slow genetic changes that take place in plants are enhanced under certain conditions to produce polyploidy or an increase in chromosome number. Polyploidy is found to be favored in perennial plants, which have a vegetative means of propagation and grow in wet soils or meadows. The incidence of polyploidy is found to vary with latitude (Grant, 1981; Hieter and Griffith, 1999). Das and Mallick (1991) carried out a comparative chemical analysis of 10 ecotypes of *Centella asiatica* to correlate their asiaticoside content with their genomic diversity. Higher amounts of asiaticoside were observed in two ecotypes collected from the subtemperate Himalayan region (northern Uttar Pradesh and Meghalaya). Two newly reported B-chromosomes were noted in these two races besides their normal karyotype of $2n = 18$ chromosomes.

2.4.2 Chemical races

From a detailed biochemical analysis of plants, it has been observed that many plants of the same species, having the same morphological characteristics, have different chemical constituents. Such plants are called *chemical varieties*, *chemotypes*, *chemodemes*, or *chemical races*. There are four chemotypes in *Acorus calamus*, which is widely used in traditional medicine. One of them does not contain the known carcinogen isoasarone. Similarly, seven chemotypes of thyme are known in the Mediterranean region alone (Bruneton, 1995). Chemotypes are indicated by the name of the compound most characteristic for that particular race, for example, *Thymus vulgaris* linalool and *T. vulgaris* thymol (Pengelly, 2004).

2.4.3 Hybridization

Hybrids of plants can be formed by a natural combination of the characteristics of related species. Theuns et al. (1991) have discussed this aspect in depth and have observed that the Turkish *Papaver bracteatum* and *P. orientale*, growing in isolated, remote locations, developed hybrids naturally. Nowadays, many of the ayurvedic herbs are collected from remote, wild areas of India and it is quite possible that some of them might exist as hybrids.

2.4.4 Ecological factors

Several ecological factors can influence the quality of medicinal plants by influencing the biomass or the amount of dry matter produced, the ratio of the different plant parts, and the level of active substances. It is the location of the plant (wild or cultivated) that determines the climate and thereby the chemistry.

2.4.4.1 Light

Cosson (1966, 1969, 1976, 1981) and Cosson et al. (1966, 1978) studied the influence of several ecophysiological factors on alkaloid production in *Datura* species. They simulated climatic conditions using the phytotron at the Centre National de la Recherche Scientifique, Gif-sur-Yvette, France. Light could modify the general metabolism of *Datura metel* and *D. tatula* with a consequent alteration in the production of scopolamine. These observations were confirmed by Nandi and Chatterjee (1976).

The stimulatory effect of light on alkaloid biosynthesis in *Papaver somniferum* has been reported by Mika (1955), who recorded a high morphine content in capsules 98 days after germination with a day length of 18 h.

2.4.4.2 Temperature

A rise in the environmental temperature can increase the general metabolism by increasing the level of active constituents. However, excessive heat can cause a loss of volatile oil from leaf surfaces. The linseed oil from plants grown in cooler climates has a high level of unsaturated fatty acids, whereas the linseed oil from plants grown in warmer areas is poor in these compounds. Mika (1955) reported a higher alkaloid content when *P. somniferum* plants were grown at 100°C than at 210°C.

2.4.4.3 Altitude

The alkaloid content of plants can be influenced by altitude, as is the case with *D. metel.* Total alkaloids were relatively lower at sea level than at high altitude (Karnick and Saxena, 1970). However, an opposite effect on alkaloids was reported earlier by Pelt et al. (1967a, 1967b). Planting density also influences the growth of herbs under cultivation. This is probably due to the modification of these plants' exposure to light (Yamada et al., 1983). Plants rich in anthracene derivatives, especially rhubarb, seem to be richer in active constituents when cultivated in high altitudes (Lebeau and Janot, 1956). However, it is also possible that the effect of altitude on plants can be species specific.

2.4.4.4 Latitude

The biosynthesis of many secondary metabolites in several plant species is influenced by light intensity and photoperiod. The long days and cool night temperatures of the northern latitudes reportedly increase the production of aromatic compounds compared with the same plant species in the south (Davik et al., 2006). The carotene, ascorbic acid, and sugar content and the flavor and aroma of some vegetables are more pronounced in the northern latitudes compared with plants grown in southern parts (Hårdh et al., 1977). Similarly, the red color (due to anthocyanin content) of strawberries is more intense in the north than in the south (Hårdh and Hårdh, 1977; Jaakola and Hohtola, 2010).

2.4.4.5 Water

Plants receive water as rain, dew, humidity, or irrigation. Water stress affects secondary metabolism. The content of alkaloids in the leaves of *D. metel* was found to decrease with rain (Gupta et al., 1974). Similar observations have been reported for some other solanaceous plants such as *Datura stramonium*, *Hyoscyamus niger*, and *Duboisia myoporoides* (Vanhaelen et al., 1991).

A report on the possible effect of rainfall on the content of secondary metabolites was published by Hofman and Menary (1979). High rainfall and high relative humidity encouraged heavy fungal growth and a 28% decrease in the dry weight of capsules of *P. somniferum*, leading to a 35% loss in morphine content.

2.4.4.6 Soil chemistry

Soil should be free from harmful levels of pathogenic organisms, pests, and toxic compounds. It has been observed that a fungal disease of the opium poppy can result in a 10% decomposition of alkaloids in just 24 h. Similarly, fungal blight on *Solanum laciniatum* causes a 20% loss of solasodine, which is used as the starting material in steroid hormone synthesis.

Soil chemistry can easily be altered by man. Therefore, soil chemistry is the most important factor that influences the production of secondary metabolites. Sodium potassium, calcium, sulfur, and phosphorus are generally known as macronutrients and soil usually receives them through fertilizers. The requirements of each species need to be studied and standardized, as plants vary in their macronutrient requirement.

Elements such as boron, zinc, copper, cobalt, iron, magnesium, and so on are required only in trace amounts and are therefore called *micronutrients*. The addition of extra amounts of these micronutrients does not have harmful effects. However, their deficiency can cause serious repercussions on the plant's chemical constitution.

Chatterjee et al. (1984) reported that manganese, molybdenum, magnesium, and boron could significantly enhance essential oil production in *Cymbopogon winterianus*. Brachet et al. (1981), on the other hand, observed that the tropane esters in *Datura innoxia* were very much sensitive to sodium chloride (NaCl) stress. Thus, monitoring of the soil chemistry is important in the cultivation of medicinal plants.

2.4.5 Ontogenic stage of harvest

The stage of growth at which a plant is harvested has a bearing on its chemical constituents. Therefore, a plant needs to be harvested at the right ontogenic stage when the desired active compounds are present in significant amounts. For example, the roots of *Withania somnifera* have a fairly high content of withanolides when they are harvested from plants that are 6–7 months old and the bark of *Holarrhena antidysenterica* has a high content of alkaloids when the tree is in full bloom (Tyagi, 2005). The vasicine content of the *Adhatoda vasica* leaves is highest at full bloom stage and the xanthotoxine content of *Heracleum candicans* is highest when the aerial portions dry up (Shiva and Mahtolia, 1998). *Ocimum sanctum* is to be harvested at full bloom stage (Bahl et al., 2000).

Schratz and Qadry (1966) reported that during the fruit development of coriander, borneol and aldehydes (trans-2-tri-decanal and n-decanal) diminished, whereas linalool, geranyl acetate, and geraniol increased. Once ripened, the fruit will have low aldehyde and borneol content, but a maximum content of linalool, which is responsible for the characteristic aroma of coriander. The right time to harvest is judged by the color of the umbels. *Terminalia chebula* fruits are to be collected near maturity, but before they turn yellow, and *Aegle marmelos* fruits are to be harvested at full ripe stage (Shiva and Mahtolia, 1998).

Very often, the stage of collection is unknown in the case of wildcrafted herbs or herbs collected from the wild. Similarly, the young leaves of peppermint contain the toxic monoterpene pulegone, while the older leaves contain the more desirable menthol and menthone (Rabak, 1917; Maroti et al., 1993; Scavroni et al., 2005). Likewise, the morphine content of the opium poppy is highest 2–3 weeks after flowering. In contrast, the solasodine content of many solanaceous berries disappears once the berries mature. Commonly, leaf drugs are collected as the flowers begin to open, the flowers are harvested just before they are fully open, and rhizomes and roots are collected as soon as their aerial parts wither away.

2.4.6 Cultivation

Considering the continuous and uniform supply of medicinal herbs and the fast depletion of natural resources, the cultivation of medicinal plants is considered to be an important way to meet the growing demand (Uniyal et al., 2000). However, not many species are cultivated. One reason is that cultivated plants are sometimes considered to be of inferior quality when compared with specimens collected from the wild (Schippmann et al., 2006). Wild ginseng roots are 5–10 times more valuable than the roots produced by cultivation (Robbins, 1998). In Botswana, local healers do not favor cultivated herbs, as they are believed to lack the potency of the material collected from the wild (Cunningham, 1994).

The medicinal properties of plants are due to the secondary metabolites needed by them to counter stress and competition in their natural environment. These compounds can be present in lesser amounts in monoculture conditions. While photosynthesis appears to be high in open conditions, enzymatic reactions that decide the quality of the herb seem to be a function of the environment (Joy, 2003). Active chemical constituents can be lower in fast-growing cultivated plants, while their wild counterparts can be older due to slow growth rates. They may have higher levels of active compounds (Schippmann et al., 2006).

Palevitch (1991) and Uniyal et al. (2000) opine that certain values can be deliberately enhanced in plants under controlled conditions of cultivation. The roots of the *Valeriana wallichii* were reported to have higher contents of valepotriates and patchouli alcohol than those reported in wild plants. There was also a marked difference in ar-curcumene, β-patchlouene, and γ-patcholuene between the cultivated and the wild plants (Singh et al., 2000).

2.4.7 Postharvest treatment

Various biochemical processes in plants continue for some time, even after the plants have been harvested. There is no control over this postharvest enzymatic activity, which may or may not be desirable. Respiration is a very active process in the living cells of freshly harvested herbs. As respiration proceeds, it releases energy that results in the heating up of the stacked crop. Fermentation also takes place. As water from the herb escapes, wilting and shriveling occur. The end result of all these reactions is the breakdown of the secondary metabolites in the herb (Boettcher and Guenther, 2005a).

In many plants, the desired compounds are formed during storage after harvest. Freshly picked vanilla beans do not have any vanillin, whereas the fermentation of the pods causes its production, involving the enzymatic hydrolysis of a glycoside. Wild cherry bark and bitter almonds derive their medicinal properties from the formation of hydrogen cyanide (HCN) and benzaldehyde produced as a result of the enzymatic cleavage of cyanogenetic glycosides. The laxative property of the senna leaf and pod is due to the presence of sennosides. Recent studies using radioimmunoassay have shown that sennosides are only formed when the leaf and pod are dried gradually at room temperature. It is known that the aromatic quality of cocoa beans is related to the cultivar, postharvest treatments (fermentation, drying, and storage), soil type, climate, geographical area of cultivation, and stage of maturity of the pod (Kattenberg and Kemmink, 1993).

Damaging enzymatic activity can be prevented by drying, as most of these enzymes are hydrolases, requiring moisture for their activity. Plants can contain up to 95% of water, as is the case with Brahmi (*Bacopa monnieri*). Thus, the removal of water from harvested herbs ensures that hydrolytic reactions do not occur. The rapid drying of the harvested material denatures enzymes and prevents decay of the herb, which can encourage the growth of microorganisms. Dry herbs having a moisture content of 5%–10% can be safely stored as enzyme activity and microbial growth do not take place at such a level.

2.4.8 Storage of herbs

From a scientific point of view, the storage of medicinal herbs is undesirable. Nevertheless, it is unavoidable. Dried herbs are very hygroscopic and can absorb up to 15% of moisture. This moisture can activate enzymes that are dormant in the stored herb with harmful effects on its chemical constituents. Moisture can cause oxidation reactions, especially of volatile oils. Lipids can go rancid as well.

Moisture encourages the growth of microorganisms in a short time. The metabolism of bacteria and fungi releases increasing amounts of moisture for microbes such as *Fusarium* and *Rhizopus*, which demand greater dampness for their growth. The microbial proliferation thus continues in the form of a cascade. The waste products emanating from these microbes render the stored herb damp or musty smelling, which are telltale signs of inferior quality. Additionally, there is also the possibility of the herb getting contaminated with mycotoxins (Boettcher and Guenther, 2005b).

Stored herbs are also prone to attack by insects, which damage the material as well as pollute it with their excreta and exoskeletons. Though described in the context of the storage of the German chamomile (*Matricaria chamomilla*), the desirable storage conditions and the characteristics of the storage room listed by Boettcher and Guenther (2005b) are equally relevant to any other herb.

Many natural products are unstable in light, especially the indole alkaloids (reserpine) from *Rauwolfia serpentina* and the ergot alkaloids (ergotamine, ergometrine). The nonnitrogenous compounds of *Cannabis sativa* are photosensitive. Tetrahydrocannabinol, a major compound, literally disappears after 2 years of careless storage (Sandberg and Corrigan, 2004).

2.5 Good cultivation practices

To encourage the scientific cultivation of medicinal plants, the European Herb Growers and Producers Association (EUROPAM) has developed *Guidelines for Good Agricultural and Wild Collection Practice (GACP) of Medicinal and Aromatic Plants* (Anonymous, 2006). Also, the WHO has published *WHO Guidelines on Good Agricultural and Collection Practices (GACP) for Medicinal Plants* (Anonymous, 2003b). These documents cover seeds and propagation material, cultivation, harvesting, drying, packing, storage, transportation, personnel, equipment, and documentation to ensure that medicinal plants are cultivated, processed, and stored with minimal negative impact.

2.6 Modern methods of extraction of bioactives

2.6.1 Essential oils

Plants contain volatile and nonvolatile secondary metabolites. The volatile compounds collectively form essential (volatile or ethereal) oils. They are invariably aromatic. They are called essential oils as they are an essence of the plant. Species belonging to Apiaceae, Asteraceae, Brassicaceae, Lamiaceae, Lauraceae, Myrtaceae, Pinaceae, Piperaceae, Rutaceae, and Zingiberaceae generally contain high amounts of essential oils (Pepeljnjak et al., 2003). About 3000 essential oils are known and among them 300 are important in the pharmaceutical, agronomic, food, sanitary, cosmetic, and perfumery industries (Bakkali et al., 2008).

Chemically, essential oils are mixtures of many different compounds. Some contain up to 60 compounds. In essential oils, two or three compounds will be present in high concentrations (20%–70%) and the others in smaller amounts. For example, carvacrol (30%) and thymol (27%) are the major components of *Origanum compactum* essential oil. Usually, these major compounds decide the biological activity of the essential oil. The chemical constituents of essential oils can be identified by thin layer chromatography, gas chromatography, and gas chromatography–mass spectrometry. The component chemicals of essential oils fall into the categories of terpenes (monoterpenes, diterpenes, triterpenes, sequiterpenes), phenylpropanes (aldehydes, phenols, phenylethers), sulfides, derivatives of anthranilic acid, and indole and acetylene derivatives (Pepeljnjak et al., 2003; Bakkali et al., 2008).

Though essential oils are produced from an endemic population, their quality differs for several reasons, including genetic, physiological, and environmental factors and processing conditions (Rohloff, 2003) (**Table 2.1**).

Essential oil-bearing plants show infraspecific and intravarietal differences in both their morphology and chemical structures. These form the basis for determining the important, chemically defined populations or chemotypes (Hay and Svoboda, 1993). **Table 2.2** gives some of the main chemotypes of important aromatic herbs.

There is no correlation between their morphological and chemical characters (Franz, 1993). This makes it difficult to characterize chemotypes on a phenotypic basis. Environmental factors exert an influence on chemotypic differentiation. Massoud (1989) reported the varying oil yield and content of bisabolol and chamazulene in chamomile and Franz (1993) has described the geographical distribution of chemotypes of chamomile throughout the Mediterranean regions. It is now believed that changes in temperature and light conditions during a period of the day may be responsible for causing distinct variations in the yield and composition of essential oils (Fahlen et al., 1997; Silva et al., 1999; Rodrigues et al., 2002).

Table 2.1 Factors Influencing the Composition of Commercial Essential Oils

Factor	Description
Genetics	
Taxon	Clone, hybrid, cultivar, population
Infraspecific	Chemotypes (distinct populations within a species)
Physiology	
Ontogenetic	Developmental changes (vegetative, generative)
Plant part	Morphological differences (root, leaf, flower, seed)
Environment	
Climate	Light, temperature, edaphic factors
Origin	Latitude, height above mean sea level, country, continent
Agriculture	Cultivation technique, fertilizer, irrigation, harvest time
Processing	
Isolation	Distillation, extraction, maceration, pressing, enfleurage
Storage	Effects of aging, storage temperature, relative humidity, etc. on raw material
Adulteration/standardization	Essential oil blending prior to distillation/directly into the essential oil

Source: Reproduced with permission from Rohloff, J., Cultivation of herbs and medicinal plants in Norway—Essential oil production and quality control. PhD thesis, Norwegian University of Science and Technology, Trondheim, 2003. With permission.

Table 2.2 Chemotypical Variation of some Economically Important Aromatic Herbs and Spices

Botanical Name	English Name	Main Chemotype
Achillea millefolium L.	Yarrow	(pro) Azulene, azulene-free, caryophyllene, germacrene D, farnescene
Artemisia vulgaris L.	Common wormwood	α-/β-Thujone, 1,8-cineole, linalool, wormwood, camphor
Anethum graveolens L.	Dill	Carvone, limonene, phellandrene
Carum carvi L.	Caraway	Carvone, limonene
Chamomilla recutita L.	Chamomile	Bisabolol, bisabololoxide A/B, bisabolonoxide A
Cuminum cyminum L.	Cumin	Cuminaldehyde, pinene
Foeniculum vulgare L. *var. dulce*	Sweet fennel	Anethole, fenchone, limonene, estragole
Lavendula angustifolia Mill.	Lavender	Linalool, linalyl acetate
Melissa officinalis L.	Lemon balm	Geraniol, citronellol
Mentha piperita L.	Peppermint	Menthol, carvone, limonene, linalool
Mentha spicata L.	Spearmint	Carvone, menthone, piperitone
Ocimum basilicum L.	Sweet basil	Estragole, linalool, eugenol, camphor, methyl cinnamate
Origanum majorana L.	Marjoram	Terpineol, sabinene hydrate
Origanum vulgare L.	Oregano	Carvacrol, thymol
Rosmarinus officinalis L.	Rosemary	Verbenone, 1,8-cineole, α-pinene
Salvia officinalis L.	Dalmatian sage	α-/β-Thujone, thujone-free, 1,8-cineole, camphor
Thymus vulgaris L.	Thyme	Thymol, carvacrol, linalool, geraniol, thujanol, terpineol
Valeriana officinalis L.	Valerian	Valeranone, valeranal, cryptofauronol

Source: Reproduced with permission from Rohloff, J., Cultivation of herbs and medicinal plants in Norway—Essential oil production and quality control. PhD thesis, Norwegian University of Science and Technology, Trondheim, 2003. With permission.

Essential oils have antibacterial, antifungal, antiviral, spasmolytic, antiphlogistic, diuretic, cytotoxic, antimutagenic, and antihelmintic properties (Pepeljnjak et al., 2003; Bakkali et al., 2008). Recently, they have also been shown to reduce anxiety and thereby improve the mood of humans (Kuriyama et al., 2005). Essential oils are extracted using several techniques.

2.6.1.1 Hydrodistillation

Most essential oils such as lavender, peppermint, eucalyptus, tulsi, vettiver, and so on are extracted by distillation. Fresh raw material is cut into pieces and put into a distillation flask containing a small amount of water. The flask is connected to a Clevenger apparatus, water circulation is started and the flask is heated, most often by a heating mantle. The steam generated inside the flask carries the essential oil, which condenses on passing through the condenser and flows into the Clevenger apparatus. The essential oil collected is drained as and when the collection chamber is full (Langenau, 1955). Detailed information on the distillation of essential oils is available in the voluminous work of Ernest Guenther (1955a).

2.6.1.2 Expression

Essential oils such as citrus oils are mechanically expressed. This method is feasible when the quantity of essential oil present is significant and the cost of the raw material is low. In a typical operation, thoroughly washed fruits are cut into halves and the juice is extracted mechanically. The leftover rind retains all the essential oil in the oil glands. The rind is then pressed in special oil presses. The resulting emulsion of water, detritus, and essential oil is passed first through a centrifuge running at a speed of 6,000–8,000 rpm and then through a clarifying centrifuge at a speed of 18,000 rpm. Clarifiers with a speed of 56,000 rpm are used to yield very clear oils (Anonymous, 2003a).

2.6.1.3 Solvent extraction

Almost all flowers contain essential oils. However, their quantities are small and they contain compounds that are very sensitive to heat. Such essential oils are extracted by solvent extraction, in which a low boiling point solvent such as hexane soaks the petals and dissolves the oil. Extracts from hexane and other hydrophobic solvents are known as *concretes*. Concretes also contain several nonfragrant waxes and resins. The essential oil portion of the concrete is later isolated using ethyl alcohol and the alcohol is finally removed by distillation. The fragrant and pure essential oil that has been isolated is called absolute (Hanna, 1999).

2.6.1.4 Enfleurage

Subtle scents that do not withstand harsh treatments are extracted by enfleurage. This method was pioneered in the French province of Grasse, the center of the French perfumery industry. Fresh flowers such as jasmine are layered between screens coated with odorless animal fat of a soft consistency. Enfleurage is carried out in cool cellars. The essential oils evaporate from the flowers and are dissolved in the animal fat. The essential oil–fat mixture is called pomade, examples being *pomade de Jasmin*, *pomade de tubereuse*, and so on. The pomade is later washed with alcohol to separate the essential oil from the fat (Hanna, 1999; Guenther, 1955b).

2.6.1.5 Supercritical fluid extraction

Supercritical fluid extraction is a new technology for extracting natural products including essential oils. Supercritical carbon dioxide is the most preferred solvent in this technology, as it is inert, noninflammable, noncorrosive, inexpensive, freely available, odorless, eco-friendly, and generally regarded as safe (GRAS). Its near ambient critical temperature makes it suitable for thermolabile natural products such as essential oils. With this technology there is no loss of the top notes or base notes of aroma compounds. There is no thermal degradation, and the products are of high purity. The problem of residual solvents does not arise, as carbon dioxide in the product evaporates completely at room temperature. Products have a longer shelf life on account of the co-extracted antioxidants and the operation is fast and economical (Shi et al., 2006) (see Chapter 4).

2.6.2 Solvent extraction of bioactives

Solvent extraction is a process that is used to recover a component from either a solid or a liquid, by keeping it in contact with a solvent that will dissolve the solute of interest. Distillation of the solvent separates the solute. Solvent extraction has several applications such as the extraction and refining of vegetable oils from oil seeds, the isolation of bioactive compounds from vegetable and animal matter, the removal of undesirable components from mixtures of organic products, the production of high-quality aromatic extracts required in the manufacture of synthetic fibers such as nylon, and the separation of metals from their ores. Due to environmental concerns, modern solvent extraction processes employ benign solvents (Wennersten, 2004).

2.6.2.1 Like dissolves like

Based on their ability to dissolve organic compounds, solvents can be broadly divided into polar (hydrophilic) and nonpolar (hydrophobic) types. The dielectric constant of a solvent is an indicator of its polarity and solvents having a dielectric constant of less than 15 are considered nonpolar. **Table 2.3** lists some of the commonly used solvents and their polarities.

The physical properties of a solvent, such as polarity, dipole movement, polarizability, and hydrogen bonding, decide which class of compounds it can dissolve and with which other solvents it is miscible. Generally, polar solvents dissolve polar compounds and nonpolar solvents dissolve nonpolar compounds. For example, strongly polar compounds such as sugars, saponins, phenolics, and tannins dissolve only in polar solvents such as water and methanol, whereas strongly nonpolar constituents such as oils, waxes, terpenoids, and glycolipids dissolve well in strongly nonpolar solvents such as hexane (Finar, 2001).

2.6.2.2 The chemistry of water extraction

Water is the ideal solvent for extracting several polar phytochemicals. Additionally, aqueous extracts are safer due to the absence of residual solvents and the extraction process is eco-friendly. Traditional medical systems employ many water extracts or decoctions. This class of medicines is prepared by boiling mixtures of crude herbs in water. During this process, many chemical compounds that are present in the crude herbs are transferred into the decoction.

Pharmaceutical chemists have always been interested in understanding the dynamics of this transfer of compounds. While studying the interaction of

Table 2.3 Commonly Used Solvents and Their Polarity

Solvent	Dielectric Constant	Boiling Point (°C)
Nonpolar Solvents		
Hexane	1.9	69
1,4-Dioxane	2.2	101
Benzene	2.3	80
Toluene	2.4	111
Diethyl ether	4.3	35
Chloroform	4.8	61
Ethyl acetate	6.0	77
Dichloromethane	8.9	40
n-Butanol	12.5	118
Polar Solvents		
Isopropyl alcohol	19.9	82
n-Propanol	20.3	97
Acetone	20.7	56
Ethanol	24.6	78
Methanol	32.7	65
Acetonitrile	37.5	82
Water	80.2	100

Source: Based on data from Smallwood, I.M., *Handbook of Organic Solvent Properties,* Arnold, London, 1996. With permission.

phytochemicals during the preparation of *Gegen Tang,* a decoction from Chinese medicine, Takaishi and Torii (1969) and Takaishi and Watanabe (1971) made a curious observation. Starch from the pueraria root (arrowroot starch) moved from the herb and dissolved in the decoction. This water-soluble starch formed a conjugate with ephedrine, the major alkaloid from the ephedra herb. It was observed that subsequent to the formation of this complex, the solubility of ephedrine in *Gegen Tang* increased 5–10 times. Yata and Tanaka (1983) reported that the saponin of bisdesmoside, which is freely soluble in water, increased the dissolution of monodesmosides, which are sparingly soluble in water. While working with the Kampo decoction *Maoo-to,* Noguchi et al. (1978b) noted that ephedrine, which was earlier detected in high quantities in the hot-water decoction of the ephedra herb alone, was later readsorbed when the ephedra was decocted in the presence of the other ingredients of *Maoo-to,* namely, licorice root, cinnamon bark, and armeniac seeds. The authors inferred that this readsorption was probably because of the presence of some "negative compounds" from the other herbs.

Chemical conversions are also reported to take place during the process of decoction. Saikosaponins a and d, which are constituents of the *Bupleurum* root, are converted into saikosaponin b (Arichi et al., 1979; Yamaji et al., 1984). Berberine-like alkaloids present in the *Coptis* rhizome or the *Phellodendrum* bark and glycyrrhizin present in licorice root cross-react and the resultant product precipitates (Tomimori and Yoshimoto, 1980; Noguchi et al., 1978a). A similar precipitation reaction occurs between alkaloids and tannins (Nakajima et al., 1994b).

It is generally believed that only compounds that are readily soluble in water are extracted in decoctions. However, a significant amount of water-insoluble compounds have also been shown to be transferred into decoctions (Arichi et al., 1979; Kano et al., 1986; Huang, 1986; Akahori and Kagawa, 1977). Nakajima and his group observed that 6,7-dimethylesculetin, a poorly water-soluble constituent of *Artemisiae capillaris spica*, is extracted in significant amounts into the Kampo formula *Inchinko-to* (Taguchi et al., 1975; Imazeki et al., 1980).

From their extensive studies on the transfer of the components of crude drugs into Kampo decoctions, Nakajima et al. (1994a, 1994b, 1994c) concluded that some of the ordinary low molecular weight components are transferred with dissolution in water, whereas others are transferred without dissolution. The relative solubility of these compounds in water, crude drug particles, insoluble substances, and the quantity of water used in a decoction are the factors that decide transfer with dissolution. On the other hand, transfer without dissolution is caused by adsorption and or partition of the compounds to the insoluble phase such as suspension or emulsion resulting from the deposition of other insoluble components and fine particles of the crude drugs. Transfer without dissolution can also happen with the formation of an emulsion or a suspension of the compounds themselves.

2.6.2.3 The various extraction modes

2.6.2.3.1 Soxhlet extraction The solvent extraction of biological materials began with the path-breaking invention by Franz Ritter von Soxhlet (1848–1926). Born in Moravia of Belgian parents, Soxhlet obtained doctorate degrees in chemistry (1872) from the University of Leipzig and medicine (1894) from the University of Halle. While studying the chemistry of milk, he invented the now-famous Soxhlet extraction apparatus to isolate lipids from milk (Soxhlet, 1879) (**Figure 2.1**).

The Soxhlet extractor (**Figure 2.2**) was originally designed to extract lipids from a solid matrix. Nevertheless, the extractor can be used to extract materials other than lipids. It is the equipment of choice when the desired compound has limited solubility in a solvent and its impurity is insoluble in that solvent. Solid material containing the substance to be extracted is put in a porous cellulose thimble and placed inside the main chamber of the Soxhlet extractor, which is then fitted to a flask containing the extraction solvent. A condenser is connected to the extractor.

Figure 2.1 *Franz Ritter von Soxhlet. (Reproduced with permission from www.cyber-lipid.org.)*

The flask with the solvent is heated and the solvent vapor moves into the condenser. It cools and trickles into the main chamber containing the material. The chamber is slowly filled with the warm solvent, which dissolves the desired substance present in the material. When the chamber is almost full, it is automatically emptied by a side arm and the solvent (menstruum) flows back into the distillation flask. This cycle is repeated till the extraction is completed. The advantage of the Soxhlet extractor is that instead of using several portions of a solvent, one batch of the solvent is recycled many times. On completion of the extraction, the solvent is removed from the extractive using a Büchi rotary evaporator (**Figure 2.3**).

In 1950, L.C. Craig invented a simple rotary evaporator system (Craig et al., 1950). This invention was first commercialized by the Swiss company Büchi Labortechnik AG in 1957 and patented in 1964 (Anonymous, 2009). The Rotavapor designed by Büchi continues to be the most widely used rotary evaporator.

The miscella (the solution of extract in the solvent) is poured into a flask that is lowered into a reservoir and is immersed in warm water of a desired temperature. The condenser is connected to a vacuum and a supply of chilled water. The flask containing the miscella is allowed to rotate at optimum speed. The hot water in the reservoir warms up the solvent, which evaporates, moves into the condenser aided by the vacuum and trickles down into a round-bottomed flask. Once the solvent is completely removed, the operation is stopped and the extract (solid or paste) is collected.

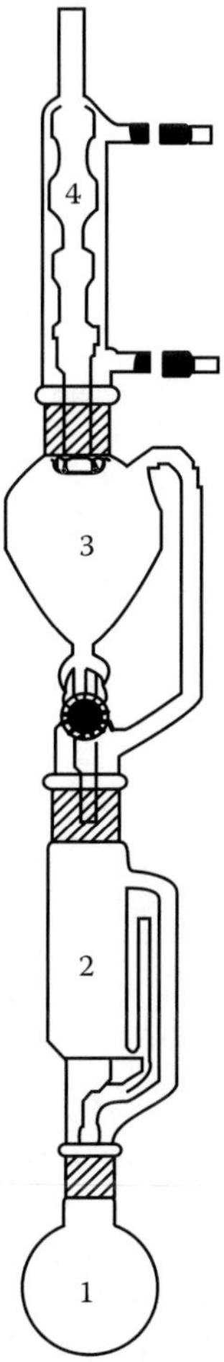

Figure 2.2 *The Soxhlet extractor. 1 = Flask with extraction solvent, 2 = extraction chamber, 3 = funnel with side arm, 4 = condenser. (Reproduced with permission from www.cyberlipid.org.)*

Based on the Soxhlet extractor, several solvent extraction systems are now available in the market. Many of them are automated or semi-automated and extract lipids efficiently.

2.6.2.3.2 Maceration This is the simplest procedure for preparing an extract. It is suitable for small and industrial quantities of a herb. A solvent is poured over powdered herb that has been placed in an appropriate vessel (extractor). After a specific period of time, the solvent will have absorbed a considerable amount of the active compounds. The miscella is now drained and the herb powder is washed with a fresh solvent. Maceration is repeated and the combined washings are used as the fluid extract or as the dry extract after the removal of the solvent. Maceration is the ideal method for preparing tinctures and the process can be static or dynamic.

Large, closed containers (extractors) are used in the industrial extraction of herbs using the maceration procedure. Powdered herb is loaded into the containers and is allowed to soak in the solvent for a stipulated period of time. After the required period of time, the miscella is drained or filtered. This type of extractor is mostly outdated now and has been replaced by extractors with built-in agitators. In the industrial version of dynamic maceration, the herb powder is loaded with solvent and the extractor is set in continuous motion

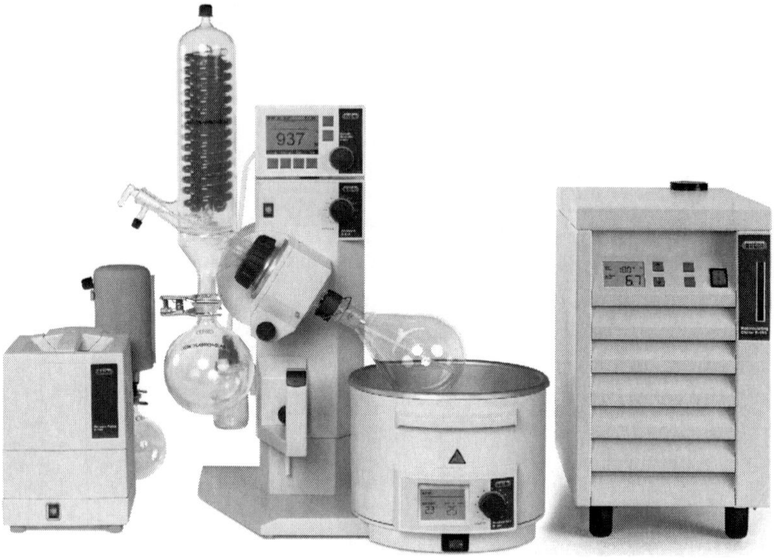

Figure 2.3 *Büchi rotary evaporator ROTAVAPOR R-210. (Photo courtesy of BUCHI India Pvt. Ltd.)*

for a specific period of time using agitators. After extraction, the miscella is drained through the aperture for this purpose.

In the present-day industrial maceration procedure, large extractors with mechanical agitators are employed. The extractors are fitted with jackets to facilitate the heating of the vessel using steam. Powdered herb is loaded into the extractor through a special aperture. Fresh solvent is pumped in and maceration under agitation is carried out for a certain length of time, whereafter the miscella is collected through a shaft at the bottom of the extractor. A fresh quantity of solvent is pumped in and the process is repeated till exhaustion of the herb is achieved. The solvent that is retained by the spent herb (marc) can be removed by heating followed by distillation.

2.6.2.3.3 Percolation In simple percolation, the herb powder is soaked well with the solvent and put in a vessel. After a specific period of time, the percolate (miscella) is drained and poured back into the vessel. The process is repeated till the herb is exhausted. To save solvent, sometimes only percolates that are less rich in extractives are poured back into the percolator.

In industrial-scale percolation, it is necessary to moisten the herb powder with a solvent before charging it into the percolator. This is to prevent a sudden swelling of the herb powder in the closed container. If water is the extraction medium, there is the possibility of the herb swelling to nearly three times its original volume causing the percolator to burst or making percolation impossible (Bombardelli, 1991). The uniform moistening of the herb powder prevents the formation of preferential channels, thereby increasing the passage of the solvent. As the porosity of the cell wall in the powder is increased, the extractive diffuses from the cell into the solvent (Bombardelli, 1991).

In industrial percolation, moistened herb powder is put in a series of percolators and the solvent is serially circulated through them. Fresh solvent is used when the herb powder reaches exhaustion stage. Fluid and soft extracts can be easily manufactured through this process. By calculating the number of percolators and the quantity of the solvent used, it is possible to prepare some extracts directly, avoiding the concentration of the extract step.

There are, however, two disadvantages. Firstly, the herb powder needs to be moistened before charging it into the percolators. Secondly, depending on the height of the percolator, the herb powder gets compacted after some time, preventing the passage of the solvent and thereby slowing down the extraction process.

2.6.2.3.4 Continuous extraction The concept of continuous extraction was conceived in Germany during the end of World War I. Faced with a shortage of oils and fats, the Germans explored better ways to make use of the soybeans that they had started to import from China. As a result, two continuous extractors based on countercurrent principles were invented. The Bollman or basket-type extractor was patented in 1919, followed by the Hildebrant or U-type extractor in 1934. Jean-Albert De Smet, a Belgian engineer, improved on these and, in 1946, he produced a patented version manufactured by his newly established company, Extraction Continue De Smet (De Smet, 1949; Anonymous, 1977). In continuous extraction, the herb powder moves against the solvent. It enters the percolator, coming into contact with solvent enriched with extractive and soon afterward encounters fresh solvent. Complete exhaustion of the herb is achieved by properly setting the movement of the herb powder toward the solvent. This method is very similar to continuous percolation. However, this extraction process is more suitable for the large-scale manufacture of extracts.

Extractors of the screw, carousel, U, or radial pressure type or their combinations are used in industrial continuous extraction. Screw extractors work on the principle of absolute countercurrent extraction, in which the herb powder and the solvent move in opposite directions.

The extractor is a long, pressure-tight, steel cylinder with provision for heating and cooling. Inside, it is equipped with a system of screw feeders. The herb powder, which is charged into the extractor through an aperture at one end of the cylinder, is driven by the screw to the opposite end. Solvent is poured into the extractor at the opposite end to the point of herb entry. Problems arise from poor solvent contact, when fine powders of herbs are used. These can be due to clumping of the powder, formation of preferred channels in the bed of soaked powder, and poor circulation of the solvent. These problems can be overcome by using pellets of the powdered herb. Ground herb is wetted with steam or water and pressed into pellets of 4–10 mm diameter, using a pellet mill (Gallagher et al., 2006). Herbs such as *Centella asiatica*, *Coleus forskohlii*, and *Kaempferia galanga* are extracted using such pellets. The herb powder and solvent are mixed together and this process can be controlled to optimum level by regulating the number of revolutions of the screw, the slope of the cylinder, the temperature of the solvent, and the pressure. Because of the harmonization of these factors, the herb powder is virtually exhausted by the time it reaches the other end, meeting continuously poured solvent (Bombardelli, 1991). The carousel and the U extractors work on the principle of relative countercurrent extraction.

To ensure the complete exhaustion of the herb, the following parameters need to be defined and strictly observed for each herb: particle size of herb, flow rate of solvent, contact times, temperature, and pressure (Bombardelli, 1991). Continuous extractors are usually of large capacity. While the industrial production of a batch of extract takes 4–6 days in the dynamic maceration process, a batch can be completed in 24 h using continuous extraction. Continuous extraction of several natural products has been reported (Wagner and Parma, 1988; Gunasekaran et al., 1989; Wiesenborn et al., 1999; Kim et al., 2001, 2002).

2.6.2.4 Concentration of extracts

The miscella obtained from solvent extraction undergoes concentration, so that final products in the form of fluid, soft, or dry extracts can be prepared. In some cases, the extract will be further purified with other solvents to prepare enriched extracts. In special cases, the miscella will be further extracted with solvents to isolate particular compounds. When concentrated at high temperatures, there is a possibility of the denaturation of compounds. To avoid this, the equipment and the temperature at which the miscella is to be concentrated, need to be chosen carefully (Bombardelli, 1991). Various kinds of equipment are employed for the concentration of extracts.

2.6.2.4.1 Jacketed evaporators Jacketed evaporators are used when the product is very viscous and the volume is small. Concentration is carried out in a small cylindrical evaporator, with or without jackets for the circulation of steam (Minton, 1986a). Jacketed evaporators are not suitable for temperature-sensitive products, as the residence time is long. In addition to this, the static head of the liquid raises the boiling point of the product at the bottom of the tank. This can be overcome to some extent by operating the tank under vacuum, to reduce the boiling point of the batch. Jacketed evaporators are still used for manufacturing processes that involve small batches of products, which are clean, viscous, and not heat sensitive (Glover, 2004).

2.6.2.4.2 Coil evaporators Metallic coils of tubes inside the cylindrical tank come in contact with the miscella that is to be concentrated. Heat radiating from steam circulating inside the coils causes the evaporation of the solvent, which moves into the condenser. Agitation of the miscella can be achieved with stirrers. Evaporation can be carried out inside the tubes as well. In such evaporators, the heating medium will be outside the coil. Coil evaporators are ideal for small quantities of miscella and when the product is difficult to handle (Minton, 1986a).

2.6.2.4.3 Falling film evaporator The miscella is fed into the top of long tubes and is allowed to fall down the walls as a thin film. Steam is used to heat the tubes. The quantity of miscella falling down in these film evaporators is small and the time spent by the fluid inside the equipment is very short. The system also operates under vacuum; therefore, it is ideal for heat-sensitive products (Minton, 1986a) (**Figure 2.4**).

2.6.2.4.4 Tray dryer Miscella is poured into trays and stacked in a jacketed chamber. The inner surface of the chamber is hot and the rising solvent vapors are removed under vacuum. Tray dryers are especially suitable for thermolabile products (Minton, 1986a).

2.6.2.4.5 Thin film evaporator Essentially, the equipment consists of a hollow, cylindrical body, heated externally and fitted with a rotor inside. The rotor agitates the miscella, which is spread on the inside of the heated cylinder. The vapors are removed under vacuum (Bombardelli, 1991). The various aspects of evaporation technology such as vapor compression, vacuum-producing equipment, and so on are treated exhaustively by Minton (1986b) (**Figure 2.5**).

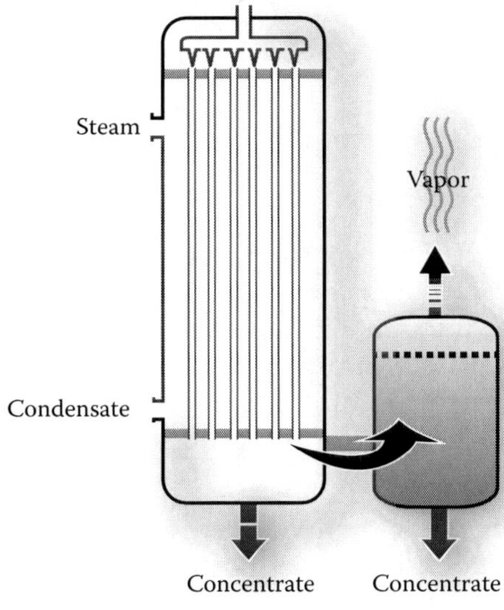

Figure 2.4 *Falling film evaporator. (Courtesy of Novasep, Pompey, France.)*

Figure 2.5 *Thin film evaporator. (Reproduced with permission from Elettronica Veneta S.p.A., Italy.)*

2.6.2.5 Clarification of extracts

Extracts that are intended for pharmaceutical or nutraceutical use should not contain solid particles extraneous to the extract. Very fine particles of the herb often escape into the extract even under stringent filtration procedures. During hot maceration of herbs, many inert substances that are present in them are extracted and become solidified on subsequent cooling. All these particles need to be removed from the extract (Bombardelli, 1991).

The clarification of extracts is achieved by simple or pressure filtration, separators, and decanters. In pressure filtration, the extract is force-pumped into several compartments, which have pores that are adequate to filter fine suspended particles. Horizontal decanters, super centrifuges, and perforated or closed-drum centrifuges are also used for clarifying extracts.

2.6.2.6 *Drying of extracts*

Various kinds of extracts such as fluid extracts, tinctures, soft extracts, and dry extracts are industrially manufactured. However, dry extracts are often preferred to the others, considering their chemical stability and ease of transportation. One added advantage is that in case of microbial contamination, dry extracts can be sterilized by gamma radiation. But this technique is not feasible in the case of liquid/semisolid extracts as the process can start a series of chemical reactions (Bombardelli, 1991).

Various types of equipment are available for drying extracts. Drum dryers, belt dryers, cabinet dryers, spray dryers, and lyophilizers are employed, depending on the quantity and the nature of the extract to be dried (Bombardelli, 1991; Cheremisinoff and Cheremisinoff, 1995).

2.6.2.6.1 Spray dryer This is the equipment of choice, if the extract that is to be dried is a water extract. Filtered and heated air is blown into the drying chamber and a slurry of the water extract is sprayed into the chamber through an atomizer. The water droplets present in the sprayed extract evaporate immediately on coming into contact with the hot air. The dried extract settles at the bottom of the chamber or it can be discharged along with the air into cyclones, from where it can be collected. As the extract is exposed to hot air for only a short time, there is no thermal damage to the phytochemicals (Cheremisinoff and Cheremisinoff, 1995) (**Figure 2.6**).

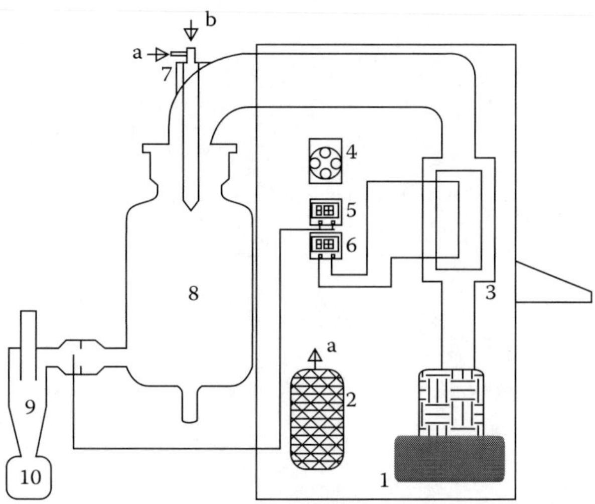

Figure 2.6 *Schematic diagram of a spray dryer. (1) Blower + air filter; (2) air compressor; (3) heater; (4) peristaltic pump; (5) temperature controller; (6) inlet thermocouple; (7) atomizer: (a) compressed air, (b) feed suspension; (8) drying chamber; (9) cyclone; (10) receptacle for dry product. (Reproduced from Souza, C.R.F. et al., 2008,* Braz. J. Chem. Eng. *25: 59–69. With permission.)*

2.6.2.6.2 Cabinet dryer Shallow pans containing concentrated extract are placed on a rack inside a large cabinet with tight walls. Hot air (60°C–80°C) passed into the cabinet circulates around the pans and leaves through an outlet. The extract is dried thoroughly during the process. Such dryers are often equipped with a vacuum (Cheremisinoff and Cheremisinoff, 1995).

2.6.2.6.3 Lyophilizer The term *lyophilization* was first coined by Rey (1977) because of the porous nature of the dried material and its ability to reabsorb the solvent, thereby attaining its original state. Lyophilization is also known as *freeze drying* or *cryodesiccation*. Lyophilization is now defined as a stabilizing process in which the substance is first frozen and the solvent is reduced first by sublimation (primary drying) and later by desorption (secondary drying) to values that no longer support biological growth or chemical reactions (Jennings, 1999). Some herbal extracts need to be lyophilized to retain their biological activity. Lyophilization is carried out at very low temperatures (–50°C) and pressure (microbars). Commonly, aqueous solutions are lyophilized (**Figure 2.7**).

2.6.2.6.4 Spray chilling Spray chilling, also known as *spray congealing*, is a drying process similar to spray drying. The only difference is the temperature

Figure 2.7 *Lyophilizer. (Reproduced with permission from Sub-Zero Lab Instruments, Chennai, India. http://www.subzeroindia.com.)*

of the air that is used and the nature of the coating. A slurry is dispersed into a chamber containing air, cooled to refrigeration temperatures. Droplets of the extract solidify and settle down. The common carriers are fatty acids, alcohols, waxes, stearates, and polyethylene glycols. Spray chilling is commonly used in marine lipid encapsulation (Jin et al., 2007).

2.7 Concluding remarks

As mentioned previously, the chemical content of a herb can vary for several reasons. Therefore, it is essential to ensure the therapeutic efficacy of an extract, based on its content of active compounds. In olden days, the concept of the herb:extract ratio was used for standardizing extracts. However, as the herb:extract ratios do not indicate the chemical content of the extracts and as sophisticated analytical instruments have become available, the chemical standardization of extracts is now accepted as a means to ensure the quality of extracts. Using such techniques, it is possible to manufacture herbal products with standard levels of particular compounds or classes of compounds (Talbott, 2002). An extract of a plant can be standardized to contain a specific chemical compound when its therapeutic efficacy can be directly related to that compound. An example is an extract of *Silybum marianum* standardized to silymarin, as silymarin is proven to be the active compound responsible for hepatoprotection (Wagner et al., 1974).

However, in many cases the therapeutic activity of an extract is due to several compounds within a class of compounds. An example is the extract of red yeast rice (*Monascus purpureus*), popular as a dietary supplement for reducing blood cholesterol. The hypolipidaemic property of *M. purpureus* is attributed to about 14 monacolins, which have the ability to inhibit the enzyme 3-hydroxy-3-methyl-glutaryl (HMG)-CoA reductase, involved in the synthesis of cholesterol (Yong-Guo et al., 2004). Red yeast rice is commonly standardized to 0.4% of total monacolins (Talbott, 2002).

In many cases, the exact chemical compound responsible for the therapeutic effect is either unknown or it is difficult to measure. In such cases, the extract can be standardized to a marker compound that can be easily quantified. The rationale is that when the extract is properly standardized to the marker compound, the other chemical constituents will also be present in the required quantities (Talbott, 2002). An example of such a marker compound is hypericin, on the basis of which an extract of St. John's wort (*Hypericum perforatum*) is standardized. The extract is standardized to contain 0.3% hypericin, although hypericin is not the compound responsible for the antidepressant action of the extract (Hobbs, 1989).

The major argument against the standardization of extracts is that a single constituent cannot heal effectively without the rest of the molecules from the herb in question. It was earlier believed that the antidepressant activity of

St. John's wort was due to hypericin (Suzuki et al., 1984). Nevertheless, it is now known that this effect is brought about by a complex mixture of flavonoids, hyperoside, rutin, and biflavones (Hobbs, 1989; Pedec and Klodziejczyk, 2000). While this is true in many cases, the standardization of extracts offers consistency in dosage, a reduction in the toxic components of plants, and predictability in therapeutic response.

A common misunderstanding is that the standardization of a herbal extract can be achieved by controlling a small portion of the chemical constituents. A ginger extract "standardized" to 5% gingerols does not say anything about the content of the individual gingerols or the composition of the other 95% of the extract. It can be reasonably inferred that the 5% designation is a statement of the strength of the extract and not its biological effect. Two different ginger extracts having the same 5% gingerol strength could vary drastically in their biological effects because of the constitution of the individual gingerols in the 5% and the biological effect that could arise from some compound in the other 95% (Dentali, 2009). Therefore, standardization encompasses a detailed description of the starting material and the entire extraction process. This is necessary, as agricultural practices and manufacturing processes can affect the outcome of the final product. As Bonati (1991) has observed, standardization signifies the body of information and the controls that are necessary to guarantee constancy of composition.

References

Akahori, A., Kagawa, K. 1977. Proceedings of Symposium, Wakan–Yaku 10, Toyama, p. 61. Cited by Nakajima et al. 1994a. *Chemical and Pharmaceutical Bulletin* 42: 1977–1983.

Anonymous. 1977. Continuous solvent extraction. *Journal of American Oil Chemists' Society* 54: 202A–206A.

Anonymous. 2003a. Essential oils of the genus citrus. In *The Complete Technology Book of Essential Oils (Aromatic Chemicals)*, 1–47. New Delhi: National Institute of Industrial Research.

Anonymous. 2003b. *WHO Guidelines on Good Agricultural and Collection Practices (GACP) for Medicinal Plants*. Geneva: World Health Organization.

Anonymous. 2004. Papaver oil. In *Handbook of Herbs Cultivation and Processing*, 313–315. New Delhi: National Institute of Industrial Research.

Anonymous. 2006. *Guidelines for Good Agricultural and Wild Collection Practice (GACP) of Medicinal and Aromatic Plants*. http://www.europam.net/documents/gacp/EUROPAM_GACP_7.3.pdf. Accessed May 23, 2015.

Anonymous. 2009. http://www.buchi.com/Where-everything-started 1939.891.0.html?id=71. Accessed June 6, 2009.

Arichi, S., Tani, T., Kubo, M. 1979. Studies on Bupleuri radix and Saikosaponin. (3) Quantitative analysis of Saikosaponin of commercial Bupleuri radix. *Medical Journal of Kinki University* 4, 235–241.

Bahl, J.R., Garg, S.N., Bansal, R.P., Naqvi, A.A., Singh, V., Kumar, S. 2000. Yield and quality of shoot essential oil from the vegetative, flowering and fruiting stage crops of *Ocimum basilicum* cv *kusumohak*. *Journal of Medicinal and Aromatic Plants* 22: 743–746.

Bakkali, F., Averbeck, S., Averbeck, D., Idaomar, M. 2008. Biological effects of essential oils—A review. *Food and Chemical Toxicology* 46: 446–475.

Boettcher, H., Guenther, I. 2005a. Raw plant material and postharvest technology. In *Chamomile: Industrial Profiles*, ed. R. Franke and H. Schilcher, 173–186. Boca Raton, FL: CRC Press.

Boettcher, H., Guenther, I. 2005b. Storage of the dry drug. In *Chamomile: Industrial Profiles*, ed. R. Franke and H. Schilcher, 211–220. Boca Raton, FL: CRC Press.

Bombardelli, E. 1991. Technologies for the processing of medicinal plants. In *The Medicinal Plant Industry*, ed. R.O.B. Wijesekera, 85–98. Boca Raton, FL: CRC Press.

Bonati, A. 1991. How and why should we standardize phytopharmaceutical drugs for clinical validation. *Journal of Ethnopharmacology* 32: 195–197.

Brachet, J., Cosson, L., Ducourtioux, D., Scheidecker, D. 1981. Effect of sodium chloride on the tropane alkaloid content in *Datura innoxia* Mill. cultivated in a controlled environment. *Physiologie Vegetale* 19: 77.

Brielman, Jr., H.L. 1999. Phytochemicals: The chemical components of plants. In *Natural Products from Plants*, ed. P.B. Kaufman, L.J. Cseke, S. Warber, J.A. Duke, and H.L. Brielman, Jr., 1–36. Boca Raton, FL: CRC Press.

Brundrett, M.C. 2002. Coevolution of roots and mycorrhizas of land plants. *New Phytologist* 154: 275–304.

Bruneton, J. 1995. Compounds of primary metabolism. In *Pharmacognosy, Phytochemistry Medicinal Plants*, 1–118. Paris: Lavoisier.

Chatterjee, S.K., Nandi, R.P., Sarkar, D.P. 1984. Effect of mineral nutrients on growth and essential oil formation in *Cymbopogon winterianus*. *Science and Culture* 50: 26.

Cheremisinoff, P.N., Cheremisinoff, N.P. 1995. Drying. In *Process Engineering Data Book*, 167–180. Boca Raton, FL: CRC Press.

Cosson, L. 1966. Effect of light on alkaloid content in *Datura*: Effect on the biogenesis of alkaloids. *Herba Hungarica* 5: 157.

Cosson, L. 1969. Effect of light on the ontogenic variations in scopolamine and hyoscyamine levels of *Datura metel* leaves. *Phytochemistry* 8: 2227.

Cosson, L. 1976. Importance of climatic factors and development stages on the production of tropane alkaloids. In *Etudes de Biologie Vegetale, Hommage au Professeur P. Chouhard*, 483. Paris: Roger Jacques.

Cosson, L. 1981. Some aspects of the metabolism of alkaloids in *Datura tatula* L. *Plantes Médicinales et Phytothérapie* 2: 269.

Cosson, L., Chouhard, P., Paris, R. 1966. Influence of light on ontogenic variations of alkaloids in *Datura tatula*. *Journal of Natural Products* 29: 19.

Cosson, L., Escudero-Morales, A., Cougol, N. 1978. Ecophysiological regulation of the metabolism of tropane alkaloids (hyoscyamine and scopolamine). *Plantes Médicinales et Phytothérapie* 12: 319.

Craig, L.C., Gregory, J.D., Hausmann, W. 1950. Versatile laboratory concentration device. *Analytical Chemistry* 22: 1462.

Cseke, L.J., Kaufman, P.B. 1999. How and why these compounds are synthesized by plants. In *Natural Products from Plants*, ed. P.B. Kaufman, L.J. Cseke, S. Warber, J.A. Duke, and H.L. Brielman, 37–90. Boca Raton, FL: CRC Press.

Cseke, L.J., Kaufman, P.B., Kirakosyan, A. 2007. The biology of essential oils in the pollination of flowers. *Natural Product Communications* 2: 1317–1336.

Cunningham, A.B. 1994. Management of medicinal plant resources: An Africa-wide overview. In *Proceedings of the 13th Plenary Meeting of AETFAT*, Zomba, Malawi, 2–11 April, 1991, Vol. 1, Montfort, Limbe, 173–189.

D'Amelio, Sr., F.S. 1999. Phytochemistry. In *Botanical: A Phytocosmetic Desk Reference*, 1–8. Boca Raton, FL: CRC Press.

Das, A., Mallick, R. 1991. Correlation between genomic diversity and asiaticoside content in *Centella asiatica* (L.) Urban. *Botanical Bulletin of Academia Sinica* 32: 1–8.

Davik, J., Bakken, A.K., Holte, K., Blomhoff, R. 2006. Effects of genotype and environment on total anti-oxidant capacity and the content of sugars and acids in strawberries (*Fragaria* × *ananassa* Duch.). *Journal of Horticultural Science & Biotechnology* 81: 1057–1063.

Dentali, S. 2009. Botanical preparations: Achieving quality products. In *Botanical Medicine: From Bench to Bedside*, ed. R. Cooper and F. Kronenberg, 1–12. New York: Mary Ann Liebert.
De Smet, 1949. Improvements relating to the continuous extraction of substances from solid matter by means of a solvent. UK Patent No. GB 688690, Application No. GB 19490032944 19491223.
Dudareva, N., Negre, F. 2005. Practical applications of research into the regulation of plant volatile emission. *Current Opinion in Plant Biology* 8: 113–118.
Fahlen, A., Welander, M., Wennersten, R. 1997. Effects of light-temperature regimes on plant growth and essential oil yield of selected aromatic herbs. *Journal of the Science of Food and Agriculture* 73: 111–119.
Finar, I.L. 2001. Physical properties and chemical substitution. In *Organic Chemistry*, Vol. 2, 1–68. Delhi: Pearson Education Asia.
Franz, C. 1993. Genetics. In *Volatile Oil Crops: Their Biology, Biochemistry and Production*, ed. R.K.M. Hay and P.G. Waterman, 63–96. Harlow: Longman Scientific & Technical.
Funk, C., Lewinsohn, E., Vogel, B.S., Steele, C.L., Croteau, R. 1994. Regulation of oleoresinosis in grand fir (*Abies grandis*): Coordinate induction of monoterpene and diterpene cyclases and two cytochrome p450—Dependent diterpenoid hydroxylases by stem wounding. *Plant Physiology* 106: 999–1005.
Gallagher, R.T., Howe, C.D., Desimone, III, E.M., et al. 2006. Harvesting and palletizing yew biomass for extraction of taxanes and other natural products. US Patent Application No. US 2006/0127510A1.
Gianiazzi-Pearson, V. 2002. Plant cell responses to arbuscular mycorrhizal fungi: Getting to the roots of the symbiosis. *The Plant Cell* 8: 1871–1883.
Glover, W.B. 2004. Selecting evaporators for process applications. *Chemical Engineering Progress*, December, 26–33.
Grant, V.P. 1981. *Polyploidy*, 283–352. New York: Columbia University Press.
Guenther, E. 1955a. *The Essential Oils*, Vols. 1–6. Toronto: Van Nostrand.
Guenther, E. 1955b. The production of essential oils: Methods of distillation, enfleurage, maceration, and extraction with volatile solvents. In *The Essential Oils*, Vol. 1, ed. E. Guenther, 189–198. Toronto: Van Nostrand.
Gunasekaran, S., Fisher, R.J., Casimir, D.J. 1989. Predicting soluble solids extraction from fruits in a reversing, single screw counter current diffusion extractor. *Journal of Food Science* 54: 1261–1265.
Gupta, S., Prabhakar, W.S., Madau, C.L. 1974. Study on the distribution of total alkaloids and major components in the different organs of *Datura metel* var. *fastuosa* at various stages of growth. *Planta Medica* 23: 370.
Hanna, T.P. 1999. These are roses scents. In *The Microscopic and Chemical Parts of Plants*, 168–170. New York: Nova.
Hårdh, K., Hårdh, J.E. 1977. Studies on quality of vegetables and strawberries at different latitudes in Finland. *Annales Agriculturae Fenniae* 16: 19–26.
Hårdh, J.E., Persson, A.R., Ottoson, L. 1977. Quality of vegetables cultivated at different latitudes in Scandinavia. *Acta Agriculturae Scandinavica* 27: 81–96.
Hay, R.K.M., Svoboda, K.P. 1993. Botany. In *Volatile Oil Crops: Their Biology, Biochemistry and Production*, ed. R.K.M. Hay and P.G. Waterman, 5–22. Harlow: Longman Scientific & Technical.
Hieter, P., Griffith, T. 1999. Polyploidy—More is more or less. *Science* 285: 210–211.
Hobbs, C. 1989. St. John's wort (*Hypericum perforatum* L.): A review. *Herbalgram* 18/19, 24–33.
Hofman, P.J., Menary, R.C. 1979. Variations in morphine, codeine and the baine in the capsules of *Papaver somniferum* L. during maturation. *Australian Journal of Agricultural Research* 31: 313.
Holton, T.A., Cornish, E.C. 1995. Genetics and biochemistry of anthocyanin biosynthesis. *The Plant Cell* 7: 1071–1083.

Huang, T.K. 1986. Chinese Traditional Herbal Drugs 17: 323. Cited by Nakajima et al. 1994a. *Chemical and Pharmaceutical Bulletin* 42: 1977–1983.

Imazeki, I., Taguchi, H., Nakajima, K., Aburada, M., Takeda, S.1980. Shoyaku- Bunseki No Giho, ed. M. Noguchi, 142–156. Cited by Nakajima et al. 1994a. *Chemical and Pharmaceutical Bulletin* 42: 1977–1983.

Jaakola, L., Hohtola, A. 2010. Effect of latitude on flavonoid biosynthesis in plants. *Plant, Cell & Environment* 33: 1239–1247.

Jennings, T.A. 1999. Introduction. In *Lyophilization: Introduction and Basic Principles*, 1–14. Boca Raton, FL: Interpharm/CRC.

Jin, Y., Perrie, C., Zhang, W., van Diepen, C., Curtis, J., Barrow, C.J. 2007. Microencapsulation of marine lipids as a vehicle for functional food delivery. In *Marine Nutraceuticals and Functional Foods*, ed. C. Barrow and F. Shahidi, 127–128. Boca Raton, FL: CRC Press.

Johnson, M., Coroteau, R. 1987. Biochemistry of conifer resistance to bark beetles and their fungal symbionts. In *Ecology and Metabolism of Plant Lipids*, ed. G. Fuller and W.D. Nes, 76–91. Washington, DC: American Chemical Society.

Joy, P.P. 2003. Agrotechnological practices for quality crude drug production in *Nilappana* (*Curculigo orchioides* Gaertn.). PhD thesis, Kerala Agriculture University, Thrissur.

Kano, Y., Saito, K., Sakurai, T., Kanemaki, S., Tanabe, M., Yasuda, M. 1986. On the evaluation of the preparation of Chinese medicinal prescriptions (1) 6-gingerol in Zingiberis Rhizoma. *Shoyakugaku Zasshi* 40: 333.

Karnick, C.R., Saxena, M.D. 1970. On the variability of alkaloids production in *Datura* species. *Planta Medica* 18: 266.

Kattenberg, H.R., Kemmink, A. 1993. The flavor of cocoa in relation to the origin and processing of the cocoa beans. In *Food, Flavors, Ingredients and Composition*, ed. G. Charalambous, 1–22. Amsterdam: Elsevier.

Kim, K.H., Tucker, M.P., Keller, F.A., Aden, A., Nguyen, Q.A. 2001. Continuous countercurrent extraction of hemicellulose from pretreated wood residues. *Applied Biochemistry and Biotechnology* 91–93: 253–267.

Kim, K.H., Tucker, M.P., Nguyen, Q.A. 2002. Effects of operating parameters on countercurrent extraction of hemicellulosic sugars from pretreated softwood. *Applied Biochemistry and Biotechnology* 98–100: 147–159.

Kuriyama, H., Watanabe, S., Nakaya, T., et al. 2005. Ambient odors of orange and lavender reduce anxiety and improve mood in dental office. *Physiology & Behavior* 86: 92–95.

Langenau, E.E. 1955. The examination and analysis of essential oils, synthetics and volatiles. In *The Essential Oils*, Vol. 1, ed. E. Guenther, 189–198. Toronto: Van Nostrand.

Larcher, W. 1995. The environment of plants. In *Physiological Plant Ecology*, 19–31. New York: Springer.

Lebeau, P., Janot, M.M. 1956. *Traite de Pharmacie Chimique*, 1113. Paris: Masson et Cie.

Marotti, M., Dellacecca, V., Piccaglia, R., Giovanelli, E. 1993. Effect of harvesting stage on the yield and essential oil composition of peppermint (*Mentha×piperita* L.). *Acta Horticulturae* 344: 370–379.

Martin, C., Smith, A.M. 1995. Starch biosynthesis. *The Plant Cell* 7: 971–985.

Massoud, H.Y. 1989. Quantiative Vererbung einiger Ertragsmerkmale und der Hauptkomponenten des ätherischen Öles von Kamille, *Matricaria chamomilla* L. (syn. *Chamomilla recutita* L.). PhD thesis, Technical University of Munich.

Mika, E.S. 1955. Growth, development and morphine content of opium poppy. *Botanical Gazette* 116: 323.

Minton, P.E. 1986a. Evaporator types and applications. In *Handbook of Evaporation Technology*, 70–97. Fairview, NJ: Noyes Publications.

Minton, P.E. 1986b. *Handbook of Evaporation Technology*, 172–247. Fairview, NJ: Noyes Publications.

Nakajima, K., Horiuchi, T., Taguchi, H., Hayashi, K., Okada, M., Maruno, M. 1994a. Physicochemical studies on decoctions of Kampo prescriptions. III. Effect of the volume ratio of the crude drug vs. extractant on transfer ratio of crude drug components into a model decoction. *Chemical and Pharmaceutical Bulletin* 42: 1991–1997.

Nakajima, K., Takeuchi, Y., Taguchi, H., Hayashi, K., Okada, M., Maruno, M. 1994b. Physicochemical studies on decoctions of Kampo prescriptions. I. Transfer of crude drug components into the decoctions. *Chemical and Pharmaceutical Bulletin* 42: 1977–1983.

Nakajima, K., Takeuchi, Y., Taguchi, H., Hayashi, K., Okada, M., Maruno, M. 1994c. Physicochemical studies on decoctions of Kampo prescriptions. II. Relationship between the hydrophobic parameter of the crude drug components and their transfer ratio into the decoctions. *Chemical and Pharmaceutical Bulletin* 42: 1984–1990.

Nandi, R.P., Chatterjee, S.K. 1976. Photoperiodic regulation of growth and alkaloid formation in *Datura innoxia* Mill. *Indian Biologist* 7: 31.

Noguchi, M., Kubo, M., Hayashi, T., Ono, M. 1978a. Studies on the pharmaceutical quality evaluation of the crude drug preparations used in oriental medicine kampoo part 1 precipitation reaction of the components of coptis rhizome and these of glycyrrhiza root or rheum rhizome in decoction solution. *Shoyakugaku Zaashi* 32: 104.

Noguchi, M., Kubo, M., Naka, Y. 1978b. Studies on the pharmaceutical quality evaluation of crude drug preparations used in Orient medicine "Kampo" IV. Behavior of alkaloids in ephedra herb mixed with other crude drugs under decoction processes. *Yakugaku Zaashi* 98: 923–928.

Ohlrogge, J., Chrispeels, M.J. 2002. Plants as chemical and pharmaceutical factories. In *Plants, Genes, and Crop Biotechnology*, ed. M.J. Chrispeels and D.E. Sadava, 500–501. Boston: Johns and Bartlett.

Palevitch, D. 1991. Agronomy applied to medicinal plant conservation. In *The Conservation of Medicinal Plants: Proceedings of an International Consultation*, ed. O. Akerele, V. Heywood, and H. Synge, 168–178. Cambridge: Cambridge University Press.

Pate, D.W. 1994. Chemical ecology of *Cannabis*. *Journal of the International Hemp Association* 1: 32–37.

Pedec, P., Klodziejczyk, P.P. 2000. Quality assurance and control for the herbal and tea industry. In *Herbs, Botanicals and Teas*, ed. G. Mazza and B.D. Oomah, 377–398. Boca Raton, FL: CRC Press.

Pelt, J.M., Younos, C., Hayon, J.C. 1967a. On the alkaloid constitution of some solanaceae from Afghanistan. I. *Annales Pharmaceutiques Françaises* 25: 59.

Pelt, J.M., Younos, C., Hayon, J.C. 1967b. On the alkaloid constitution of some solanaceae from Afghanistan. II. *Annales Pharmaceutiques Françaises* 25: 101.

Pengelly, A. 2004. Essential oils and resins. In *The Constituents of Medicinal Plants*, 2nd edn, 86. Oxford: Oxford University Press.

Pepeljnjak, S., Kosalec, I., Kalodera, Z., Kustrak, D. 2003. Natural antimycotics from Croatian plants. In *Plant Derived Antimycotics: Current Trends and Future Prospects*, ed. M. Rai and D. Mares, 49–79. New York: The Haworth Press.

Rabak, F. 1917. The effect of cultural and climatic conditions on the yield and quality of peppermint oil. *Bulletin of Plant Industry* 80: 450–454.

Raguso, R.A., Light, D.M., Pichersky, E. 1996. Electroantennogram responses of *Hyles lineata* (Sphingidae: Lepidoptera), to volatile compound from *Clarkia breweri* (Onagraceae) and other moth-pollinated flowers. *Journal of Chemical Ecology* 22: 1735–1766.

Rey, L.R. 1977. Glimpses into the fundamental aspects of freeze-drying. In *Freeze-Drying of Biological Products, Biological Standardization Series,* Vol. 36, ed. V.J. Cabasso and R.H. Regamey, 19–27. Basel: S. Karger.

Robbins, C.S. 1998. *American Ginseng: The Root of North America's Medicinal Herb Trade.* Washington, DC: Traffic North America.

Rodrigues, R.A.F., Queiroga, C.L., Rodrigues, M.V.N., Foglio, M.A., Sartoratto, A., Montanari, I. 2002. Study of the variation of the composition of the essential oil of leaves and flowers of *Achyrocline alata* (DC) along a period of the day. *Journal of Essential Oil Research* 14: 280–281.

Rohloff, J. 2003. Cultivation of herbs and medicinal plants in Norway—Essential oil production and quality control. PhD thesis, Norwegian University of Science and Technology, Trondheim.

Sandberg, F., Corrigan, D. 2004. The quality control of herbal medicinal products. In *Natural Remedies: Their Origins and Uses*, 11–22. New York: Taylor & Francis.

Scavroni, J., Boaro, C.S.F., Marques, M.O.M., Ferreira, L.C. 2005. Yield and composition of the essential oil of *Mentha piperita* L. (Lamiaceae) grown with biosolid. *Brazilian Journal of Plant Physiology* 17: 345–352.

Schippmann, U., Leaman, D., Cunningham, A.B. 2006. A comparison of cultivation and wild collection of medicinal and aromatic plants under sustainability aspects. In *Medicinal and Aromatic Plants*, ed. R.J. Bogers, L.E. Cracker, and D. Lange, 75–95. Heidelberg: Springer.

Schratz, E., Qadry, S.M.J.S. 1966. Die Zusammensetzung des aetherischen Oeles in *Coriandrum sativum*. III Oelkomposition im Verlaufe der Ontogenese. *Planta Medica* 14: 436–442.

Shi, J., Kassama, L.S., Kakuda, Y. 2006. Supercritical fluid technology for extraction of bioactive components. In *Functional Food Ingredients and Nutraceuticals: Processing Technologies*, ed. J. Shi, 3–44. Boca Raton, FL: CRC Press.

Shiva, M.P., Mahtolia, D.C. 1998. Strategies to overcome problems for utilization of medicinal plants. In *Prospects of Medicinal Plants*, ed. P.L. Gautam, R. Raina, U. Srivastava, S.P. Raychaudhari, and B.B. Singh, 131–136. New Delhi: Indian Society of Plant Genetic Resources.

Silva, M.G.D., Craveir, A.A., Matos, F.J.A., Machado, M.I.L., Alencar, J.W. 1999. Chemical variation during daytime of constituents of the essential oil of *Ocimum gratissimum* leaves. *Fitoterapia* 70: 32–34.

Singh, R.D., Ahuja, P.S., Nagar, P.K., et al. 2000. Effect of manuring and shade on yield and quality of *Valeriana wallichii*. *Journal of Medicinal and Aromatic Plants Science* 22: 669–670.

Smallwood, I.M. 1996. *Handbook of Organic Solvent Properties*. London: Arnold.

Souza, C.R.F., Schiavetto, I.A., Thomazini, F.C.F., Oliveira, W.P. 2008. Processing of *Rosmarinus officinalis* Linne extract on spray and spouted bed dryers. *Brazilian Journal of Chemical Engineering* 25: 59–69.

Soxhlet, F. 1879. Die gewichtsanalytische Bestimmung des Milchfettes. *Dinglers Polytechnisches Journal* 232: 461–465.

Suzuki, O., Katsumata, Y., Oya, M., Bladt, S., Wagner, H. 1984. Inhibition of monoamine oxidase by hypericin. *Planta Medica* 50: 272–274.

Taguchi, H., Endo, T., Nakajima, K., Aburada, M. 1975. *Proceedings of the Symposium Wakan Yaku* 9, Toyama, p. 85. Cited by Nakajima et al. 1994a. *Chemical and Pharmaceutical Bulletin* 42: 1977–1983.

Takaishi, K., Torii, Y. 1969. Studies on the decoction of Chinese medicines. I. On the interaction of some chemicals with starch in aqueous solution. *Yakugaku Zasshi* 89: 538–543.

Takaishi, K., Watanabe, Y. 1971. Studies on the decoction of Chinese medicines. II. The extraction of some drugs by starch aqueous solution and the property of Gegen Tang. *Yakugaku Zasshi* 91: 1092–1097.

Talbott, S.M. 2002. Critical evaluation of dietary supplements. In *A Guide to Understanding Dietary Supplements*, 28–33. New York: Haworth.

Theis, N., Lerdau, M. 2003. The evolution of function in plant secondary metabolites. *International Journal of Plant Sciences* 164: S93–S102.

Theuns, H.G., Janssen, R.H.A.M., Salemnik, C.A. 1991. The alkaloids of the *Papaver* Section *Oxytona* Bernh. In *Herbs, Spices and Medicinal Plants*, ed. L.E. Cracker and J.E. Simon, 57–110. Phoenix, AZ: Oryx Press.

Tomimori, T., Yoshimoto, M. 1980. Quantitative variation of glycyrrhizin in the decoction of Glycyrrhizae Radix mixed with other crude drugs. *Shoyakugaku Zaashi* 34: 138.

Tyagi, D.K. 2005. Pharma industry. In *Pharma Forestry: Field Guide to Medicinal Plants*, 28–30. New Delhi: Atlantic Publishers & Distributors.

Uniyal, R.C., Uniyal, M.R., Jain, P. 2000. *Cultivation of Medicinal Plants in India: A Reference Book*. New Delhi: Traffic.

Vanhaelen, M., Lejoly, J., Hanocq, M., Molle, L. 1991. Climatic and geographical aspects of medicinal plant constituents. In *The Medicinal Plant Industry*, ed. R.O.B. Wijesekera, 59–76. Boca Raton, FL: CRC Press.

Wagner, H., Diesel, P., Seitz, M. 1974. The chemistry and analysis of silymarin from *Silybum marianum* Gaertn. *Arzneimittelforschung* 24: 466–474.

Wagner, J.P., Parma, D.G. 1988. Continuous solvent extraction process for recovery of natural rubber from guayule. *Polymer-Plastics Technology and Engineering* 27: 335–350.

Wennersten, R. 2004. Extraction of organic compounds. In *Solvent Extraction Principles and Practice*, ed. J. Rydberg, M. Cox, C. Musikas, and G.R. Choppin, 415–454. New York: Marcel Dekker.

Wiesenborn, D.P., Wang, J., Chang, K.C., Schwarz, J.G. 1999. Comparison of continuous and batch processes for pectin extraction from sunflower heads. *Industrial Crops and Products* 9: 171–181.

Yamada, H., Ota Y., Tsujiwaki, Y., Ikemasu, M., Mizukami, H., Ikenaga, T., Ohashi, H. 1983. Ecology and cultivation of *Swertia* plant. XII. Study on planting density in cultivation of *S. japonica*. *Shoyakugaku Zaashi* 37: 115.

Yamaguchi, S. 2008. Gibberellin metabolism and its regulation. *Annual Review of Plant Biology* 59: 225–251.

Yamaji, A., Mareda, Y., Oishi, M., et al. 1984. Determination of saikosaponins in Chinese medicinal extracts containing Bupleuri radix. *Yakugaku Zaashi* 104: 812–815.

Yata, N., Tanaka, O. 1983. *Journal of Traditional Sino-Japan Medicine* 4: 68. Cited by Nakajima et al. 1994a. *Chemical and Pharmaceutical Bulletin* 42: 1977–1983.

Yong-Guo, L., Fang, Z., Zheng-Tao, W., Zhi-Bi, H. 2004. Identification and chemical profiling of monacolins in red yeast rice using high-performance liquid chromatography with photodiode array detector and mass spectrometry. *Journal of Pharmaceutical and Biomedical Analysis* 35: 1101–1112.

3

Extraction of the Bioactives

[N]ever talk about science, *show* it to them.

Lawrence Bragg (1890–1971)
Nobel Laureate in Physics (1915)

3.1 Introduction

Plants produce a multitude of chemical substances, most of which are biologically active in human beings. Although primarily intended for use by the plants themselves, humans can also benefit from them by making useful products such as medicines, nutraceuticals, cosmetics, fortified foods and beverages, fragrances, flavors, personal care products, and various consumer goods. The extraction of the chemical substances is a *sine qua non* for the development of such products. A high-quality herbal extract is manufactured using a good-quality herb, an appropriate solvent and equipment, good manufacturing technology, and valid analytical methods. This chapter attempts to identify and describe the steps that contribute to the production of superior-quality extracts, at the laboratory and manufacturing plant levels.

3.2 Selection of plant species

The success of an R&D project on product development rests mainly on the choice of the herb. Therefore, the focus should be on a herb that has interesting properties, as evidenced by traditional use. Indian systems of medicine such as Ayurveda, *Unani*, and Siddha are veritable treasure houses of knowledge on medicinal plants. *Caraka Samhita* (Sharma and Dash, 2004), *Suśruta Samhita* (Bhishagratna, 1916), *Aṣṭāṅgasamgraha* (Mitra and Sharma, 2006), and *Aṣṭāṅgahṛdaya* (Murty, 2007) are the most important Sanskrit sources of ayurvedic information. There are many works on *Unani*, all written in either

Persian or Arabic. Nevertheless, plenty of valuable information is available in the English language (Said, 1997; Anonymous, 1994, 2001b, 2006a, 2006b, 2007c, 2007d, 2007e, 2007f, 2007b, 2008b, 2008c, 2009b, 2009c, 2010h, 2011b). Information on the medicinal plants used in the Siddha system of medicine can be obtained from Anonymous (1993) and the famous Tamil–English dictionary of T.V. Sambasivam Pillai (1931).

The World Health Organization (WHO) has published four collections of monographs on selected medicinal plants (Anonymous, 1999b, 2002b, 2007h, 2009h). These works contain descriptions of botanical features, geographical distributions, listings of major chemical constituents, traditional medical uses, pharmacology, dosage, and contraindications. In addition, two publications—*Medicinal Plants in China* (Anonymous, 1997b) and *Medicinal Plants in the South Pacific* (Anonymous, 1998a)—provide an introduction to the traditional use of numerous plants of Asia and the Far East.

In the past, very little information about medicinal plants was published in standard medical or pharmacy journals in the United States. Access to information was difficult and the information that was available in the European scientific press was not considered to be reliable. Consequently, many healthcare professionals were under the false notion that there was no scientific evidence in support of the safety and efficacy of herbal dietary supplements. This situation has now undergone a sea change. In the past 10 years, great progress has been made in the United States in disseminating the available scientific and clinical information on the quality, safety, and efficacy of medicinal herbs. The *United States Pharmacopoeia* has published many official botanical monographs and many more are in preparation (Mahady, 2003).

In addition to these publications, databases such as the International Bibliographic Information on Dietary Supplements (IBIDS) and Computer Access to Research on Dietary Supplements (CARDS) and many websites disseminate authentic information on medicinal plants (Mahady, 2003).

While selecting plant species for R&D work, it needs to be ascertained whether the work can generate any intellectual property for the organization. Patents issued by the United States Patent and Trademark Office and patent applications published by them are available on their official website (www.patft.uspto.gov). Information on patents issued by the European Patent Office (EPO) is available at www.ep.espacenet.com and www.epo.org. The EPO provides access to materials covering a wide range of patent-related topics. There is also a facility for e-learning (www.e-learning.epo.org).

To summarize the process of selecting the plant species, the R&D team should zero in on a medicinal plant that can form the basis of an innovative and efficacious product, which can be protected by a patent. The plant should be freely available in tons in any given season. There is no point in working on a rare herb that has extraordinary properties. Nevertheless, to begin with, only a small quantity of the herb (10–15 kg) should be procured, preferably from four different geographical regions of the country, and if possible, it should be collected in different seasons of the year.

3.3 Procuring initial samples of the herb

Before embarking on the collection of plant materials, the R&D team should target collection locations and probable dates for collection. This is necessary to procure useful specimens. In some cases, it is also essential to obtain collection permits and establish contacts with officials, as in the case of reserve forests and protected areas. The Convention on International Trade in Endangered Species of Wild Fauna and Flora (CITES), an international agreement between nations, calls on member countries to ensure that international trade in species of wild animals and plants does not threaten their survival. In accordance with this agreement, the Ministry of Commerce, Government of India, has banned the export of 29 plants, plant parts, and their derivatives and extracts *vide* notification No. 24 (RE-98)/1997-2002, dated October 14, 1998. The list includes plants such as *Kaempferia galanga*, *Picrorhiza kurroa*, and *Pterocarpus santalinus*. Similarly, there exists in the United States a list of Chinese herbs that cannot be exported from China or imported into the United States (Anonymous, 2015). It needs to be confirmed, therefore, that the plant species to be collected is not included in such lists.

Soon after collecting the desired plant material, a voucher herbarium specimen of the same should be prepared. According to Miller and Nagorsen (1992), voucher specimens are representative specimens that are collected in biological field surveys and research and are preserved to permit independent verification of results and to allow further study. A voucher herbarium specimen is a pressed sample of the plant, deposited in a recognized herbarium, where it will be maintained on a long-term basis. Herbaria can also include seeds, wood sections, pollen, microscope slides, frozen DNA extractions, and fluid-preserved flowers or fruits (Anonymous, 2010c, 2010g). Such samples support research work and can be examined to verify the identity of the plant species used in a study. The preservation of voucher specimens is necessitated by the constantly changing plant systematics and shifts in species alignments and groupings.

A good specimen of the plant material that has been collected should be pressed, dried, and mounted on herbarium paper. Pressing must start immediately on collection of the material. When pressed, wilted plants will yield inferior specimens. Plants should be carefully arranged on a newspaper or blotting paper and then placed inside a wooden plant press, which is then tightened using straps or ropes. Plants are usually arranged on the paper in a lifelike arrangement, so that the dried specimen will resemble a live counterpart. The blotting paper or newspaper sheet is changed daily to facilitate the complete drying of the material (Anonymous, 2010g).

The pressed and dried specimen is affixed with glue to a sheet of heavy paper that provides physical support and allows the specimen to be handled and stored without damaging it. The botanical identity of the specimen is ascertained with the help of dichotomous keys, published descriptions of plants, illustrations and photographs, comparison with properly identified herbarium specimens, and discussions with experienced plant taxonomists (Anonymous, 2010g). In the case of unknown plants, the help of recognized herbaria, such as the Central National Herbarium, Howrah, Calcutta, can be sought. Equally resourceful are the Royal Botanic Garden, Kew, England; Brooklyn Botanic Garden, New York; and Fairchild Tropical Botanic Garden, Florida.

On completion of the identification of the plant specimen, a specimen label should be pasted on the space beside the specimen and an institutional seal affixed to the sheet. The label should have the following components: the Latin binomial with author citation, detailed location of collection, other plants growing in association, plant habitat, frequency (frequent, rare, or occasional), description of the plant (color of flower or fruit, fragrance, leaf orientation, etc.), names of the collector and other collectors present on the expedition, collection number, and date of collection (Anonymous, 2010g) (**Figure 3.1**). Detailed procedures for preparing herbarium specimens are described by Anonymous (1999a, 2010g) and Carter et al. (2007).

3.4 Procuring larger consignments of herbs

Procuring herbs in large quantities is a crucial aspect that needs to be addressed by a group that has linguistic and negotiation skills. Needless to say, the herb purchase team should communicate effectively with the suppliers. As a preliminary step, attempts should be made to prepare an extensive inventory of herb suppliers with their contact details. A vendor approval questionnaire can be circulated among them to gather information such as the number of years in operation, major herbs traded, warehousing facilities, major clients, etc. Herb suppliers with integrity and efficiency can be identified from business transactions. As far as possible, herbs should only be procured from such suppliers. Suppliers should not be changed without a valid reason.

BRITISH COLUMBIA
CONSERVATION DATA CENTRE
-Flora of British Columbia-

Alisma gramineum **Lej.**

LOCALITY:	Carnation Creek estuary, ca. 14 km NE of Bamfield
HABITAT:	Dominant on tidal mud flats with Plantago maritima, Honkenya peploides, Spergularia, and Salicornia virginica; slope 1%; asp W
UTM:	10U 353300 5419600 **NAD** 83
LAT_LONG:	48°10′/125°00′ **ELEV** 0 m
COLLECTOR:	G.W. Douglas, J.L. Penny & N. Alexander
COLL NO:	13298 **COLL. DATE:** 98-06-30
PLOT NO.:	_____ **DET.:** GWD/98
NOTES:	ca. 500 plants/1 ha.; flowers reddish-tinged

Figure 3.1 *Specimen label on herbarium sheet. (Reproduced with permission of British Columbia Conservation Data Centre, Ministry of Environment, Canada.)*

The herb purchase team should obtain and retain voucher specimens of all consignments of herbs supplied in bulk. It is sometimes possible that a single commercial batch of a herb may have been collected by different herb collectors in different localities. In such cases, the voucher material should include samples taken from the entire purchased lot, so that there is a greater likelihood of including multiple species (Hildreth et al., 2007). The WHO insists on sampling 10% of containers from a consignment. Samples should be collected from the top, middle, and bottom of each container, then mixed homogeneously and reduced to a sample of an appropriate volume (Anonymous, 1998b). Hildreth et al. (2007) have described in detail a standard operating procedure for collecting voucher specimens of herbs to be used in the nutraceutical industry.

3.5 Quality control of raw material

A common problem encountered in the crude herb trade is the presence of morphologically allied and geographically co-occurring species (Srirama et al., 2010). This can arise from accidental or intentional adulteration of crude herbs and can lower the trade value of the herb in question and threaten the safety of medicines prepared with them (Wieniawski, 2001; Song et al., 2009). Therefore, the identity of the crude herb procured from a supplier needs to be ascertained. This is achieved by studying the macromorphology, micromorphology, ultraviolet/visible (UV/VIS) spectroscopy, infrared (IR) spectroscopy, and thin layer chromatography (TLC).

3.5.1 Macromorphology

The pharmacognosist in charge of the identification of raw materials should be able to classify the incoming raw material into one of the following groups, namely, powders, woods, barks, leaves, flowers, seeds, fruits, whole plants, rhizomes, or roots, and unorganized drugs such as beeswax, guggul gum, olibanum, vegetable oils, agar, and opium, which have a generally uniform structure, but are not composed of cells. By studying the pharmacognostical characteristics of the samples, a clear distinction can be made between the powders of minerals, starches, and herbs (D'Amelio, 1999).

3.5.2 Micromorphology

Microscopy techniques are invariably used to study the finer aspects of the herb. For effective identification, the pharmacognosist relies on characteristics such as crystals, trichomes, palisade ratios, vessels, vein-islet number, starch granules, fibers, sclerenechyma, etc. Chemical solutions such as chloral hydrate solution, potassium hydroxide solution, ether–alcohol mixture, and a solution of chlorinated soda are used for dissolving the chemical inclusions in cells and isolating the tissue elements so that the botanical identity of the herbal material can be established unambiguously (D'Amelio, 1999) (**Figure 3.2**).

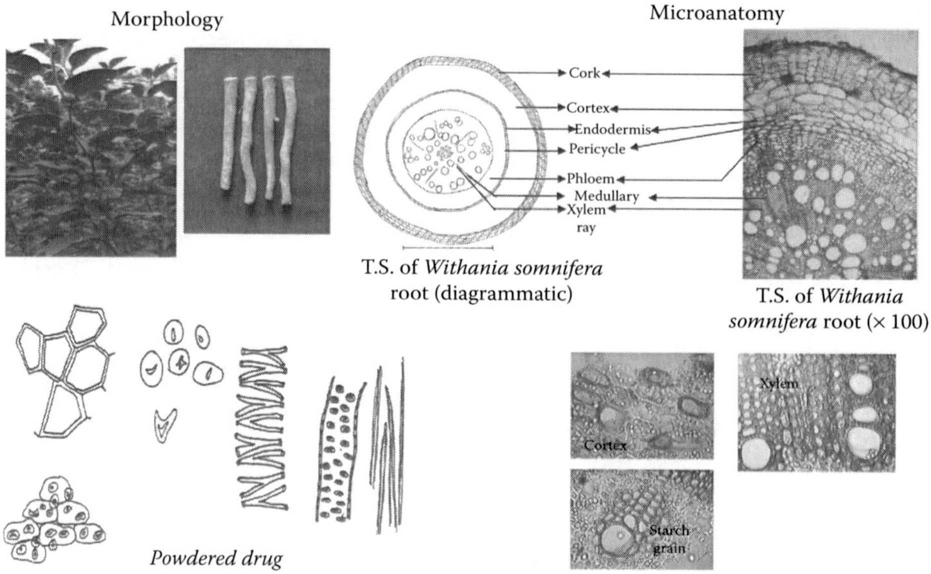

Figure 3.2 *Morphology and microanatomy of* Withania somnifera. *(Reproduced from Chemistry and Manufacturing Control of Botanical Products, Herbal International Summit and Expo on Medicinal, Aromatic products and Spices (Hi-Maps), April 4, 2008, with permission of Bedi, 2008.)*

3.5.3. Ultraviolet and visible spectroscopy

The identity of the plant material can be established by studying the absorption spectra of the constituent compounds. A dilute solution of the herb in an appropriate solvent such as ethanol, methanol, water, hexane, ether, and petroleum ether is studied in a UV/VIS spectrophotometer at 200–400 nm for colorless compounds or 200–700 nm for colored compounds (D'Amelio, 1999).

3.5.4 Infrared spectroscopy

The herbal material is mixed with KBr and pressed into a disc, which is placed inside a cuvette and introduced into the IR spectrophotometer. The IR spectrum obtained is used as a reference spectrum against which spectra of subsequent batches of herbs can be matched (D'Amelio, 1999).

3.5.5 Thin layer chromatography

Developed in 1938 by Nikolai Izmailov and Maria Schreiber of the Institute of Experimental Pharmacy of the State University of Kharkov, Ukraine, TLC has now become a vital tool in the analysis of herbs, drugs, and cosmetic preparations (Izmailov and Schreiber, 1938; Ettre, 2002). This is an inexpensive technique to obtain semiquantitative information on the major active compounds in a herb. Glass plates and aluminum or plastic sheets are coated with an absorbent such as silica gel 60 F_{254} or aluminum oxide. An extract of the herb is spotted at one end of the plate, which is then placed in a TLC chamber containing an appropriate solvent system that rises up on the TLC plate carrying the constituent compounds. When the solvent rises up to the required height, the plate is taken out of the chamber, dried and sprayed with a specific reagent. Distinct spots that appear on the plate can be visualized and identified in daylight or under UV light (254 or 365 nm). The spots are identified on the basis of appropriate reference standards spotted alongside the herb extract. TLC can be used for any class of compounds. An improved version of TLC, known as high-performance TLC (HPTLC), permits the quantification of compounds present in the spots (Wagner and Bladt, 1996; D'Amelio, 1999; Stahl, 2005; Vasisht, 2008).

In addition to the aforementioned tests, several other tests are carried out to evaluate the quality of the raw material. These parameters include extractive values, ash values, bulk density, physical constants such as specific gravity, optical rotation, viscosity, refractive index, and so on in the case of oils, fats, oleoresins, and balsams, microbial status and assay of active compound or categories such as alkaloids, saponins, tannins, etc. (Lohar, 2010). On the basis of tests related to quality control (QC), certificates of analysis (COA) of the herb collected from at least four different geographical regions are prepared. After careful study of these COAs, limits are set for the major parameters, so that the raw material specification of the herb in question can be formulated. Raw material specifications for the herb *Coleus forskohlii* and the various tests and methods to be used are provided in **Tables 3.1** through **3.3** to serve as examples. Only consignments that comply with this specification should be accepted for solvent extraction and subsequent product development activities.

3.5.6 Loss on drying

Not more than 10%. Determine on 10 g of powdered sample. Dry at 105°C in an oven until constant weight is reached.

Table 3.1 Raw Material Specifications of *C. forskohlii* Herb: Sampling and Handling

Standard packing	50–100 kg of the herb packed in gunny bags.
Storage requirement	To be stored in well-closed bags.
Sampling level	Sample all bags if the number of bags is three or less. If the number of bags is more than three, use the formula $\sqrt{n} + 1$, where n is the number of bags. Thus, determine the minimum number of bags to be sampled. Sample from randomly selected bags, a minimum quantity of 50 g, from each bag into a clean and dry polythene bag.
Quantity of composite sample for analysis	1 kg.
Handling and precaution	Use gloves and a nose mask while handling the herb.
Retest period	One year. If not tested after 1 year, carry out retest prior to extraction.

Table 3.2 Raw Material Specifications of *C. forskohlii* Herb: Standards and Limits

Description	Pale brown slices of tuber with characteristic odor. The material should be physically examined for the presence of fungal growth, insects, and adulterants.
Identification	
Macroscopy	To pass the test
Microscopy	To pass the test
TLC	To pass the test
HPLC	Positive for forskolin
Foreign Matter	
Sand and silica (% w/w)	Not more than 1
Physicochemical Parameters	
Moisture content (% w/w)	Not more than 12
Total ash (%)	Not more than 8
Acid-insoluble ash (% w/w)	Not more than 2
pH of 5% (w/v) suspension	4.0–7.5
Alcohol soluble extractives (% w/w)	Not less than 15
Water soluble extractives (% w/w)	Not less than 20
Microbial Limits	Less than 10^5/g
Total fungal count	Less than 10^3/g
Total Enterobacteriaceae	Less than 10^3/g
E. coli	To be absent
Salmonella	To be absent
Heavy Metals	
Lead	Less than 10 ppm
Arsenic	Less than 3 ppm
Cadmium	Less than 1 ppm
Mercury	Less than 0.1 ppm
Aflatoxins	
B1	Less than 5 ppb
Sum of B1 + G1 + B2 + G2	Less than 10 ppb
Pesticide residues	To meet the requirements
Phytochemical analysis	To meet the requirements of Grade A, B, or C
Grade A	>1% forskolin (superior)
Grade B	>0.8% forskolin (good)
Grade C	>0.5% forskolin (average)

Source: Reproduced with permission from Agarwal, A. and Murali, B., *Quality Assessment of Selected Indian Medicinal Plants*, Vol. 1, 1–252. National Medicinal Plants Board & Natural Remedies, Bangalore, New Delhi, 2010.

Table 3.3 Raw Material Specifications of *C. forskohlii* Herb: Thin Layer Chromatography

Description

Pale brown slices of tuber with characteristic odor. The material should be physically examined for the presence of fungal growth, insects, and adulterants.

Identification: To comply with TLC.

The Rf values of the spots obtained from the sample preparation should be comparable with that of the reference material and forskolin standard.

Mobile Phase

Toluene:ethyl acetate

85:15

Stationary Phase

Precoated silica gel $60F_{254}$ plate

Sample and Reference Material Preparation

Accurately weigh a 2 g sample of coarse powder and *Coleus* reference material and transfer to separate 250 mL beakers. Extract each with 50 mL of acetonitrile by boiling on a water bath for about 20 min and transfer the extracts to two other 250 mL beakers. Repeat the process four or five times till the extracts are colorless. Concentrate the extracts to 80–90 mL, cool to room temperature, transfer to 100 mL volumetric flasks, make up the volume to 100 mL with acetonitrile, mix well, and filter.

Standard Preparation

Weigh 2.5 mg of forskolin reference standard in a 25 mL volumetric flask, dissolve in 15 mL of acetonitrile by warming on a water bath for 10–15 min, cool and make up with acetonitrile.

Procedure

Separately apply equal volumes (10 μL) of the sample, reference material, and forskolin standard. Develop the chromatogram to three-quarters of the plate, in the chamber previously saturated with mobile phase. Dry the plate under a current of air. Spray the plate with vanillin–sulfuric acid reagent. The reagent is prepared by mixing equal quantities of 5% w/v vanillin in glacial acetic acid and 10% v/w sulfuric acid in water. The reagent is colorless. Discard the reagent if color develops. After spraying, heat the plate at 60°C–70°C for 5–10 min and observe in visible light. The TLC profile of the sample should be similar to that of the reference material and also with a spot at a retention factor (Rf) corresponding to that of forskolin, indicating that the sample is genuine.

Source: Reproduced with permission from Agarwal, A., and Murali, B., *Quality Assessment of Selected Indian Medicinal Plants*, Vol. 1, 1–252, National Medicinal Plants Board & Natural Remedies, Bangalore, New Delhi, 2010.

3.5.7 Yield of extract (user test)

Not less than 15–20%. Accurately weigh about 100 g of *Coleus* root powder from the representative sample (W1), transfer into a 1 L round-bottomed (RB) flask and reflux with 300 mL of methanol for 3 h. Cool to room temperature, filter the supernatant solution through polypropylene cloth into a filtration flask. Continue the refluxing two more times using 300 mL of methanol each time.

Combine the extracts and filter again through polypropylene cloth into a filtration flask. If some sediment appears, use light vacuum. Distill the methanol and dry completely using a rotary evaporator and a 2 L RB flask (W2). Cool and take the final weight of the RB flask (W3). Calculate the yield as follows:

$$\frac{(w_3 - w_2)}{w_1} \times 100 \tag{3.1}$$

3.5.8 Assay by high-performance liquid chromatography (HPLC)

The content of forskolin is estimated by HPLC. The raw material should contain not less than 1% of forskolin on a dry basis. The raw material is examined by isocratic liquid chromatography using a HPLC system equipped with a dual/quaternary pump, a manual/autoinjector, and a photodiode array or UV detector supported by suitable software. The details are given in **Table 3.4**.

3.6 Traceability of herbs

The herb procurement personnel should also obtain from suppliers vital information helpful to achieve the traceability of the material. The concept of the traceability of herbs originated from tracking bioengineered products. Following the events of September 11, 2001, there was increased demand for assuring the traceability of all food, herbs, and nutraceuticals brought into the United States and Europe, not only for QC, but also to prevent bioterrorism (Shorab, 2008; Lachance, 2008). The following kinds of information help in the traceability of herbs: location of collection, name and address of farmer in the case of cultivated herb, pesticides and fertilizers used, time of harvest, age of herb, mode of harvesting, postharvest treatments, number of days taken to transport the produce from site of harvest to the extraction unit, and mode of transport and size, in the case of fruits (average diameter and weight, number of fruits per kilogram).

Instrumental techniques can be successfully utilized to trace the source of herbal raw materials. This was demonstrated by Cambrai et al. (2010) in their study on the geographical origin of cocoa beans used in the commercial production of chocolates by Valrhona (Tain-l'Hermittage, France). Using gas chromatography–mass spectrometry (GC-MS), the authors were able to identify two or three characteristic (height) volatile compounds in chocolates manufactured using cocoa beans procured from Africa, the Caribbean Islands, and Madagascar (**Table 3.5**).

Table 3.4 Estimation of Forskolin by HPLC

Chromatographic Conditions

Mobile phase	Degassed mixture of 55 volumes of water and 45 volumes of acetonitrile
Column	Stainless steel column (250 × 4.6 mm) packed with octadecylsilane bonded to porous silica
Detector	Photodiode array or UV detector
Wavelength	220 nm
Flow rate	1.8 mL/min
Run time	60 min
Injection volume	20 μL

Standard Preparation

Accurately weigh 10 mg of forskolin reference standard and transfer to a 10 mL volumetric flask. Dissolve in 5 mL of acetonitrile by mild warming on a water bath, cool, make up to 10 mL with acetonitrile and mix well.

Sample Preparation

Accurately weigh a 3 g sample of coarse powder and transfer to a 250 mL beaker. Extract with 50 mL of acetonitrile by boiling on a water bath for about 20 min and transfer the extract to a 250 mL beaker. Repeat the process four or five times till the extract is colorless.

Pool the filtrates and concentrate to 80–90 mL, cool to room temperature, transfer to a 100 mL volumetric flask, make up the volume to 100 mL with acetonitrile, mix well and filter through 0.45 membrane.

Procedure

Chromatograph the standard preparation by injecting 20 μL. Calculate the relative standard deviation (RSD; not more than 2%) for at least three replicate injections and a tailing factor of not more than 1.5. Chromatograph the sample preparation and calculate the percentage of forskolin from the peak responses.

Calculation

The content of forskolin on a dry basis can be calculated using the following formula:

$$\frac{A_s}{A_{std}} \times \frac{W_{std}}{stdd} \times \frac{sd}{swt} \times \frac{P_{std}}{100} \times 100$$

where:

A_s = area of the sample
A_{std} = area of the standard
W_{std} = weight of standard in milligrams
$stdd$ = standard dilution
sd = sample dilution
swt = sample weight in milligrams
P_{std} = purity of standard

Source: Reproduced with permission from Agarwal, A. and Murali, B. *Quality Assessment of Selected Indian Medicinal Plants*, Vol. 1, 1–252, National Medicinal Plants Board & Natural Remedies, Bangalore, New Delhi, 2010.

Table 3.5 Volatile Compounds Characteristic of Cocoa Beans from Three Geographical Regions

Characteristic Compound	Geographical Group
Benzaldehyde	Africa
Linalool	Africa
Phenylacetaldehyde	Africa
(*E,E*)-2,4-Nonadienal	Caribbean
(*E,E*)-2,4-Decadienal	Caribbean
2-Phenyl-2-butenal	Madagascar
4-Methyl-2-phenyl-2-pentenal	Madagascar

Source: Adapted with permission from Cambrai et al., 2010, *Journal of Agricultural and Food Chemistry* 58: 1478–1483. Copyright 2015 American Chemical Society

3.7 Cleaning herbs

Consignments of crude herbs received from suppliers or cultivators need to be cleaned before utilizing them for extraction. The US Food and Drug Administration sponsored a 3-year study to develop standards on insect, bird, rodent, and other animal contamination levels in crude and ground spices. The study analyzed 1000 samples and found that a cultivated herb such as thyme contained insect fragments, rodent hairs, feather barbules, mites, thrips, and aphids (Gecan et al., 1986). Cleaning herbs undoubtedly improves the quality of the extracts derived from them. Metal pieces in crude drugs are removed to protect the end user from hazards. Metal pieces and stones can also damage pulverizing equipment.

There are well-defined procedures for cleaning spices. For example, black pepper is first passed through mechanical sifters to remove pinheads, vegetable seeds, dust, and similar contaminants. Thereafter, the spice is sent to multiple sieve-cum-air classifiers and gravity separators for the removal of dust, light foreign matter, and stones. The material passes from gravity separators to mechanical washers fitted with brushes for removing dust, dirt, and mold. The washed pepper is then centrifuged to remove water, dried, and sent for sterilization, either by steam or gamma radiation (Ravindran and Kallupurackal, 2001).

Tainter and Grenis (2001) have described in detail the various methods and equipment used in the cleaning of herbs. All the cleaning equipment is based on the principle of the physical difference between the herb and the foreign material that is being removed. The various cleaning equipment used in the herbal extraction industry are magnets, sifters (vibrating screens), air tables, destoners, cyclone dust collectors, air separators, aspirators, indent separators, and spiral separators. Metal pieces, other extraneous matter, weeds, stones, lumps of mud, and dust are efficiently removed in this way (Venskutonis, 2002).

Tainter and Grenis (2001) have described a microanalytical method for determining dirt present in the crude herb. A small amount of herb is floated in water and the heavy and light portions of dirt are studied. The light and heavy particles are examined under a microscope and particles of dirt such as insect fragments, rodent hairs, feather barbules, and others are counted. However, only a trained analyst can identify these tiny particles under a microscope. The European Spice Association (ESA) and the American Spice Trade Association (ASTA) have set standards for the cleanliness of herbs and spices (Muggeridge, 2001; Anonymous, 2011a).

3.8 Comminution of herbs

Only dried herbal materials are considered in modern solvent extraction. Comminution or size reduction is a process by which raw materials such as leaves, roots, bark, heartwood, or seeds are reduced to powders that can pass through sieves of different mesh sizes. Pulverizing equipment is selected on the basis of the material that is to be powdered.

Chunks of heartwood need to be minced before being fed into pulverizers. These mincing machines cut the wood lengthwise or crosswise, reducing it to small pieces. Such pieces are fed into hammer mills. An array of hammer-like solid metal blocks fixed to the shaft of the machine swing at high speed and powder the pieces of wood dropped into the chamber that houses them. The number of hammers varies with different models.

The particle size of the powder is dependent on the perforated screen that is fixed to the wall of the chamber (**Figure 3.3**). Another piece of equipment, called the *knife mill*, has sets of sharp knives fixed to a rotor (**Figure 3.4**). These knife mills are suitable for powdering leaves, roots, and barks.

The pulverization of herbs can also be carried out using a pin mill (teeth mill, impact mill). This equipment has a circular row of pins (teeth) fixed inside the chamber. Bits of herbs that are dropped into it are powdered through the beating, shearing, crushing, and colliding that the material undergoes during the process (**Figure 3.5**).

Large-sized pin mills have a screw feeder with a variable speed drive, ensuring the uniform feeding of the herb into the mill. There are other types of comminution equipment such as the shredding mill, ball mill, slow-speed attrition mill, micromill, and air-swept mill.

Two major problems are encountered during the pulverization of herbs. One is the disruption of the oil-bearing parts of the herb and the subsequent release of essential oils, which become prone to oxidative changes. Secondly, the heat generated during the operation causes vaporization of the aromatic components. It is thus evident that the temperature needs to be controlled in some grinding operations (Venskutonis, 2002). Cryogenic grinding advocated

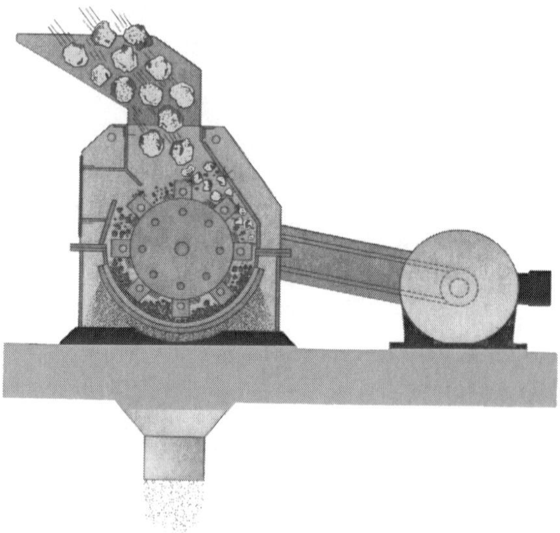

Figure 3.3 *Hammer mill. (Reproduced with kind courtesy of Schutte-Buffalo Hammermill, LLC., Buffalo, New York. www.hammermills.com.)*

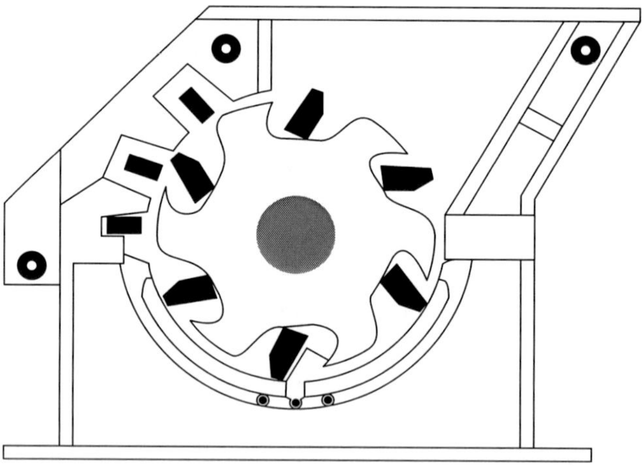

Figure 3.4 *Knife mill. (Reproduced with permission of PALLMANN Maschinenfabrik GmbH & Co. KG, Zweibrücken, Germany. www.pallmannindustries.com.)*

by Cohodas (1969) is an effective way to counter the loss of essential oils during grinding. In this method, the herb and a gaseous refrigerant such as nitrogen are fed into the milling head at a controlled rate. The cryogenic blanket provided by nitrogen prevents the loss of aromatic compounds. The risk of fire hazard is also eliminated.

Figure 3.5 *Pin mill. (Reproduced with permission of PALLMANN Maschinenfabrik GmbH & Co. KG, Zweibrücken, Germany. www.pallmannindustries.com.)*

Different comminution processes are adopted for different herbs. Herbs with more stems and stalks need shredding or cutting mills, whereas hammer mills are suitable for hard and brittle materials such as resins, and leafy plants such as senna and *Melissa*. Hard materials such as cinnamon bark and dry ginger rhizomes can be pulverized in a two or three stage process involving cutting, shredding, and pulverization in hammer or pin mills. Pulverization should be carried out with care, so that the herbal material is not damaged due to the generation of heat (Anonymous, 2001c).

3.8.1 Sieves

The pulverized material is passed though sieves with a definite mesh size. Powders are usually classified as coarse powder, moderately coarse powder, moderately fine powder, fine powder, and very fine powder. They are obtained in degrees of coarseness or fineness, after sieving the comminuted material through sieves of the appropriate nominal mesh aperture sizes. The nominal mesh aperture size and the corresponding sieve numbers are as follows: 1.70 mm (#10), 1.00 mm (#16), 710 μm (#22), 500 μm (#30), 355 μm (#44), 250 μm (#60), 180 μm (#85), 150 μm (#100), and 125 μm (#120). The particles of a *coarse powder* pass through a sieve with a nominal mesh aperture of 1.70 mm and not more than 40% through a sieve with a nominal mesh aperture of about 355 μm. The particles of a *very fine powder* pass through a sieve with a nominal mesh aperture of 125 μm (Anonymous, 1980). Coarse powders are generally used in solvent extraction work. A small quantity of the pulverized herb should be properly labeled and stored as a voucher sample.

3.9 Selection of extraction solvent

To get a high yield of target compounds, the extraction solvent should have several desirable properties. It should solubilize the compounds selectively, it should be stable at the extraction temperature, and it should not react with the desired compounds. Its viscosity should be low to enable easy passage through the particles of herb. Usually, low boiling point solvents are preferred, as they can be profitably recovered. The extraction solvent should be safe, be available in large quantities, have low flammability, and be reasonably priced (Takeuchi et al., 2009).

3.9.1 Selectivity of solvents

As mentioned earlier (Section 2.6.2.1), the selectivity or the ability of the solvent to extract the desired compound is an important factor that decides the success of the extraction process. Considering the selectivity and the ability to extract natural products in a stable form, devoid of undesirable chemical transformations, aliphatic alcohols or their mixtures with water are known to have great extractive power. Usually, ethyl alcohol of various strengths is used. The literature indicates that 70% or 80% of aqueous alcohol is suitable for the extraction of heartwood, bark, seeds, and root. However, aqueous alcohol with a strength of less than 50% is suitable for the extraction of leaves or aerial parts. By using 50% aqueous alcohol, one can avoid extracting chlorophyll and resinous substances, the removal of which later is a considerably difficult task (Bombardelii, 1991).

How the moistening of the herb powder can set off a series of chemical changes is exemplified by the conversion of the primary glucosides A, B, and C of *Digitalis lanata*. When moistened with water or 20% aqueous alcohol at room temperature, lanatosides A, B, and C are transformed into digoxin, gitoxin, and digitoxin, respectively. Therefore, to extract the primary glucosides, it is necessary to use a hydroalcoholic mixture of a strength that is above 50%. This is required to inhibit the action of hydrolytic enzymes. Many of the constituents of plants can behave in this manner and therefore the extraction solvent and the extraction temperature are decided judiciously (Bombardelli, 1991).

Many alkaloids can be extracted using the procedure outlined here. Powdered herb is extracted with boiling methanol. The solvent is distilled out and the extract is treated with inorganic acids, thereby forming the salts of alkaloids. This mixture is partitioned with chloroform. The chloroform layer contains many impurities present in the mixture and it is discarded. The aqueous layer is treated with sodium carbonate, which causes the liberation of free bases that need to be extracted with chloroform. The resultant mixture of alkaloids can be separated into individual compounds by various separation techniques (Finar, 2001).

Flavonoids and terpenes can be extracted at neutral pH with ethyl acetate or acetone. For other classes of chemical compounds, it is very difficult to assign specific solvents, on account of the diversity of the chemical structures and polarities of the compounds. Extraction solvents are decided to suit the herb in question (Bombardelli, 1991).

3.9.2 Analysis of solvent

All organic solvents, including deionized water, should be analyzed in a QC laboratory before being utilized in extraction work. A COA of the solvent in question is prepared and only those lots that conform to the specifications are accepted for extraction. Batches of solvents that fail in analysis should be rejected. The parameters for deionized water and rectified spirit are given as examples in the following subsections.

3.9.3 Deionized water

Color, clarity, odor, taste, pH, acidity, alkalinity, ammonium content, residue on evaporation, heavy metals, calcium, magnesium, chloride, nitrate, sulfate, and oxidizable substances are determined (**Table 3.6**).

3.9.4 Rectified spirit

Color, clarity, odor, miscibility, moisture, specific gravity, acidity, alkalinity, boiling range, assay by distillation, and content of miscible and immiscible solvents by GC need to be estimated in the case of rectified spirit (**Table 3.7**).

Table 3.6 Certificate of Analysis of Deionized Water

Product: **Deionized water**
Batch No.: -----

Parameter	Specification	Certificate of Analysis
Description	Clear, colorless liquid, odorless and tasteless	Clear, colorless liquid, odorless and tasteless
pH	Between 4.5 and 7.0	6.9
Acidity/alkalinity	To comply with the test for acidity/alkalinity	Complies
Ammonium	To comply with the test for ammonium	Complies
Residue on evaporation	Not more than 0.001%	0.0005%
Heavy metals	Not more than 0.1 ppm	Complies
Calcium and magnesium (hardness as $CaCO_3$)	Not more than 5 ppm	Complies
Chloride	To comply with the test for chloride	Complies
Nitrate	Not more than 0.2 ppm	Complies
Sulfate	To comply with the test for sulfate	Complies
Oxidizable substances	To comply with the test for oxidizable substances	Complies
Total microbial count	Not more than 100 cfu/mL	Complies
Total yeast and mold count	Should be absent	Absent
E. coli	Should be absent	Absent
Salmonella	Should be absent	Absent
Staphylococcus aureus	Should be absent	Absent
Pseudomonas aeruginosa	Should be absent	Absent

Table 3.7 Certificate of Analysis of Rectified Spirit

Product: **Rectified spirit**
Batch No.: -----

Parameter	**Specification**	**Certificate of Analysis**
Description	Clear, colorless, flammable liquid, free of suspended matter and rust particles, with characteristic odor	Clear, colorless, flammable liquid, free of suspended matter and rust particles, with characteristic odor
Miscibility	Miscible with water, chloroform, ether, and glycerin	Complies
Moisture content by KF method	Not more than 5%	4.5%
Specific gravity	Between 0.8084 and 0.8104 at 25°C	0.8084
Acidity or alkalinity	Not more than 0.2 mL of 0.1 M NaOH/HCl is required for 20 mL of sample	Complies
Boiling range	Boils at about 78°C	Complies
Assay by distillation	Not less than 94.7% and more than 95.2% v/v	Complies
Methanol	Should be absent	Complies
Purity by GC	Not less than 99.5%	99.69%

3.10 Extraction of herbs

3.10.1 Optimization of extraction process

The act of obtaining the best possible result or the effort for achieving the optimal solution under a given set of circumstances is known as *optimization*. The goal of this effort is either to minimize the production cost or to maximize profit, by way of enhancing product quality and operation yield. Such operations are made efficient by applying appropriate optimization methods and adopting suitable technological and managerial decisions (Tzia, 2003).

Optimization is defined as the process of finding the conditions that produce the optimum (maximum or minimum) value of a function of a set of decision variables in the wake of restrictions imposed (Edgar et al., 2001). The conditions (the processing variables) that produce the desired optimum result are called *optimum conditions* and the best possible design is called the *optimal design* (Tzia, 2003).

The optimization of any manufacturing process can be achieved by empirical or statistical methods. The empirical method makes use of the one-factor-at-a-time strategy, in which one factor varies at a time, while all others remain constant. The major disadvantage of the empirical method is that it fails to consider any interaction between the factors. An interaction is the failure of one factor to produce the same effect on the response at different levels of another factor. The right approach to deal with several factors is to adopt a factorial experiment strategy, in which the factors are varied altogether, instead of just changing one at a time (Montgomery, 2009).

3.10.2 Response surface methodology

Process optimization has been made easier with the adoption of response surface methodology (RSM). Originally introduced by Box and Wilson (1951), RSM is a collection of mathematical and statistical techniques that are widely used for developing, improving, and optimizing processes. It is also applied in the design, development, and formulation of new products and for improving existing product designs. RSM makes use of a series of designed experiments to obtain an optimal response. RSM is extensively used in industrial processes in which several variables influence some performance measure or quality characteristic of the product, or the process itself. This performance measure or quality characteristic is called the *response*. The *input variables* are called *independent variables*. These are controlled by the engineer or the R&D scientist (Myers et al., 2009). RSM is applied to the optimization of process technologies in diverse disciplines such as biotechnology (Kawaguti et al., 2006), edible oil technology (Mani et al., 2007), food technology (Sharma et al., 2006; Kong et al., 2010), microbiology (Iyer et al., 2010), mining (Aslan 2007; Aslan et al., 2008; Aslan and Ünal, 2010; Karimi et al., 2011), molecular biology (Maldonado et al., 2007), pharmaceutics (Palamakula et al., 2004), pollution control (Ye et al., 2010), and synthetic chemistry (Lee et al., 2010). Its usefulness is valued in economics (Burdick and Naylor, 1969), psychology (Meyer, 1963), and even warfare (Grier et al., 1997).

Unlike the empirical method of process optimization, RSM takes into account the possible interaction among the independent variables and minimizes the number of experiments. RSM has been successfully applied to the optimization of the solvent extraction of herbs (Mani et al., 2007; Cacace and Mazza, 2003; Karacabey and Mazza, 2008).

The end result of the extraction of a natural product is influenced by several operational parameters such as the extraction method, particle size, type of solvent, solid to solvent ratio, extraction temperature, residence (static) time (duration of extraction), extraction cycle (number of extractions), and pH (Luthria et al., 2007; Luthria, 2008; Karacabey and Mazza, 2008). These are the input variables or independent variables of the process. These variables influence the end result or response, which consists of the extraction yield, the content of the active substance in the extract, the unwanted substances in the extract, and the cost of the extract.

3.10.2.1 Selection of independent variables

3.10.2.1.1 Solvent The influence of each independent variable on the extraction process is studied first with the help of single-factor experiments. The selection of the ideal solvent is the most crucial among them. Sadeghi-aliabadi et al. (2009) optimized the solvent extraction of taxol from *Taxus baccata* needles. The appropriate solvent was selected from a series of normal soaking extraction (NSE) experiments. Defatted herb powder (1 g) was put in a test tube with 5 mL of a solvent such as acetone (100%, 50%, and 20% in water) or ethanol (97%, 50%, and 20% in water). The test tubes with the herb powder and solvent were kept at room temperature for 24 h, with occasional shaking. The content of each test tube was filtered and the miscella was saved. The herb powder was extracted three times with the same solvent in this way. The combined miscella from each sample of herb powder was dried in a rotary evaporator. The content of taxol in each sample was analyzed using HPLC and the results were plotted on a graph. Acetone (100%) was selected as the ideal solvent, as it was found to be the solvent with the highest efficiency to extract taxol.

Another example of solvent selection is provided by Wang et al. (2010), who optimized the microwave-assisted extraction of flavonoids from *Radix puerariae*. A dry powder of the herb (500 mg) was extracted with 15 mL of ethanol of varying concentrations (50%, 60%, 70%, 80%, and 90% in water). The samples of miscella were filtered and freeze-dried, and their flavonoid content was estimated spectrophotometrically. The yield of flavonoids increased sharply with the increase in ethanol concentration up to 70%, whereafter the yield decreased. Therefore, 70% ethanol was selected for further studies (**Figure 3.6a**). Similar results have been reported by Duan et al. (2001) and Ma ct al. (2005).

3.10.2.1.2 Solid to solvent ratio (solvent volume) Experiments were performed by increasing the volume of the solvent from 20 to 60 mL/g, while keeping the solvent concentration and residence time constant. **Figure 3.6b** shows that the yield of the extraction reached its maximum at 30 mL/g, after which it showed a downward trend. Therefore, 30 mL/g was selected as the basis for fixing the solid to solvent ratio.

3.10.2.1.3 Microwave power To evaluate the effect of microwave power, extractions were carried out at varying microwave power, while keeping the other independent variables constant. The results show that the yield of extraction increased with increasing microwave power up to 255 W, whereafter it remained steady. Therefore, 255 W microwave power was selected for the subsequent experiments (**Figure 3.6c**).

3.10.2.1.4 Extraction time Extraction experiments were conducted for different durations of time, while keeping the other independent variables constant. **Figure 3.6d** shows that the rate of extraction increased gradually up to 6 min, after which it remained unchanged. Therefore, 6 min was chosen as the appropriate residence time (Wang et al., 2010).

These single-factor preliminary experiments showed that the operational parameters for microwave-assisted extraction of flavonoids from *Radix puerariae* are ethanol concentration (70%), solid to solvent ratio (30 mL/g), microwave power (255 W), and extraction time (6 min).

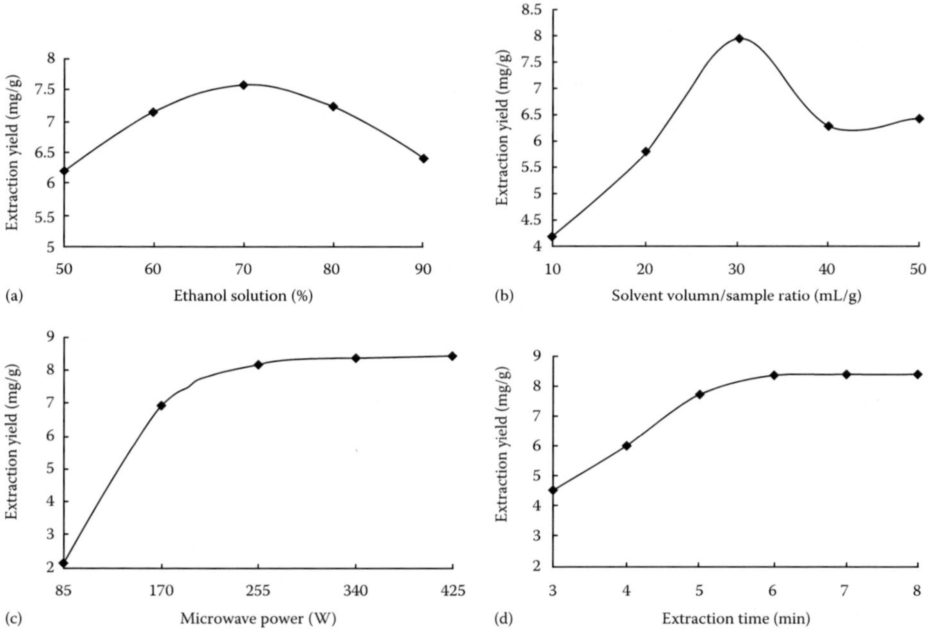

Figure 3.6 *Effects of ethanol concentration (a), solid to solvent ratio (b), microwave power (c), and extraction time (d) on the yield of flavonoids from* Radix puerariae. *(Reproduced with permission from Wang et al.,* Journal of Medicinal Plants Research *4, 304, 2010.)*

3.10.2.2 Central composite rotatable design

Central composite rotatable design (CCRD) is a very efficient statistical approach to obtain information on the effects of the independent variables on the response variable and overall experimental error, using a minimum number of experiments. Usually, the Taguchi method of robust parameter design is employed. Using this method, it is possible to design products or processes that are robust to environmental conditions and component variation, with minimum variability in the output response of the products and the target value (Myers et al., 2009).

In CCRD, each independent variable is represented at five levels. The experimental range for each independent variable is selected on the basis of the results of literature data and preliminary experiments (Juntachote et al., 2006). The real value of each level of the independent variable in question (X_i) is converted into a coded value (x_i). The independent variables are coded according to the following equation:

$$x_i = \frac{\left(X_i - X_0\right)}{\Delta X_i} \qquad i = 1, 2, 3, 4, \text{ or } 5 \tag{3.2}$$

where:

x_i and X_i are the coded (dimensionless) and the actual value of the independent variable i, respectively

X_0 is the actual value of the independent variable i at the center point

ΔX_i is the step change of X_i corresponding to a unit variation of the coded value (see **Table 3.8**)

The conversion of the actual value to a coded value brings the variables to set limits. The coefficients of regression therefore become comparable. For example, let us take the real values of particle size and extraction temperature in **Table 3.8**. The real values of these two variables have a difference of 10 orders of magnitude. If these real values are used in regression analysis, one cannot find out which of these two variables—the particle size or the extraction temperature—has a greater influence on the extraction process. Both variables come to the same order of magnitude on scaling and regression analysis becomes fruitful.

A design of the experimental protocol based on the CCRD is planned as soon as the real values of the independent variables at desired levels are converted into coded values. Twenty combinations of the coded values are required when three independent variables at five levels are considered. The experimental matrix is designed using specialized software. Actual experiments are thereafter performed on the basis of each of these 20 combinations using a small quantity of pulverized herb (20 g) and solvent (250 mL) (Mani et al., 2007). Usually, duplicate experiments are carried out at all design points except for the center point (0,0,0) for which triplicate extractions are performed. Combinations of the five levels of the three independent variables considered by Mani et al. (2007) for optimizing the extraction of Moringa oil are provided in **Table 3.8**.

Data from the CCRD experiment design are subjected to a second-order polynomial regression analysis using least square regression methodology. This exercise furnishes the parameters of the mathematical models that explain

Table 3.8 Real and Coded Levels of Three Independent Variables Related to the Extraction of Moringa Seed Kernel Oil

Particle Size (mm)		Extraction Temperature (°C)		Residence Time (h)	
Real Value X_1	Coded Value x_1	Real Value X_2	Coded Value x_2	Real Value X_3	Coded Value x_3
2.2	1.682	76.8	1.682	7.4	1.682
1.7	1	70	1	6.0	1
1.0	0	60	0	4.0	0
0.30	−1	50	−1	2.0	−1
0.18	−1.682	43.2	−1.682	0.6	−1.682

Source: Data from Mani et al., *Food and Bioproducts Processing* 85, 328, 2007.

the main effect and two-factor interactions. An analysis of variance (ANOVA) is performed on the experimental data to quantitatively estimate the relative contribution of each independent variable on the response variable. The relative significance of the factors is represented in terms of the F-ratio. A greater F-ratio indicates that the process parameter exerts greater significance on the response variable (Aslan and Ünal, 2010). Response surfaces are generated by statistical analysis software. A representative flowchart for process optimization is provided in **Figure 3.7**.

3.10.2.3 Response surfaces

Response surfaces are visualized graphically. The graph allows one to observe the shape of a response surface in the form of hills, valleys, and ridgelines (**Figure 3.8**). In this graph, each value of x_1 and x_2 generates a y value. This three-dimensional graph shows the response surface from the side and is called a *response surface plot* (**Figure 3.8**). The dots in the graph indicate the location of y with respect to x_1 and x_2 in space. The area passing through them will be the response surface (Bradley, 2007).

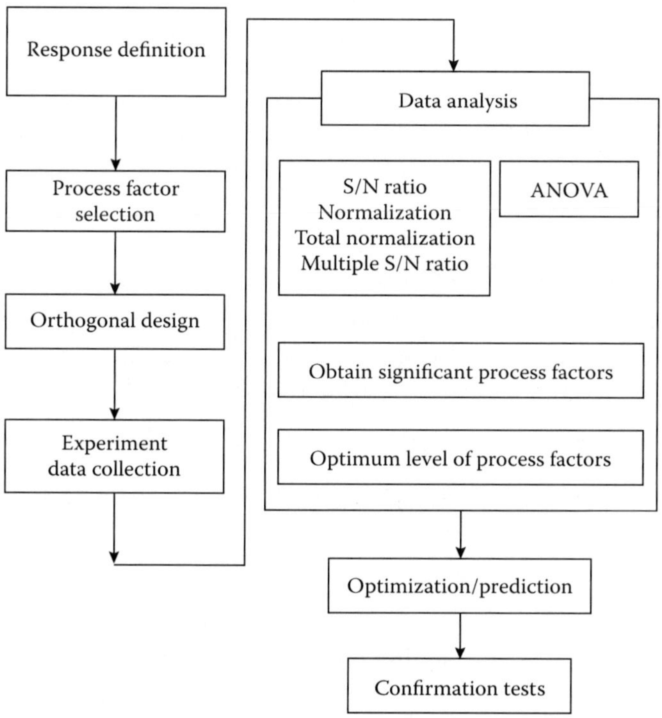

Figure 3.7 *Flowchart for process optimization. S/N = signal to noise ratio. (Reproduced from* Fuel Processing Technology, *92, Aslan, N. and Ünal, I., Multi-response optimization of oil agglomeration with multiple performance characteristics, 1157–1163, 2010 with permission from Elsevier.)*

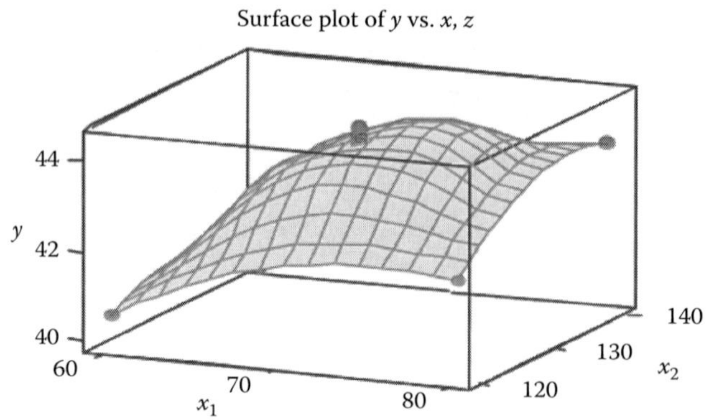

***Figure 3.8** A response surface plot. (Reproduced from Bradley, N., The response surface methodology. Master of Science Thesis, Indiana University of South Bend, 2007.)*

3.10.2.4 Interpretation of response surfaces

Mani et al. (2007) reported the optimization of the solvent extraction of *Moringa oleifera* seed oil using RSM. The influence of the particle size, extraction temperature, and residence time (duration of extraction) on the oil yield was studied using hexane, petroleum ether, and acetone as the extraction solvents. **Figures 3.9** through **3.11** show the response surface plots with each of these solvents. The circles in the plots indicate the yield of oil at various operating conditions.

3.10.2.5 Extraction of oil with hexane

For extractions with hexane, the regression equation (Equation 3.3) indicates that the yield of oil was negatively correlated with the particle size and extraction temperature and positively correlated with the residence time.

$$Yhx = 30.81 - 3.64x_1 - 0.57x_2 + 1.93x_3 - 1.83x_1^2 + 0.87x_1x_2$$
$$+ 1.5x_1x_3 - 0.72x_2^2 + 0.32x_2x_3 - 2.4x_3^2 \qquad r^2 = 0.93 \qquad (3.3)$$

where

Yhx is the percentage of oil extracted with hexane

x_1, x_2, and x_3 are the coded values of the independent variables X, namely, particle size (X_1), extraction temperature (X_2), and residence time (X_3), respectively (Mani et al., 2007).

An increase in the particle size and the extraction temperature, therefore, decreased the oil yield, while an increase in the residence time increased the oil yield. The equation in **Table 3.3** indicates that the magnitude of the coefficients for the particle size was larger than the coefficients for the extraction

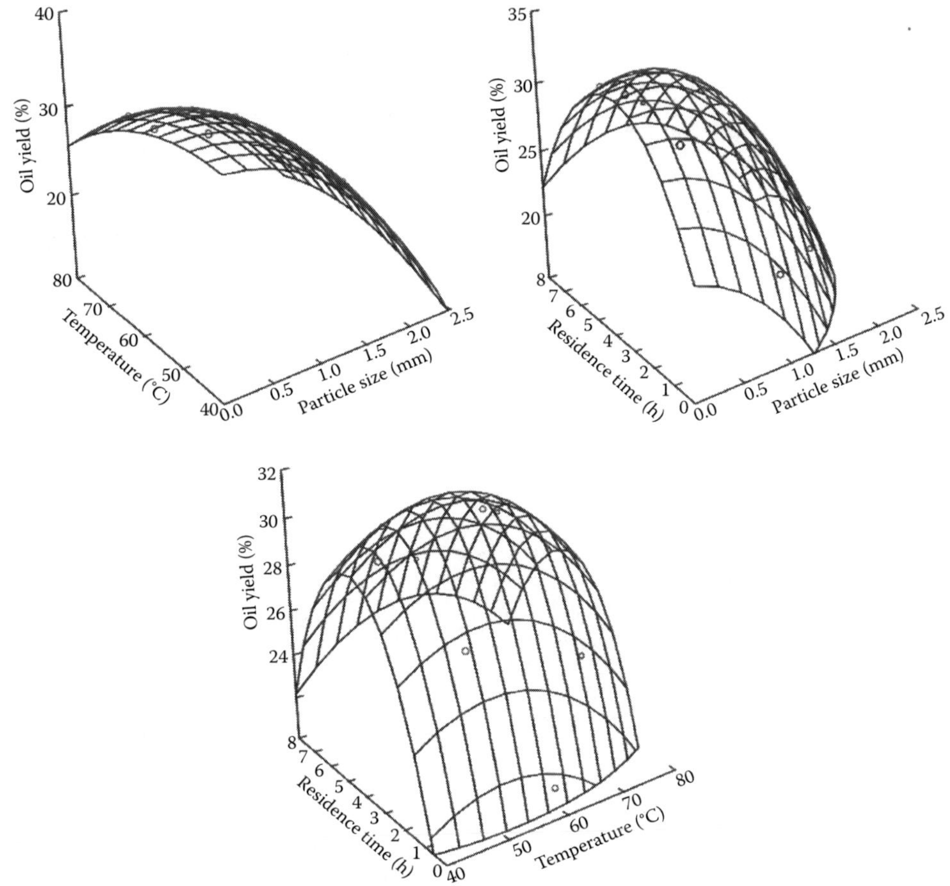

Figure 3.9 *Response surface plots of percentage of oil extracted with hexane. (Reproduced from* Food and Bioproducts Processing, *85, Mani, S., et al., Optimization of solvent extraction of Moringa* (M. oleifera) *seed kernel oil using response surface methodology, 328–335, 2007 with permission from Elsevier.)*

temperature and residence time, suggesting that the particle size had a more significant effect on the percentage of oil yield than the other two independent variables (**Figure 3.9**).

The lack of fit and pure error associated with ANOVA were not significant, indicating that the number of experimental combinations in CCRD experiments was sufficient to determine the effect of the independent variables on the percentage of oil yield.

The low yield of oil associated with the large particle size is due to the resistance offered by the particles to the penetration of the solvents (**Figure 3.9**). A rise in the extraction temperature causes an increase in the diffusivity of the solute and solvent. This results in a higher yield. Nevertheless, at higher

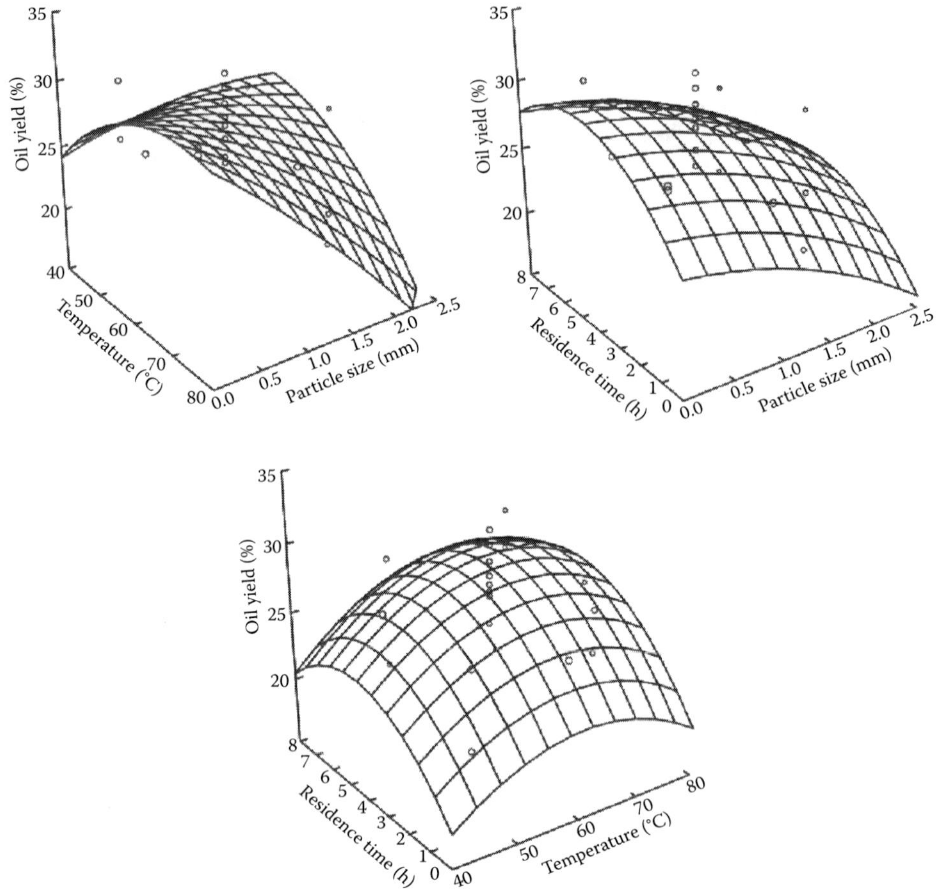

Figure 3.10 *Response surface plots of percentage of oil extracted with petroleum ether. (Reproduced from Food and Bioproducts Processing, 85, Mani, S., et al., Optimization of solvent extraction of Moringa (M. oleifera) seed kernel oil using response surface methodology, 328–335, 2007 with permission from Elsevier.)*

extraction temperatures, the oil yield decreases due to lowering of the solvent density in the particles of the herb (Treybal, 1980). Heating solvents above their boiling point does not enhance the yield of oil. The yield of oil is constant up to 60°C. However, it decreases as the temperature increases. In a similar way, the yield of oil is constant for particle sizes below 1.0 mm. Thereafter, it reduces significantly with an increase in the particle size. The yield of oil increases with the residence time and reaches maximum value at 6 h residence time. A further increase in the residence time does not increase the oil yield (**Figure 3.9**). A decrease in the oil yield at higher extraction temperatures has also been observed in celery and castor seeds (Akaranta and Anusiem, 1996; Papamichali et al., 2000).

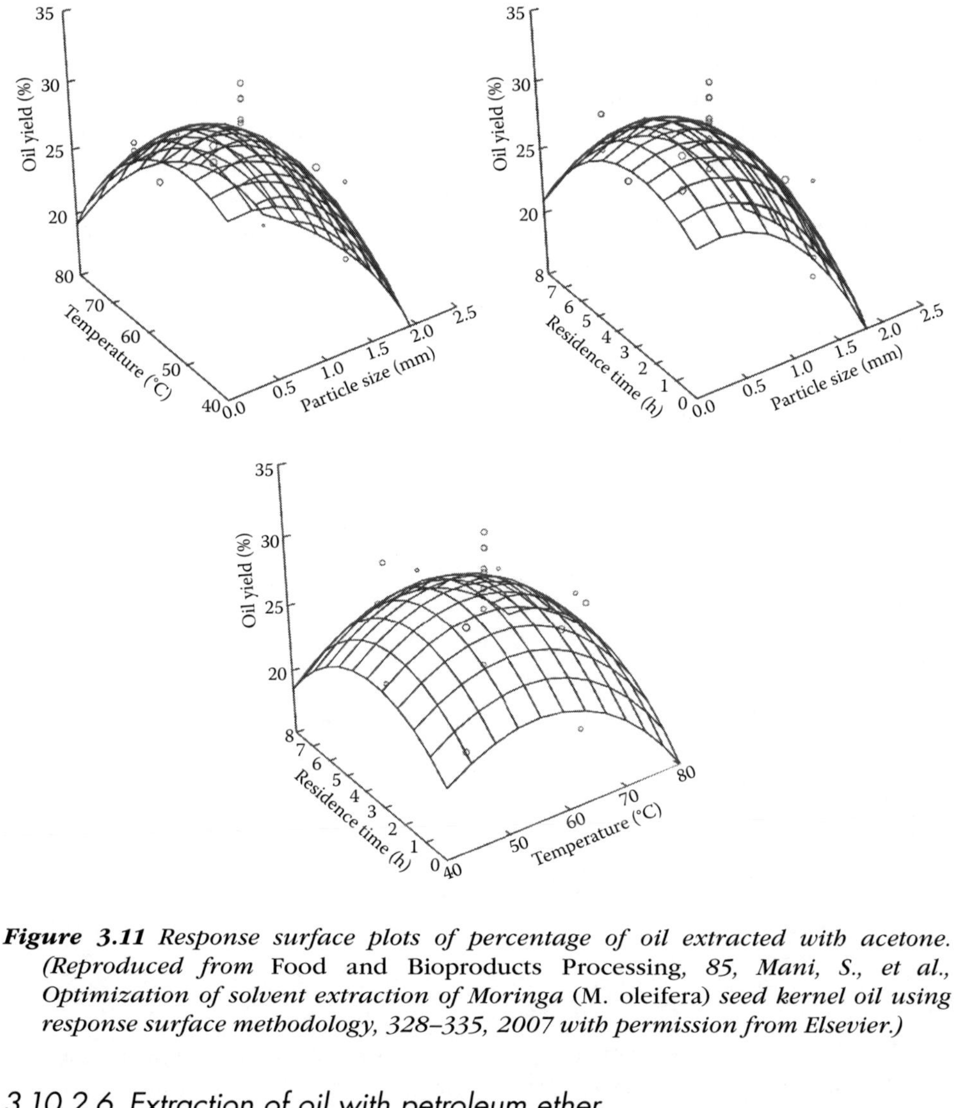

Figure 3.11 *Response surface plots of percentage of oil extracted with acetone. (Reproduced from* Food and Bioproducts Processing, *85, Mani, S., et al., Optimization of solvent extraction of Moringa* (M. oleifera) *seed kernel oil using response surface methodology, 328–335, 2007 with permission from Elsevier.)*

3.10.2.6 Extraction of oil with petroleum ether

The regression equation (Equation 3.4) shows that the percentage of oil extraction was negatively correlated with the particle size and positively correlated with the extraction temperature and residence time. This means that an increase in the particle size decreased the oil yield. However, an increase in the extraction temperature and residence time caused an increase in the oil yield (**Figure 3.10**).

$$Ype = 28.9 - 1.47x_1 + 0.81x_2 + 1.28x_3 - 0.83x_1^2 + 0.99x_1x_2 - 0.16x_1x_3 - 0.91x_2^2 + 0.32x_2x_3 - 1.18x_3^2 \qquad r^2 = 0.77 \qquad (3.4)$$

where *Ype* is the percentage of oil extracted with petroleum ether.

In experiments using petroleum ether as the extraction solvent, the percentage of oil was high at small particle size. The yield reduced at large particle size. The extraction temperature also exerted a similar effect. Nevertheless, the particle size had a greater effect than the extraction temperature. Lack of fit and pure error were insignificant, indicating that the experimental design was sufficient to study the effect of the independent variables on the yield of oil.

3.10.2.7 Extraction of oil with acetone

The yield of oil was negatively correlated with the particle size and extraction temperature. However, it was positively correlated with the residence time (Equation 3.5). From the coefficient for residence time, it can be inferred that particle size is the most significant parameter affecting oil yield with acetone.

$$Yac = 29.25 - 2.82x_1 - 0.56x_2 + 0.31x_3 - 2.37x_1^2 + 0.77x_1x_2 + 0.87x_1x_3 - 1.37x_2^2 + 0.49x_2x_3 - 1.34x_3^2 \qquad r^2 = 0.90 \quad (3.5)$$

where *Yac* is the percentage of oil extracted with acetone.

The response surface plots for acetone extraction show that a smaller particle size with a high extraction temperature produced a higher yield of oil (**Figure 3.11**). The plots also show that the oil yield can be improved at a shorter residence time, if the particle size is small. It is also evident that the yield of oil is high when the extraction temperature is nearer to the boiling point of acetone (60°C).

3.10.2.8 Contour plots

In addition to a response surface plot, the results of a CCRD experiment can also be represented as a two-dimensional contour plot, in which contours of equal response are plotted on a grid of values of the independent variables (Freund et al., 2003). A contour plot helps in the identification of an optimum point on the response surface. Contour plots can show the contour lines of x_1 and x_2 pairs, having a common response value y (Bradley, 2007) (**Figure 3.12**).

The contour plot shows the shading becoming increasingly intense, as the rings go from the periphery to the center. The color code in the inset indicates a proportional increase in the value of y. The lines of x_1 and x_2 intersecting the central area with the most intense shading indicate their optimum values; y maximizes for x_1 between 70 and 77 and x_2 between 126 and 135.

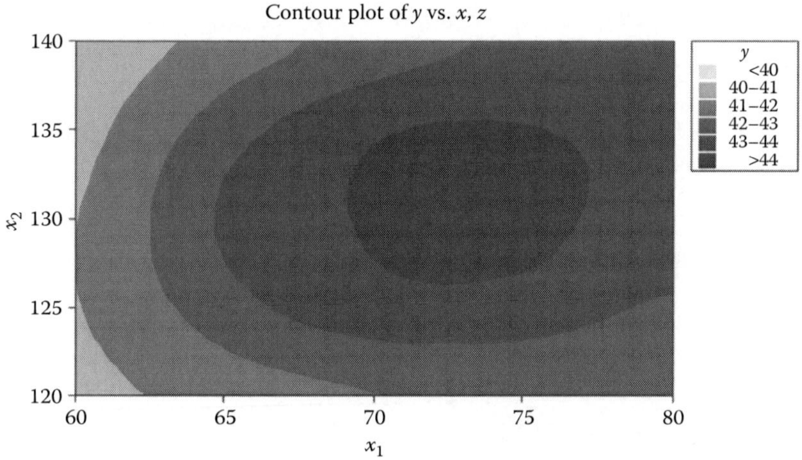

Figure 3.12 *Contour plot. (Reproduced from Bradley, N., The response surface methodology. Master of Science Thesis, Indiana University of South Bend, 2007.)*

3.10.2.9 Optimization of independent variables

Mani et al. (2007) optimized the independent variables for the maximum percentage of oil extracted from the seeds of *M. oleifera*, using the procedure reported by Jaya and Das (2004) and Uddin et al. (2004). They solved the second-order regression equations using Microsoft Excel Solver (**Table 3.9**).

The results show that the optimum combination for the maximum yield of oil for hexane and acetone is within the range of levels of the independent variables. The optimum values for the extraction temperature and residence time are out of the selected range in the case of petroleum ether. When comparing the optimum values of the independent variables for extraction with the three solvents, hexane and acetone gave comparable values. Nevertheless, as hexane extracted the maximum oil and as it is a cheaper solvent, it can be considered to be the solvent of choice (Mani et al., 2007).

Table 3.9 Optimum Value of Independent Variables for Maximum Yield of Oil

Independent Variables	Hexane	Petroleum Ether	Acetone
Particle size (mm): X_1	0.62	1.01	1.36
Extraction temperature (°C): X_2	56.47	77.77	69.41
Residence time (h): X_3	7.09	9.47	6.75
Calculated maximum yield (%)	33.47	29.98	30.33

Source: Data from Mani et al., *Food and Bioproducts Processing* 85, 328, 2007.

3.10.2.10 Verification of optimal extraction conditions

To verify the predictive capacity of the combination of independent variables, optimum conditions are established by RSM. The predicted results are then compared with actual experiments carried out in the optimum conditions. Uma et al. (2010) optimized the process parameters for extracting total phenolic compounds from henna leaves. Under the optimum conditions, the model predicted a maximum response of 7231.34 mg gallic acid equivalents (GAE)/100 g henna extract. The real experiments yielded a value of 7203.74 ± 197.8 mg GAE/100 g of extract. The difference between the predicted and the actual values was less than 10%, reflecting the adequacy of the predicted optimization (**Table 3.10**).

3.10.2.11 Software for RSM

Several specialized software packages are used in the design, graphical representations, and interpretation of the results of response surface studies. Design-Expert® and Minitab are the two widely used software. Minitab's response surface design capabilities and sound advice on the use of the designs are available at: http://www.minitab.com/support/docs/rel14/14helpfiles/DOE/ResponseSurfaceDesigns.pdf. Design-Expert is considered to be the best software for response surfaces. While some software cannot go beyond 7 factors, Design-Expert can work with 30 factors. It is also considered to be more user-friendly (Ryan, 2007). D.o.E. Fusion, EChip, JMP, Matlab, SAS, Sigma Plot, Stat-Ease, Statgraphics Plus, Statistica, Statistical Graphic Plus, and Systat are some of the other software packages utilized in RSM studies of the solvent extraction of herbs.

3.10.2.12 Influence of independent variables on the extraction process

3.10.2.12.1 Extraction solvent The extraction solvent affects the composition of extracts. Du et al. (2004) observed that the ratio of neutral to malonyl ginsenosides in an aqueous ethanol extract of American ginseng increased with the proportion of ethanol in the solvent mixture. The maximum extraction of neutral ginsenosides was obtained with 70% alcohol and the highest yield of malonyl ginsenosides was achieved with 40% ethanol.

Table 3.10 Optimal Extraction Conditions and the Predicted and Experimental Values for Total Phenolic Compounds

Optimum Conditions		Total Phenolic Compounds (mg GAE/100 g extract)		
		Experimental	Predicted	Difference (%)
Acetone concentration	48.07%	—	—	—
Extraction temperature	39.57°C	7203.74 ± 197.8	7231.34	0.38%
Extraction time	73.78 min	—	—	—

Source: Data from Uma et al., *Sains Malaysiana* 39, 119, 2010.

Working with a range of aqueous ethanol, Shimoyamada et al. (1993) observed that the solubility of soyasaponin Bb (soyasaponin-I) was maximum in 60% ethanol. The best solvent found for microwave-assisted extraction of glycyrrhizic acid from licorice roots was 50%–60% ethanol (Pan et al., 2000).

When red ginseng was extracted at 80°C (5 × 8 h), the yield of solids decreased and the recovery of ginsenosides increased with ethanol concentration. The optimum composition was found to be 70% (Sung and Yang, 1985). Pure water, 10% ethanol, and 0.5% ammonia in water could extract more or less the same quantity of glycyrrhizic acid from licorice roots (Wang et al., 2004).

Cacace and Mazza (2003) optimized the extraction of anthocyanins from black currants using aqueous ethanol as the solvent. Total phenolics increased with the concentration of ethanol, up to a maximum of 60% and thereafter decreased with a further increase in the solvent concentration.

It is known that the extraction solvent can affect the physicochemical properties of saponin extracts. These include the particle size, size distribution, morphology, water uptake profiles, sorption isotherms, densities, flow properties, and compaction profiles. All of these have great significance in pharmaceutics (Endale et al., 2004; Güçlü-Üstündağ and Mazza, 2007).

3.10.2.12.2 Solvent:solid ratio Wang et al. (2004) reported that the glycyrrhizic acid content of an ethanolic extract of licorice decreased with an increasing solvent:feed ratio, from 339 mg/mL at 6 mL/g to 245 mg/mL at 10 mL/g. However, the yield of extractive remained in the range of 75%–83%. Contrary to this, an increase in the recovery of glycyrrhizic acid with microwave-assisted extraction was reported with a solvent:feed ratio of 1.88% at 5:1 to 2.58% at 20:1 (Pan et al., 2000). Muir et al. (2002) reported that the optimum ratio for the extraction of quinoa saponin was found to be 10:1–15:1.

3.10.2.12.3 Temperature The saponin yield of ethanol–water extracts is known to increase with temperature (Kerem et al., 2005). The yield of the total extract from red ginseng increased, but the recovery of saponin decreased with temperature, especially at 100°C (Sung et al., 1985). The solubility of gypsophia saponin in water was found to increase with temperature from 7.4 g/100 mL at 30°C to 18.0 g/100 mL at 70°C (Biran and Baykut, 1975).

Cacace and Mazza (2003) reported that the maximum extraction of anthocyanins was obtained at 30°C. Increasing the temperature caused further degradation of the anthocyanins and a loss of yield. The yield of a multistage countercurrent extractive of licorice root and its content of glycyrrhizic acid increased with temperature in the range of 30°C–70°C (Wang et al., 2004).

Temperature affects the composition of licorice extract, reflected in its flavor characteristics. Extraction at 65°C–82°C yielded an extract with a significantly high content of glycyrrhizic acid, sugar, and inorganic salt. The extract had a mild sweet flavor. However, high temperature yielded a stronger licorice character and balanced sweetness (Vora and Testa, 1997).

3.10.2.12.4 Particle size Particle size reduction (pulverization) is generally done to increase the mass transfer efficiency of the extraction (Güçlü-Üstündağ and Mazza, 2007). Nevertheless, the influence of the particle size on the efficiency of the extraction is often overlooked (Luthria, 2008). Mukhopadhyay et al. (2006) demonstrated that the recovery of phenolics from the black cohosh plant matrix was strongly influenced by variations in the matrix particle size. Extraction efficiency increased nearly threefold, as the particle size decreased from >2 to <0.25 mm. This is because the surface area per unit mass increases with a decrease in particle size.

3.10.2.12.5 Number of extractions Mukhopadhyay et al. (2006) extracted black cohosh root powder with methanol:water (60:40 v/v). Nearly 97% of the phenolics were extracted in the first extraction cycle. An additional 8.8% were obtained with the second extraction and 4.2% of the phenolics were extracted in the third extraction.

Using acetone:water (50:50 v/v) as the extraction solvent, Luthria (2009) extracted dried parsley flakes. Nearly 88% of the phenolics were extracted in the first extraction cycle. Nevertheless, when the composition of the solvent mixture was changed to either 90:10 (v/v) or 10:90 (v/v), the efficiency of the extraction cycle was reduced by 30%. **Table 3.11** provides examples of some independent variables that have significant effects on the extraction process.

3.11 Pilot-scale extraction

A pilot plant is a processing system that is operated to generate information about the behavior of the system for use in the design of larger facilities. Pilot plants are less expensive than large production plants and are used to test ideas for new products. Bruhn and Sandberg (1991) remark that well-equipped pilot plants manned by experienced and knowledgeable staff should be an essential feature of organizations engaged in the development and optimization of herbal extraction processes. Pilot plants of various volumes are in use. The well-known Büchi extractor handles materials in the 10–25 L scale. Larger versions of pilot plants of 80 or 250 L capacities are also commonly used. The engineers and operators of such a pilot plant efficiently define the parameters for the scale-up of the extraction process (Glasgow, 1997).

After the optimization of the extraction process, the R&D team and the pilot-plant personnel will carry out three pilot-scale extraction operations, based on the optimized process. Comminuted herb (10–20 kg) is extracted with the appropriate volume of solvent. The miscella should not be stored for long periods at room temperature or in sunlight, as there is a risk of artifact

Table 3.11 Independent Variables with Significant Effect on Extraction of Actives

Sl No.	Extractive	Most Significant Parameter(s)[a]	Reference
1	Resveratrol from *Vitis vinifera* fruits	Ethanol concentration (54% ethanol/water) Temperature (83.6°C)	Karacabey and Mazza (2008)
2	Anthocyanins from black currants	Solvent to solid ratio (19 L of 60% ethanol/water per kilogram of berries)	Cacace and Mazza (2003)
3	Anthocyanins from sweet purple potato (*Ipomoea batatas*)	Extraction temperature (80°C) Solid to solvent ratio (1:32)	Fan et al. (2008)
4	Phenolic acids from black cohosh (*Cimicifugaracemosa*)	Particle size (between 0.25 and 0.425 mm)	Mukhopadhyay et al. (2006)
5	Phenolics and flavonoids from pink guava (*Psidium guavajava*) fruits	Extraction time (5 h) Temperature (60°C)	Kong et al. (2010)
6	Phenolic compounds from *Cympopogon citratus, Alpinia galanga, Ocimum sanctum,* and *Rosmarinus officinalis*	75% ethanol/water	Juntachote et al. (2006)
7	Phenolic compounds from *Cocos nucifera* shell	Extraction time (>50 min)	Rodrigues and Pinto (2007)
8	Polysaccharides from *Glycyrrhiza glabra* roots	Extraction time (4.3 h) Solid to solvent ratio (1:35)	RenJie (2008)
9	Oleanoic acid from *Lantana camara* roots	Particle size (0.5 mm) Solid to solvent ratio (1:55) Solvent composition (52.5% methanol/ethyl acetate)	Banik and Pandey (2008)
10	Oil from Moringa (*M. oleifera*) seeds	Particle size (0.62 mm)	Mani et al. (2007)
11	Quercetin and kaempferol from *Coriandrum sativum* herb	Concentration of HCl in the solvent (55% methanol/water containing 1.85 M HCl)	Hadjmohammadi and Sharifi (2009)
12	Betacyanin from red beetroots (*Beta vulgaris*)	Solvent temperature (70°C) Solid to solvent ratio (1:5)	de Azeredo et al. (2009)
13	Juice of sapodilla (*Achras zapota*)	Extraction temperature (60°C)	Sin et al. (2006)

[a] Figures in parentheses indicate optimum values.

formation and the decomposition of the constituents of the extract. The solvent in the miscella is distilled off in a rotary evaporator with the water bath kept at a temperature below 40°C to prevent the decomposition of thermolabile compounds. Sufficient precautions should be taken to minimize the risk of fire and explosion while using and storing highly inflammable solvents and solvents such as diethyl ether, which tend to form explosive peroxides (Jones and Kinghorn, 2006). Samples of the three batches of extracts are analyzed and corresponding COAs prepared. Unit recovery of the active is also calculated (*vide infra*). Based on these COAs, a pilot-scale specification is prepared. A sample of 5 g of extract from each pilot-scale extraction should be packed and stored as a voucher sample for future use.

3.12 Enrichment of extract

3.12.1 Example of a process

Extracts obtained by solvent extraction sometimes undergo a process of enrichment so that the content of the active ingredient is enhanced. This point can be illustrated by the example of a *C. forskohlii* extract.

The roots of *C. forskohlii* are coarsely powdered and extracted with toluene at 80°C–90°C. The miscella is concentrated and the toluene is completely removed *in vacuo*. The brown-colored extract obtained will have a forskolin content of 8%–9%. This extract is boiled with hexane at 50°C for 2 h, with stirring. The yellow product obtained is filtered through a nutsche filter and dried *in vacuo*. This extract will have a forskolin content of 60%–65%.

The extract with 60%–65% of forskolin is extracted with toluene at 60°C–70°C while stirring and filtered (first extract). The insoluble portion is re-extracted twice with toluene at 60°C–70°C and filtered (second and third extracts). The insoluble substance after the third extraction has an assay greater than 80%. The first extract is concentrated and used for making *Coleus* extract 10% and 20% products. The second and third extracts are combined, concentrated, and dried. The product above 80% assay is combined with the insoluble residue.

The extract containing more than 80% of forskolin is refluxed with methanol and then decolorized with activated charcoal. The hot solution is filtered and the volume is reduced to one-fourth *in vacuo*. The concentrated extract is allowed to crystallize and the crystals are recovered by filtration. The crystalline material will have a forskolin content of 95%.

The extract with 95% of forskolin is refluxed with anhydrous methanol, filtered and concentrated to one-fourth. The concentrated fraction is allowed to crystallize at first for 6 h at room temperature and then at 5°C for 2 h. The crystals are washed with cold methanol and dried completely *in vacuo*. The crystalline material will have a forskolin content of 98% (Saksena et al., 1985; Ammon and Muller, 1985).

3.12.2 Enrichment of saponins

The herb powder is defatted with n-hexane, followed by extraction with methanol. The methanol extract is concentrated, suspended in deionized water, presaturated with n-butanol, and partitioned with n-butanol. Diethyl ether is added to the butanol fraction and this causes the saponins to precipitate (Hostettmann et al., 1991; Jones and Kinghorn, 2006).

3.12.3 Enrichment of alkaloids

Alkaloids containing basic amines are extracted according to the following scheme. Ground plant material is defatted with hexane or petroleum ether and extracted with methanol. The defatted methanolic extract is suspended in tartaric acid and partitioned with ethyl acetate presaturated with water. The aqueous acid phase is made alkaline with Na_2CO_3 solution, titrating to pH 11 and partitioned with ethyl acetate. The aqueous extract contains quaternary amines and alkaloid *N*-oxides. The second ethyl acetate partition contains 1°, 2° and 3° amines (Jones and Kinghorn, 2006; Cordell, 1981). Alkaloids can also be extracted with 10% acetic acid in ethyl alcohol, followed by concentration and precipitation by the dropwise addition of NH_4OH (Harborne, 1998).

3.13 Analysis of extract

3.13.1 Certificate of analysis

On completion of the extraction process, a sample (50 g) of the extract is submitted to the quality assurance QA/QC department for detailed analysis. Another sample of 5 g should be packed and stored as a voucher sample for future use. The parameters to be considered for analysis are color, consistency, solubility, mesh size, loss on drying (LOD), pH of 1% w/v solution in water, total ash, acid-insoluble ash, heavy metals (arsenic, lead, cadmium, mercury), tapped and loose bulk density, content of actives, residual solvents, aflatoxins, pesticide residues, and microbial load (total microbial count, yeast and mold count, *Escherichia coli*, *Salmonella*, *Staphylococcus aureus*, and *Pseudomonas aeruginosa*). Analytical methods for these and several other tests are available (Cuniff, 1995a, 1995b; Anonymous, 1995, 2002a, 2006c, 2006d, 2007a, 2007g; Rajpal, 2002; Hurst, 2008; Agarwal and Murali, 2010). A COA of the extract is prepared on the basis of the analysis (**Table 3.12**).

3.13.1.1 Significance of the components of the COA

3.13.1.1.1 Bulk density Bulk density is a characteristic that is considered in the case of powders and pellets. It is defined as the weight per unit volume of the material (mass/volume). This parameter reflects the particle size and dispersion, which in turn decide the material flow consistency and packaging quantity. A powder with a low bulk density will have more volume and less mass. Two types of bulk densities are considered in pharmaceutics and food technology—loose bulk density and packed (tapped) bulk density. Loose bulk density is the weight per volume that has been measured when the sample is in a loose, noncompacted, or poured condition. Tapped bulk density is the weight per volume that has been measured when the sample has been packed or compacted, for example, in a bin or a drum, or after containerized transportation. As the material is compacted, the entrapped air is displaced and the void space is reduced. Consequently, the value of the packed bulk density is higher than that of the loose bulk density. Bulk density plays an important role in the content uniformity of solid dosage forms, drug dissolution, manufacturing of dosage forms, and food product development (Jambhelkar, 2005).

Table 3.12 COA of *Passiflora incarnata* Dry Extract

Product: ***Passiflora incarnata* dry extract**
Batch No.:

Parameter	Certificate of Analysis
Name of herb	*Passiflora incarnata*
Excipients used	None
Description	Brown powder
Identification	Complies by TLC
Solubility	
Water solubles	85%
Alcohol solubles	85%
Tapped bulk density	1.12 g/mL
Loose bulk density	084 g/mL
Loss on drying	5.30%
Assay by HPLC	1.5%
Content of vitexin	
pH (1% w/v solution in water)	6
Residual solvent	Complies
Acid-insoluble ash	1.38%
Arsenic	<0.1 ppm
Lead	1.48 ppm
Cadmium	0.92 ppm
Mercury	<0.5 ppm
Aflatoxins	Not detected
Pesticide residues	Not detected
Total microbial count	<100 cfu/g
Yeast and mold count	<100 cfu/g
Escherichia coli	Absent
Salmonella	Absent
Staphylococcus aureus	Absent
Pseudomonas aeruginosa	Absent
Remarks: Complies with respect to above tests	

Source: Adapted from Rajpal, V., *Standardization of Botanicals*, Vol. 1, 1–286, Eastern Publishers, Delhi, 2002.

3.13.1.1.2 Ash value/residue on ignition When crude herbs or their extracts are incinerated, they leave behind an inorganic ash, which is helpful in checking the purity of the material. This residue arises from the mineral matter in the sample. The determination of ash values helps in assessing the quality of herbs, extracts, and food substances. It also helps in the identification and rejection of inferior materials. For example, exhausted ginger can be detected from the content of water-soluble ash. The acid-insoluble component of ash is an indicator of dirt and sand in the sample (Trease and Evans, 1983; Sathe, 1999).

3.13.1.1.3 Loss on drying Loss on drying (LOD) is an important characteristic of dried herbs and extracts, as it indicates the efficiency of drying. LOD is usually synonymous with "moisture content." However, as herbs contain an array of complex chemical molecules, several volatile substances including essential oils may be present in a sample of a herb or a herbal extract. Therefore, the term *loss on drying* indicates loss of water and other volatile substances in the sample. LOD is crucial to the development of pharmaceutical dosage forms such as granules, tablets, and capsules. It has a direct bearing on the chemical and microbial stability of the dosage forms and on their physical properties such as size distribution and flowability (Souza et al., 2008).

3.13.1.1.4 Residual solvents Organic solvents that are used in the extraction of bioactives are usually removed by distillation, followed by drying. However, trace quantities of the solvents cling tenaciously to the particles of extracts even when very efficient equipment is used. This amount of solvent is generally referred to as a residual solvent. As these solvents can be hazardous to health, their levels need to be limited. In the past, each regulatory agency used to adopt its own guidelines for residual solvents in medicinal products, and this resulted in different acceptance limits. An international effort was made to arrive at uniform guidelines and on July 17, 1997, the ICH Harmonized Tripartite Guideline Step 4 was adopted (Anonymous, 1997a, 2010d). It is recognized by many countries (Anonymous, 2001a, 2006c).

According to the guideline, there are three classes of solvents. Class 1 includes solvents such as benzene and carbon tetrachloride, and they are to be avoided on account of their carcinogenicity and environmental hazards. Class 2 solvents have moderate toxicity and their use is to be limited. Examples of this class are acetonitrile, chloroform, methanol, hexane, etc. Class 3 solvents, which include acetic acid, acetone, ethanol, ethyl acetate, ether, and so on, have low toxicity. The content of residual solvents is determined using GC (Anonymous, 2009a).

If on analysis a sample of extract is found to contain a residual solvent above the specified limit, the solvent can be removed by drying the extract in a vacuum tray drier at an appropriately high temperature (Willis, 2003). Alternatively, the extract can be suspended in absolute ethanol and the ethanol removed by distillation. The residual solvent will mix with the ethanol azeotropically and can be removed by the distillation of the ethanol.

3.13.1.1.5 Microbial count Herbs and herbal extracts are often found to be contaminated with microbes. A nationwide outbreak of salmonellosis occurred in Germany in 1993, because of paprika-powdered potato chips contaminated by many strains. Toxigenic molds such as *Aspergillus flavus, A. parasiticus, A.*

fumigatus, A. ochraceus, Penicillium citrinum, and *P. islandicum* have been detected in some spices (Farkas, 2000). The biological structure of the herb often predisposes it to higher levels of microbial contamination, as in the case of plants having trichomes on their surfaces. Several factors, such as the condition of the environment where the herb grows and the way that it is harvested, dried, processed, and transported, contribute to microbial contamination. Microbial load is also dependent on the plant part used. For example, the rhizomes of turmeric and ginger may have more microbes than the seeds of mustard, because microbes from soil can enter the rhizomes more easily (Das, 2008).

Pathogenic microbes may also find their way into herbal extracts. The microbial content of products is estimated according to specific methods (Anonymous, 2009d). There are no official regulatory guidelines on the microbial contamination of herbs. The limits vary from product to product and from country to country. For example, the total bacterial content of the saw palmetto dietary supplement should not exceed 10^4 cfu/g, the total yeast and mold count should not exceed 100 cfu/g, and *Salmonella* species, *E. coli* and *S. aureus* should be absent (Anonymous, 2009e). The extraction of herbs with alcohol reduces their microbial content because of direct contact with alcohol (Hoffmann, 2003).

Microbial contamination of botanical raw materials decreases with the increasing level of processing, going from roots and rhizomes to flowers and leaves, to fruits and seeds, to liquid extracts and dry extracts (Katušin-Ražem et al., 2001). Crude herbs, spices, and extracts are usually decontaminated by irradiation or fumigation with ethylene oxide gas. However, irradiation is a preferable sanitation method (Katušin-Ražem et al., 2001; Koseki et al., 2002; Migdal and Chmielewski, 2005). The limits for bacterial and mold contamination in crude herbs and extracts are to be strictly adhered to.

3.13.1.1.6 Pesticide residues Through the use of pesticides, humans have achieved a more abundant supply of food and have enhanced their control of diseases that are spread by pests. However, associated with these pest control procedures is the problem of the chemical contamination of the environment. These pesticides do not disintegrate easily and therefore end up as residues in food, vegetables, and the somatic tissues of man and animals. The residue problem is associated with persistent pesticides such as dichlorodiphenyltrichloroethane (DDT), aldrin, dieldrin, gamma BHC, chlordane, and heptachlor. These pesticides have a long residence time in the soil and are capable of contaminating crops in successive years after the initial application. Endrin-containing soybeans were harvested from fields where cotton had been grown and treated with endrin. Similarly, alfalfa grown in soil where a previous crop was treated with dieldrin and heptachlor was found to contain residues of these pesticides (Kraybill, 1969). To curb the health hazards posed by pesticide residues, the US Pharmacopeia (USP) has set limits for them (Anonymous, 2009f).

3.13.1.1.7 Heavy metals Herbal extracts can be contaminated with heavy metals such as lead, mercury, bismuth, arsenic, antimony, tin, cadmium, silver, copper, and molybdenum. They are estimated according to pharmacopoeial methods. The usual limit is not more than 0.001% (10 ppm) (Anonymous, 2009g).

3.14 Specification for the extract

Three laboratory-scale extractions need to be carried out. On analysis of the products from these extractions, the range of variation in the values of the parameters can be judged. On the basis of the results of these extractions, a specification sheet for the extract can be prepared. A product specification is a written statement of the characteristics of the product to facilitate its production and acceptance.

3.15 Costing of the extract

The cost of an extract is pivotal to its marketability. A commercially viable product should have unique application, stability, and above all, affordability. It should never be prohibitively expensive. The costing of the product should be complete as soon as it is ready for marketing. The cost of manufacturing extracts is influenced by several factors that can be broadly categorized into direct costs, fixed costs, and general expenses. Direct costs consider expenses that depend directly on the production rate. Examples are the cost of the raw materials, solvents, utilities, and operating labor. The raw material cost should include the price of the herbal raw material, the cost of grinding the raw material to the required size, the cost of solvents, and the cost of transporting the raw and finished products. Utilities include power supply, cooling tower, chilling plant, steam boiler, and vacuum. Fixed costs are independent of the production rate and are to be considered even if the operation is interrupted. These include the cost of the property, preparation of the ground for construction, civil construction, cost of equipment, depreciation, taxes, and insurance. General expenses are the expenses that are needed to maintain the business and are made up of administrative costs, sales expenses, research, and development (Patel et al., 2006; Pereira et al., 2007; Rosa and Meireles, 2009).

Turton et al. (1998) have formulated a methodology to estimate the cost of manufacture, which is defined as the weighed sum of fixed capital investment, the cost of labor, the cost of raw material, the cost of waste treatment, and the cost of utilities.

A computational simulation of the processes is also adopted to predict the production costs at industrial scale (Rouf et al., 2001; Kwiatkowski et al., 2006; Takeuchi et al., 2008). The process simulator Superpro Designer is a group of software tools that are capable of predicting process and economic parameters, quantifying the process parameters that are used in the estimation of the cost of manufacture (Rouf et al., 2001).

3.16 Unit recovery

The calculation of unit recovery or compound(s) recovered from the starting material is important in the extraction and isolation process as it has multiple steps in the process. The calculation of unit recovery reflects the completeness of the extraction of the active(s).

Units = yield of the product × assay of the product

$$U_r = \frac{U_f}{U_s} \times 100 \quad (3.6)$$

where:

U_r = unit recovery
U_f = units in the final product
U_s = units in the starting material

This is illustrated by the following examples.

Example 3.1:

Bacoside was extracted with water from 1 kg of *Bacopa monnieri* leaves having a bacoside content of 1%. The extraction yielded 100 g of extract having an assay of 8%.

Units in the starting material = 1000 × 1 = 1000
Units in the extract = 100 × 8 = 800

$$U_{rb} = \frac{800}{1000} \times 100 = 80\% \quad (3.7)$$

U_{rb} = unit recovery of bacosides
20% of units remain in the marc.

Example 3.2:

C. forskohlii extract (100 g) having 10% forskolin was subjected to crystallization and purification. This yielded 50 g with an assay of 17%. Unit recovery is calculated as follows:

Units in the initial extract = 100 × 10 = 1000
Units in the final product = 50 × 17 = 850

$$U_r = \frac{850}{1000} \times 100 = 85\% \quad (3.8)$$

U_r = unit recovery
15% of units remain in the mother liquor.

3.17 Stability studies

Herbal extracts and the nutraceutical products that are derived from them are invariably subjected to stability studies, which are intended to monitor the possible changes in them over time and at different storage conditions. Stability assessment is usually done during the development of the product to ascertain its quality and shelf life. In earlier days, stability studies were confined to the detection of degradation products. However, according to the International Conference on Harmonization (ICH) guidelines, stability should be studied using a combination of analytical methods concerning the identity, purity, and biological activity of the product. Several ICH guidelines deal with various aspects of stability testing (Aubry et al., 2008).

Usually, accelerated and real-time stability studies are carried out to assess the effects of storage conditions on the physical and chemical qualities of an extract. Accelerated stability is studied by storing the product in conditions that are harsher than that of a real-time situation, in which case the product is stored in ordinary conditions. A 6 month accelerated stability trial will provide data to support a 2 year shelf life. For stability studies, a sample from one production batch or a sample from a pilot-plant batch whose batch size is at least 10% of a production batch is selected.

The packing material should represent the material that is used for the storage and distribution of the product. Extracts are packed in primary packs of 12 μPET (polyethylene terephthalate)/100μ low-density polyethylene (LDPE) bags (11.9 × 8.6 cm, three sides presealed) and heat sealed, followed by a secondary pack of 12 μPET/12 μ aluminum foil/12 μPET/100 μ food-grade LDPE (15.5 × 12 cm, three sides presealed). Finally, the pack is heat sealed.

An accelerated stability study is carried out at 40°C ± 2°C/75 ± 5% relative humidity (RH) in Chamber A and real-time stability at 25°C ± 2°C/60 ± 5% RH in Chamber B. The reference samples for the study will be packed in identical packs and stored in cold storage or a refrigerator at a temperature of less than 10°C or any other condition appropriate to the sample.

Twenty to twenty-four packets of 25–50 g each of the extract are packed and kept in both the stability chambers. Six to ten packets are stored in Chamber A for the accelerated stability study and 10–14 packets are stored in Chamber B for the real-time study. Equal numbers of packets containing sufficient quantities of the extract (depending on the parameters chosen) are kept as reference samples. Reference samples are intended for a comparison of the parameters. The labels on the packets should contain the following information: name of product, batch number, date of manufacture, quantity of sample, pack or sample number, storage conditions, date of loading into the chamber, and date of withdrawal.

The samples kept for the accelerated stability study and the reference samples are withdrawn at the initial point and after 3 and 6 months. The real-time stability samples are withdrawn at the initial point and after 6, 12, 18, 24, and 36 months. Samples from each station are withdrawn at the period mentioned on the label. Withdrawal should be on a date ±7 days of the storage period and analysis should be carried out immediately. The following parameters are evaluated in both the accelerated and the real-time stability studies.

3.17.1 Physicochemical properties

The important parameters are color, odor, appearance (powder, lumps, etc.), pH, solubility, moisture/LOD, volatile oil, total ash content, acid-insoluble ash, bulk density, tapped bulk density, total soluble solids, extractive values (alcohol, water, etc.), and sieve analysis.

3.17.2 Identity of the product by any one method

Product identity is established with IR spectrophotometry, UV/VIS spectrophotometry, TLC, HPTLC, HPLC, or GC.

3.17.3 Estimation of actives

The content of actives is estimated by HPLC, HPTLC, GC, or gravimetry.

3.17.4 Category compounds

Category compounds such as phenols, sterols, alkaloids, and so on are estimated.

3.17.5 Fingerprints

Fingerprinting is done using HPLC or HPTLC.

3.17.6 Microbiological parameters

1. Total viable aerobic count
2. Total fungal count
3. Total Enterobacteriaceae count
4. *Staphylococcus aureus*
5. *E. coli*
6. *Pseudomonas aeruginosa*
7. *Salmonella* species

The recorded stability data are evaluated and the shelf life is determined for the product based on these data (Anonymous, 2003). The type of packing required for the product and its storage conditions are decided on the basis of the stability data (Vadas, 2001).

3.18 Toxicological studies

Herbal extracts are invariably used in foods, beverages, nutraceuticals, medicines, and cosmetics. Therefore, they should be safe to use. The short-term and long-term toxicity due to ingestion, intravenous injection, and topical application are studied before the product is offered to the market. Usually acute, subacute, chronic, and subchronic toxicity are studied in the case of orally administered extracts. Dermal irritation studies are essential in the case of topically applied products (ointments, creams, lotions, cosmetic gels), and ophthalmic products require eye irritation tests.

Acute oral toxicity is usually defined as the adverse effects that occur immediately or a short time following a single or a short period of exposure *per os* (Walum, 1998). It is usually determined according to the technique of Horn (1956). Doses >1500 mg/kg are generally used. The method provides a reasonably accurate estimate of the median lethal dose (LD_{50}), although it is called the *approximate* LD_{50} (ALD_{50}). Further testing is done at ½ ALD_{50} for crude extracts and at 1/5 ALD_{50} for pure substances.

Subacute (2–3 weeks) and chronic (26–104 weeks) toxicity studies are conducted to study the toxicity of the test substance by repeated exposure. A no observable effect level (NOEL) and a no observable adverse effect level (NOAEL) can be established from these studies. Reproductive and developmental toxicity studies consider the effects of the test substance on the mating or fertility of parents, maternal or embryo toxicity, lactation, weaning, growth, and development of pups (Mutai, 2000).

The Organization of Economic Cooperation and Development (OECD) has set forth several guidelines for these toxicity studies (Anonymous, 2010a, 2010f; Jaijoy et al., 2010). However, the choice of the test model is dictated mostly by the interest of the buyer of the extracts.

3.19 Material safety data sheet (MSDS)

The MSDS is a document containing data on the properties of a substance. It is an important component of product stewardship and workplace safety. The MSDS is intended to provide workers and emergency personnel with procedures for handling or working with a substance in a safe manner and includes information such as physical data (melting point/boiling point/flash point), toxicity, health effects, first aid, reactivity, disposal, protective equipment, and spill handling procedures. The MSDS of the extract in question should be prepared on completion of a laboratory-scale extraction and an analysis of the extract. The format of an MSDS varies from country to country, depending on the regulatory requirements. A specimen of an MSDS is provided in **Table 3.13**.

Table 3.13 Material Safety Data Sheet of Green Coffee Extract

Product Name: **Green Coffee Extract**	
Batch No. ----	
1. Method of production	Water extraction
Residual solvent	Nil
Additives if any	Nil
2. Manufacturer and supplier	(Provide address)
3. Trade name and synonyms	None
4. Intended use of product	As dietary supplement
5. Physical state	Powdery extract
6. Odor and appearance	Characteristic
7. Specific gravity	Not applicable
8. Optical rotation	Not applicable
9. Refractive index	Not applicable
10. Solubility in water at 25°C	Fully soluble
11. Flammability. If flammable, specify conditions	Not flammable
12. Irritancy of product	No
13. Whether gloves are to be used	Yes
14. Whether irritant to eyes	Avoid contact with eyes
15. Storage requirements	No special requirement
16. Special shipping information	No special requirement
17. First aid measures	Flush with water

3.20 Commercial manufacture of the extract

3.20.1 Critical control points

The transition from a successful pilot-scale to a full-scale commercial process requires careful planning and implementation. This involves the identification of critical control points. These steps can contribute to the variability of the final product if not controlled. A flow diagram of the complete process can be helpful in identifying these critical process steps. The following account of some such critical control points illustrates their importance.

3.20.1.1 Turmeric extract

For obtaining a standardized mixture of curcumin, demethoxy curcumin, and bisdemethoxy curcumin, the acetone extract derived from turmeric powder is washed with hexane to remove the oily matter. This is a critical step. The hexane-soluble substances should be completely removed from the acetone extract; failure to do this can result in a loss in the yield of the final product. Similarly, care should be taken to ensure that the value of the LOD does not go beyond 2%.

3.20.1.2 Horse chestnut extract (90% aescin)

1. Coarse powder of the horse chestnut is extracted with aqueous methanol and the miscella is concentrated to 16% TDS. It should be ensured that the methanol does not evaporate from the extractor.
2. The concentrated aqueous methanol extract is fractionated with saturated n-butanol in a clean reactor. The layers are mixed and separated every 4 h. The saturated n-butanol layer is concentrated completely, dissolved in demineralized (DM) water, and spray-dried.

   The separation of the aqueous n-butanol layer is critical. At least 3 h are required for the complete separation of the layers. The n-butanol extract should be dried completely. Otherwise, the assay of aescin will be low. Care should be taken to ensure that an emulsion is not formed when the aqueous methanol fraction is agitated with n-butanol.
3. n-Butanol extract is dissolved in two volumes of methanol and poured into eight volumes of acetone. The mixture is stirred and centrifuged, and the precipitate is dissolved in DM water and spray-dried.

   The n-butanol extract should be completely soluble in methanol. Sometimes, insoluble black particles may appear, and these are removed by filtration. Similarly, while adding the methanolic solution to acetone, the rate of precipitation needs to be watched closely. The methanolic solution should be added slowly. Two to three hours are required for a 1 ton batch.

3.20.1.3 Boswellia serrata extract

The gum oleoresin extracted from *Boswellia serrata* gum is mixed with 5% aqueous KOH solution, warmed to 80°C, and agitated for 8 h. The reaction mass is cooled to room temperature and mixed with isopropyl alcohol (IPA). After stirring for 30 min, petroleum ether is added and agitated. The petroleum ether fraction is thereafter separated. The remaining aqueous alkaline solution is concentrated by removing the IPA by distillation. Boswellic acids are precipitated from the mixture by adding dilute HCl. Four critical steps influence the success of the operation.

1. The nonpolar impurities from the oleoresin are removed completely by washing with petroleum ether. Incomplete washing will result in the oily material passing forward in the process, making the final product a paste and not a free-flowing powder.
2. IPA from the alkaline layer is removed completely. Otherwise, there will be a lowering of the aescin content.
3. Organic solvents are removed completely. Otherwise, the final product will not be a free-flowing powder.

4. Dilute HCl should be added with agitation of the reaction mass. Failure to do so can result in the formation of lumps causing subsequent discoloration of the final product and a lowering of the aescin content.

3.20.1.4 Lecithin

Lecithin is a family of products based on naturally occurring phospholipids. A number of factors in oilseed processing can affect the composition of phospholipids in crude oil and therefore have a significant effect on the isolated lecithin. The critical control points that affect the crude oil are bean quality, stripper temperature, residual solvent, and particulate contamination. Similarly, drying of the final product can be affected by the temperature, residence time, vacuum, temperature out of cooler, and final moisture (Dashiell, 1990).

3.20.2 Master formula

The commercial extraction of herbs is carried out as a batch processing operation. A batch is a defined quantity of starting material processed in a single process or a series of processes so that it could be expected to be homogeneous. The full procedure for the extraction process is detailed in a master formula (MF), which describes the preparations to be made, the kind of equipment to be used, and the method to be followed (Chaloner-Larsson et al., 1997). **Table 3.14** lists some of the elements of an MF. The original approved MF, with the signatures of the production and QA officials, is filed in a safe place. An approved copy of the MF is made available to the production department for every batch production run.

The entire process of the solvent extraction of every batch is recorded in a batch manufacturing record (BMR). A distinctive combination of numbers and/or letters is assigned to each batch and is used on labels and all documents pertaining to the batch. All aspects of the operation, such as the preparation of raw materials, the analysis of solvents, the cleaning of extractors, the removal of solvents from the miscella *in vacuo*, the analysis of raw materials and extracts, and so on are carried out according to standard operating procedures (SOP), which are detailed written instructions that specify how a test or an administrative procedure is to be performed (Chaloner-Larsson et al., 1997). The BMR should contain the name of the product, dates and times of commencement and completion of production, name of the person responsible for each stage of production, initials of the operator of different steps of production and of the person who checked each of these operations, batch number, a record of in-process controls and the initials of the person who carried them out, the yield of extract obtained at different and pertinent stages of production, and detailed notes on special problems encountered, with authorization for any deviation from the MF (Anonymous, 2008a). On completion

Table 3.14 Some Elements of the Master Formula for *Kaempferia galanga* Ester

Steps of the Operation	
Kaempferia galanga tubers: 1700 kg	
Hexane: 7000 L	
1.1	Check and certify that the equipment is clean as per SOP.
1.2 **Extraction I.** Charge 1700 kg of *Kaempferia galanga* tuber powder into the extractor.	
1.3	Charge 3000 L of hexane into the extractor. Ensure soaking.
1.4	Heat to reflux and maintain for 3 h.
1.5	Filter and collect the extract.
1.6 **Extraction II.** Charge 2000 L of hexane into the extractor and ensure soaking.	
1.7	Heat to reflux and maintain for 3 h.
1.8	Filter and collect the extract.
1.9 **Extraction III.** Charge 2000 L of hexane into the extractor and ensure soaking.	
1.10	Heat to reflux and maintain for 3 h.
1.11	Filter and collect the extract.
1.12	Combine all the filtrates and send for concentration.
2.0 **Concentration**	
2.1	Check and certify that the equipment is clean as per SOP.
2.2	Charge the combined filtrates and distill off hexane completely.Expected yield of crude oil = 32–35 kg.
3.0 **High vacuum distillation**	
3.1	Check and certify that the equipment is clean as per SOP.
3.2	Distill off under high vacuum at 120°C–170°C.
3.3	Cool the distillate to 0°C–4°C.
3.4	Filter and dry. Expected yield: Galanga ester—9 to 11 kg Galanga oil—15 to 16 kg

of the production process, the BMR is returned to the QA, where it is stored in an appropriate place for a period of 5 years.

3.20.3 Formulation and finishing

The extracts obtained from solvent extraction are generally formulated with pharmacologically inactive or inert excipients. These are intended to give the desired properties for stability, aesthetics, drug delivery, to limit microbial growth, and to facilitate the incorporation of the extract into products. These excipients have fundamental effects on the bioavailability, bioequivalence, and stability of formulations. They possess physical properties such as boiling point, melting point, bulk and tapped density, compression characteristics, hygroscopicity, flowability, particle size distribution, refractive index, specific surface area, and stability. Small variations in the physical properties of an excipient produce differences in the behavior of the formulated products. Excipients are therefore chosen with care (Wade and Weller, 1994).

One major reason for using excipients is to dilute extracts to the desired strength. For example, using dicalcium phosphate, magnesium carbonate, and talc, the dry extract of *C. forkohlii* having a forskolin content of 60%–65% is diluted to *C. forskohlii* extract 10%. However, the drug–excipient compatibility is to be studied, as was done in the case of sennosides A and B. Verloop et al. (2004) observed that in the presence of water, a sennoside extract is incompatible with sodium carbonate, stearic acid, citric acid, polyethylene glycol (PEG), lactose, glucose, and sorbitol. The common excipients used in the formulation of extracts are calcium carbonate, calcium stearate, cetyl alcohol, dicalcium phosphate, tribasic calcium phosphate, microcrystalline cellulose, dextrin, dextrose, lactose, lecithin, magnesium carbonate, magnesium stearate, maltodextrin, peanut oil, sesame oil, soybean oil, and talc (Wade and Weller, 1994).

3.20.4 Packing of extracts

Free-flowing extracts are packed into food-grade LDPE bags (457 × 355 × 432 mm, weight approximately 100 g) and the opening is closed using the twist and tie method. The bag is sealed by applying a tamperproof clip. The bag is put inside another LDPE bag and the opening is twisted and sealed with a tamperproof clip. The bag is then placed inside a fiber craft drum (125 × 75 cm, weight approximately 3.5 kg). The lid is closed and sealed with a metallic ring. Tamper-evident seals are also provided. Extracts that are paste-like in consistency (e.g., oleoresins) are packed in white pails and small-mouthed, high-density polyethylene (HDPE) drums. The packed extracts are stored in a cool, dry place.

3.20.5 Finished product specification

A finished product specification is prepared after three manufacturing plant batches (**Table 3.15**).

3.20.6 Unit calculation

The unit recovery of the actives is calculated as described in Section 3.16.

3.20.7 Voucher sample

A sample (50 g) of the commercially manufactured extract should be stored as a voucher sample for any future analysis.

Table 3.15 Finished Product Specification of *Terminalia arjuna* Extract

Specification No. ------

Product code ------

Botanical name: *Terminalia arjuna*

CAS No.: 93456-04-3

Plant part used: Bark

Solvent used in manufacture: Methanol

Other solvent used: Methylene dichloride

Final extract ratio: 10:1 to 17:1

Marker for standardization: Arjunglucoside

Excipients used: None

Parameters	Limit	Reference
Description	Dark brown powder[a]	Visual
Identification	To comply by HPLC	In-house method
Alcohol solubles (1% w/v solution in 95% v/v alcohol)	Not less than 80.0% w/w	API <191>
Loss on drying	Not more than 8.0% w/w (dried at 105°C)	USP <731>
Tapped bulk density	Between 0.80 and 1.10 g/mL	USP <616>
Loose bulk density	Between 0.50 and 0.80 g/mL	USP <616>
Sieve test (passes through) 40 mesh	Not less than 95% w/w	USP <786>
Chemical assay Content of arjunglucoside	Not less than 3.0% w/w	HPLC, in-house method
Others		
Total heavy metals	Not more than 20 ppm (μg/g)	USP <231> Method II
Lead	Not more than 3 ppm (μg/g)	ICP-MS, AOAC <990.08>
Arsenic	Not more than 1 ppm (μg/g)	ICP-MS, AOAC <990.08>
Cadmium	Not more than 1 ppm (μg/g)	ICP-MS, AOAC <990.08>
Mercury	Not more than 0.1 ppm (μg/g)	ICP-MS, AOAC <971.21C>
Residual solvents	To comply as per USP	USP <467>
Residual pesticides	To comply as per USP	USP <561>
Microbiological profile		
Total plate count	Not more than 5000 cfu/g	USP <2021>
Yeast and mold count	Not more than 100 cfu/g	USP <2021>
Escherichia coli	Negative/10 g	USP <2022>
Salmonella	Negative/10 g	USP <2022>
Staphylococcus aureus	Negative/10 g	USP <2022>
Pseudomonas aerugnosa	Negative/10 g	USP <62>
Enterobacteriaceae	Negative/10 g	USP <2021>

(*Continued*)

Table 3.15 (Continued) Finished Product Specification of *Terminalia arjuna* Extract

Additional information	
Sanitizing treatment	Nonirradiated and not treated with ETO
Genetic modification status	GMO-free
BSE/TSE status	BSE/TSE-free
Cultivated or wildcrafted	Cultivated
Storage condition	Store at room temperature
Shelf life	2 years
Name and address of manufacturer	--------

Note: AOAC = Official Method of Analysis of AOAC International; API = *Ayurvedic Pharmacopoeia of India Part 1, Vol. 2*; IP = *Indian Pharmacopoeia Vol. 1*; USP = United States Pharmacopoeia; BSE = Bovine spongiform encephalopathy; TSE = Transmissible spongiform encephalopathy.

[a] As this is a herbal product, there is a likelihood of minor color variation, on account of the geographical and seasonal variations of raw material.

3.20.8 Product master file

The product master file (drug master file) is a complete application dossier for the registration of a product intended for use in Europe, Japan, and the United States. It is prepared according to a specification known as the *Common Technical Document* (CTD) and the accepted format is that of the ICH CTD (Anonymous, 2010b). Essentially, a product master file contains information on the manufacturing process and the process controls, control of raw materials, analytical procedures, stability, etc. (Anonymous, 2010e) (**Table 3.16**).

3.20.9 *Kosher* and *halal* certification

Kosher and halal certification are two certification systems that focus on the dietary laws of Jews and Muslims, respectively. The kosher (*kashrus*) dietary laws of Jews determine which foods are kosher (fit or proper).These laws are of biblical origin, coming mainly from the original five books of the Holy Scriptures, the *Torah*, which has remained unchanged over centuries. The halal dietary laws determine which foods are "lawful' or permitted for Muslims. These laws are found in the *Quran* and the Traditions of the Prophet (*Hadith*) (Kocturk, 2002; Regenstein et al., 2003).

Kosher and halal certification of a herbal extract and nutraceutical or functional food ensures that the Jewish and Islamic dietary laws are not compromised in their manufacture. These certifications are of importance to the herbal medicine and food industry in Europe and the United States. Several agencies offer kosher and halal certification. However, the most prestigious agencies are the Orthodox Union (Union of Orthodox Jewish Congregations of America, Eleven Broadway, New York, NY 0004, USA; website: www.ou.org) for kosher certification and IFANCA International (A division of Islamic Food

Table 3.16 The CTD Format

No.	Particulars
1.	General informationNomenclature
1.1.1	International proprietary name
1.1.2	Compendial name
1.1.3	Chemical name
1.1.4	Company or laboratory code(s)
1.1.5	Other nonproprietary name (s)
1.1.6	Chemical abstracts service (CAS) registry number
1.2	Structure
1.2.1	Structural formula
1.2.2	Molecular formula
1.2.3	Relative molecular weight
1.3	General properties
1.3.1	Physicochemical characteristics
1.3.1.1	Physical characteristics
1.3.1.2	Solubility
1.3.1.3	Melting point
1.3.1.4	Polymorphism
1.3.1.5	Particle size
2.	Manufacture
2.1	Manufacturer(s)
2.2	Description of manufacturing process and process controls
2.2.1	Synthetic route
2.2.2	Description of the process
2.2.3	Manufacturing scheme
2.3	Controls of the materials
2.4	Controls of critical steps and intermediates
2.5	Process validation and or evaluation
2.6	Manufacturing process development
3.	Characterization
3.1	Elucidation of structure and other characteristics
3.1.1	IR spectrum
3.1.2	^{1}H NMR spectrum
3.1.3	^{13}C NMR spectrum
3.1.4	Mass spectrum
3.1.5	Key intermediates of synthesis
3.2	Impurities
3.2.1	By-products of synthesis
3.2.2	Intermediates of synthesis
3.2.3	Residual solvents and reagents

(*Continued*)

Table 3.16 (Continued) The CTD Format

No.	Particulars
3.2.4	Trace elements from use of catalysts
3.2.5	Degradation product
3.2.6	Analytical methods and detection limits
4.	Control of drug substance
4.1	Specification
4.2	Analytical procedures
4.2.1	Description
4.2.2	Identification A (IR)
4.2.3	Identification B (chlorides)
4.2.4	Appearance in solution
4.2.5	pH
4.2.6	Fluoroquinolonic acid
4.2.7	Heavy metals
4.2.8	Water
4.2.9	Sulfated ash
4.2.10	Related substances (HPLC)
4.2.11	Assay (titration)
4.2.12	Residual solvents (GC)
4.3	Validation of analytical procedures
4.4	Batch analyses
4.5	Justification of specification
5.	Reference standard materials
5.1	Related substances
6.	Container closure system
6.1	Packaging
6.1.1	Drums
6.1.2	Polyethylene bags
7.	Stability
7.1	Stability summary
7.2	Postapproval stability protocol and stability commitment
7.3	Stability data
7.3.1	Batches tested
7.3.2	General test methodology
7.3.2.1	Accelerated storage conditions and testing frequency
7.3.2.2	Long-term storage conditions and testing frequency
7.3.3	Analytical test procedure
7.3.4	Validation of analytical test procedures
7.3.5	Results of tests
7.3.5.1	Accelerated stability results
7.3.5.2	Long-term stability studies

and Nutrition Council of America, 5901, N. Cicero Avenue, Suite 309, Chicago, Illinois 60646, USA; website: www.ifanca.org) for halal certification. Although basically of religious significance, over the years both kosher and halal certification has assumed the status of symbols of quality, especially in Europe and the United States.

3.21 Quality assurance of extracts

A company that is manufacturing herbal extracts or products should have an efficient QA and QC department. QA is defined as the system of monitoring, inspecting, and auditing, ensuring that the product has been manufactured according to the SOPs, right from the manufacture of the product to its delivery to the end user (Pedec and Klodziejczyk, 2000).

The company must engage in the manufacture of quality products and occasionally some of the products may not conform to the set standards. The QC system ensures that such unacceptable products are detected and their manufacture is prevented. Thus, quality is maintained and wastage is prevented. Therefore, detection and prevention of lapses is the primary goal of a QC system. A QA system ensures that all batches of the manufactured product will be of the same quality (Pedec and Klodziejczyk, 2000).

QA is constantly involved in activities such as quality planning, QC, quality improvement, quality auditing, and reliability. Depending on the product line and the size of the company, QA and QC may be separate or combined. Where QA is a separate entity, it is responsible for monitoring the entire manufacturing activities of the company, to assure the company management that the production facilities, personnel, manufacturing methods, various practices in the company, records, and control mechanisms conform to good manufacturing processes. QA is also expected to be fully conversant with all the regulatory requirements related to the business (Pedec and Klodziejczyk, 2000).

QA personnel should have knowledge of and expertise in the technology used in the production; they should also have training ability, interpersonal skills, and abilities in auditing and investigation. In Canada, the QA/QC department of a herbal product manufacturing company should employ staff with a university degree in a science related to the activity being carried out. With this expertise, they will be able to identify starting materials from a botanical perspective and appreciate and understand the sources of herbal raw material, cultivation practices, the ontogenic stage of the herb, environmental factors that affect the quality of herbs, and various certification systems related to international trade (Pedec and Klodziejczyk, 2000).

The QA/QC manager provides leadership for compliance with standards and thereby maintains continuous improvements in the manufacturing processes and the quality of the products. He/she should have skills such as business knowledge, leadership, management, communication, and knowledge of manufacturing processes, quality consciousness, validation, chemistry, microbiology, botany, audits, suppliers and their strengths, and customer awareness (Pedec and Klodziejczyk, 2000).

Cremer (1996) states that "a quality professional's true job is to learn, drive the quality transition and take the organization where it cannot take itself." Therefore, staff in QA/QC should receive sufficient training for their functions and very specially good manufacturing practices (GMPs), SOPs, MSDSs, and record maintenance.

3.22 Interaction with marketing team

The development and marketing of nutraceuticals is a complex process (Van Kleef et al., 2002, 2005), requiring significant research efforts in the identification of functional compounds and assessing their physiological effects through suitable food matrices (Siró et al., 2008). It also calls for close interaction between the R&D and marketing teams.

One of the first steps in product development is an exploration of the diseases with which consumers are concerned. The marketing team gleans this information from market research and communicates it to the R&D, which initiates the entire R&D program. The outcome of this activity is a standardized herbal extract that can be incorporated into a variety of products.

In order to market the standardized extract, the marketing team establishes contact with manufacturers of fortified foods and beverages. At this point, the R&D team lends its support with product information. Product literature essentially consists of information on the botany and chemistry of the medicinal herb from which the extract is derived, the method and flow chart of the manufacture of the extract, the excipients used, the nutraceutical applications of the extract and possible food matrices that can be fortified with it, the mechanism of action of the major compounds in the extract, the finished product specification, and the stability data and toxicological information.

Consumers have limited knowledge and awareness of the health benefits of newly developed nutraceuticals. Therefore, there is a strong need for specific information and communication activities. The general public, including physicians and nutrition advisers, need to be informed about the medicinal value of the product in a language that speaks of the benefits and not the "science" (Siró et al., 2008). Interaction between R&D and marketing is vital in this context.

3.23 Concluding remarks

There is growing interest in the extraction of biologically active compounds from medicinal herbs. Experts from several scientific disciplines are engaged in this activity. A freshly harvested herb is carefully dried, comminuted to the desired particle size, and extracted with the appropriate solvent. Care should be taken during the extraction to minimize the interference from compounds that may coextract with the desired compounds and to prevent degradation of the compounds. After the optimization of the extraction conditions at the laboratory scale, the process is scaled-up to the pilot scale and later to the production scale. Effective scale-up is achieved through the logical application of well-characterized biological, chemical, and engineering principles (Martin et al., 2006).

Innovation should continue in the sphere of the solvent extraction of herbs. The R&D team should constantly strive to develop environment-friendly processes, apply novel processing methods, and enhance the existing process technology. It is also important to acquire expertise in the optimization of the process through a reduction in the use of utilities, an increase in the yield of extractives, a reduction in the quantity of solvents, and a reduction in the processing time (Aziz et al., 2003).

Some recent examples highlight these points. After utilizing cocoa beans, the pod shells of cocoa fruits are generally discarded. They decompose in the cocoa plantations, generating foul odors and causing environmental pollution. Chan and Choo (2013) investigated whether the pod shells could be put to better use. Their studies show that commercial pectin can be extracted from the pod shells with citric acid as the solvent.

In addition to oil, cottonseed contains gossypol, a compound that is toxic to livestock (Alexander et al., 2008). Cottonseed triacylglycerols are nonpolar components, while gossypol is polar. On account of its high polarity, gossypol is readily extracted by polar solvents such as ethyl alcohol and IPA. Because of the different solubility characteristics between triacylglycerols and gossypol, it is impossible to isolate both types of components using a single-step extraction with one solvent such as hexane or alcohol. To overcome this problem, Kuk and Hron (1998) employed a solvent system of isohexane and either 2%–25% ethanol or IPA. Isohexane was chosen as it is not a health risk like n-hexane. This extraction system produces a cottonseed meal of superior quality.

The use of green (eco-friendly) solvents has also begun. One such solvent is ethyl lactate (ethyl 2-hydroxy-propanoate), which is fully biodegradable, noncorrosive, noncarcinogenic, and nonozone depleting. The use of this solvent has already been demonstrated in the extraction of sclareol from clary sage (Tombokan, 2008), carotenoids from different plant matrices (Ishida and Chapman, 2009; Strati and Oreopoulou, 2011), γ-linolenic acid from spirulina (Golmakani et al., 2012), caffeine from green coffee beans and green tea leaves (Bermejo et al., 2012), and thymol from thyme (Angelov et al., 2013), and the fractionation of squalene and tocopherol from edible oils (Hernández et al., 2011; Vicente et al., 2011).

d-Limonene, which is obtained from citrus peels through steam distillation followed by a deterpenation process, is shown to be a substitute for solvents such as dichloromethane, toluene, and hexane. The major advantages of using limonene are decreases in extraction time, energy consumption, solvents recycled, and environmental impact (Chemat et al., 2012a). The findings of Chemat-Djenni et al. (2010) and Angelov et al. (2013) demonstrate that d-limonene can be used in the extraction of lycopene and thymol, respectively. Toxicological studies show that d-limonene has very low toxicity and does not pose any mutagenic, carcinogenic, or nephrotoxic risk to humans (Sun, 2007).

Preliminary studies show that glycerol, α-pinene, and p-cymene can also be used as biosolvents for oilseed extraction (Chemat et al., 2012b; Fine et al., 2013; Boutekedjiret et al., 2013). It is essential to continue the quest for green solvents as more and more new products enter the global market in the form of plant extracts and essential oils (Baser, 2003).

References

Agarwal, A., Murali, B. 2010. *Quality Assessment of Selected Indian Medicinal Plants*, Volume 1, 1–252. Bangalore: National Medicinal Plants Board, New Delhi & Natural Remedies.

Akaranta, O., Anusiem, A.C.I. 1996. A bioresource solvent for extraction of castor oil. *Industrial Crops and Products* 5: 273–277.

Alexander, J., Benford, D., Cockburn, A., et al. 2008. Gossypol as undesirable substance in animal feed. *The EFSA Journal* 908: 1–55.

Ammon, H.P., Muller, A.B. 1985. Forskolin—From an Ayurvedic remedy to a modern agent. *Planta Medica* 6: 473–477.

Angelov, I., Bermejo, D.V., Stateva, R.P., Reglero, G., Ibañez, E.,Fornari T. 2013. Extraction of thymol from different varieties of thyme plants using green solvents. In *III Iberoamerican Conference on Supercritical Fluids, Cartagena de Indias (Colombia)*, 1–7.

Anonymous. 1980. *British Pharmacopoeia*, Volume 2, A 195. London: Her Majesty's Stationery Office.

Anonymous. 1993. *Formulary of Siddha Medicines*. Chennai: The Indian Medical Practitioners' Co-Operative Pharmacy and Stores.

Anonymous.1994. *Unani Pharmacopoeia (Formulary of Unani Medicines)*. Chennai: The Indian Medical Practitioners' Co-Operative Pharmacy and Stores.

Anonymous. 1995. *Bacteriological Analytical Manual*, 8th edn. Silver Spring, MD: Food and Drug Administration, AOAC International.

Anonymous. 1997a. ICH harmonized tripartite guideline for residual solvents, Q3C Step 4, 16 July. http://www.nihs.go.jp/drug/ich_q3c_step4/q3cdrf_9.html. Accessed July 20, 2010.

Anonymous. 1997b. *Medicinal Plants in China*. Manila: WHO.

Anonymous. 1998a. *Medicinal Plants in the South Pacific*. Manila: WHO.

Anonymous. 1998b.*Quality Control Methods for Medicinal Plant Materials*. Geneva: WHO.

Anonymous. 1999a. *Voucher Specimen Collection, Preparation, Identification and Storage Protocol—Plants & Fungi, Standards for Components of British Columbia's Biodiversity No. 4b*. Prepared by Ministry of Environment, Lands and Parks Resources Inventory Branch for the Terrestrial Ecosystems Task Force Resources Inventory Committee, Victoria, BC, Canada. http://www.ilmb.gov.bc.ca/risc/pubs/tebiodiv/voucherb/assets/voucherb.pdf. Accessed June 27, 2009.

Anonymous. 1999b. *WHO Monographs on Selected Medicinal Plants*, Volume 1, 1–295. Geneva: WHO.

Anonymous. 2001a. Guideline for residual solvents, residual solvents test, and models for the test in monographs. In *The Japanese Pharmacopoeia*, 15th edn, 1682–1683. Tokyo: Society of Japanese Pharmacopoeia.
Anonymous. 2001b. *National Formulary of Unani Medicine*, Part III, 1st edn. Delhi: The Controller of Publications.
Anonymous. 2001c. Processing of medicinal plants—Grinding, sieving and comminution. In *A Guide to the European Market for Medicinal Plants and Extracts*, 60–61. London: Commonwealth Secretariat.
Anonymous. 2002a. *Indian Herbal Pharmacopoeia*. Revised new edition. Mumbai: Indian Drug Manufacturers' Association.
Anonymous. 2002b. *WHO Monographs on Selected Medicinal Plants*, Vol. 2, 1–358. Geneva: WHO.
Anonymous. 2003. Note for guidance on stability testing—Stability testing of new drug substances and products. http://www.emea.europa.eu/pdfs/human/ich/273699en.pdf. Accessed June 27, 2009.
Anonymous. 2006a. *National Formulary of Unani Medicine, Part I*, 1–336. New Delhi: Central Council for Research in Unani Medicine, Government of India.
Anonymous. 2006b. *National Formulary of Unani Medicine, Part IV*, 1st edn, 1–231. Delhi: The Controller of Publications.
Anonymous. 2006c. Residual solvents. In *European Pharmacopoeia*, 3rd edn, 335–344. Strasbourg: Council of Europe.
Anonymous. 2006d. *The Japanese Pharmacopoeia*, English Version, 15th edn, 1–1788. Tokyo: Society of Japanese Pharmacopoeia.
Anonymous. 2007a. *Indian Pharmacopoeia 2007*, Vols. 1–3. Ghaziabad: The Indian Pharmacopoeia Commission.
Anonymous. 2007b. *National Formulary of Unani Medicine, Part II*, Vol. I, 1–197. New Delhi: Department of Ayurveda, Yoga & Naturopathy, Unani, Siddha and Homoeopathy (AYUSH), Government of India.
Anonymous. 2007c. *The Unani Pharmacopoeia of India, Part I*, Volume I, 1–208. New Delhi: Department of Ayurveda, Yoga & Naturopathy, Unani, Siddha and Homoeopathy (AYUSH), Government of India.
Anonymous. 2007d. *The Unani Pharmacopoeia of India, Part I*, Volume II, 1–221. New Delhi: Department of Ayurveda, Yoga & Naturopathy, Unani, Siddha and Homoeopathy (AYUSH), Government of India.
Anonymous. 2007e. *The Unani Pharmacopoeia of India, Part I*, Volume III, 1–239. New Delhi: Department of Ayurveda, Yoga & Naturopathy, Unani, Siddha and Homoeopathy (AYUSH), Government of India, New Delhi.
Anonymous. 2007f. *The Unani Pharmacopoeia of India, Part I*, Volume IV, 1–222. New Delhi: Department of Ayurveda, Yoga & Naturopathy, Unani, Siddha and Homoeopathy (AYUSH), Government of India.
Anonymous. 2007g. *The United States Pharmacopoeia—The National Formulary*, Volumes 1–3. Rockville, MD: The United States Pharmacopoeial Convention.
Anonymous. 2007h. *WHO Monographs on Selected Medicinal Plants*, Volume 3, 1–390. Geneva: WHO.
Anonymous. 2008a. *Guidelines on Good Manufacturing Practice for Traditional Medicines and Health Supplements*, 1st edn. Kuala Lumpur: National Pharmaceutical Control Bureau of Malaysia.
Anonymous. 2008b. *National Formulary of Unani Medicine, Part V*, 1–207. New Delhi: Department of Ayurveda, Yoga & Naturopathy, Unani, Siddha and Homoeopathy (AYUSH), Government of India.
Anonymous. 2008c. *The Unani Pharmacopoeia of India, Part I*, Volume V, 1–229. New Delhi: Department of Ayurveda, Yoga & Naturopathy, Unani, Siddha and Homoeopathy (AYUSH), Government of India.

Anonymous. 2009a. *Second Supplement to U.S. Pharmacopoeia*, 32nd edn, 4144–4145. Rockville, MD: The United States Pharmacopoeia Convention.
Anonymous. 2009b. *The Unani Pharmacopoeia of India, Part I*, Volume VI, 1–230. New Delhi: Department of Ayurveda, Yoga & Naturopathy, Unani, Siddha and Homoeopathy (AYUSH), Government of India.
Anonymous. 2009c. *The Unani Pharmacopoeia of India, Part II*, Volume I: Formulations, 1st edn, 1–291. New Delhi: Department of Ayurveda, Yoga & Naturopathy, Unani, Siddha and Homoeopathy (AYUSH), Government of India.
Anonymous. 2009d. *The United States Pharmacopoeia USP 32, NF 27, Chapter 2021*, Volume 1. Rockville, MD: The United States Pharmacopoeial Convention.
Anonymous. 2009e. *The United States Pharmacopoeia USP 32, NF 27*, Volume 1, 1069–1070. Rockville, MD: The United States Pharmacopoeial Convention.
Anonymous. 2009f. *The United Sates Pharmacopoeia USP 32, NF 27*, Volume 1, 187–189. Rockville, MD: The United States Pharmacopoeial Convention.
Anonymous. 2009g. *The United States Pharmacopoeia USP 32, NF 27, Chapter 231*, Volume 1. Rockville, MD: The United States Pharmacopoeial Convention.
Anonymous. 2009h. *WHO Monographs on Selected Medicinal Plants*, Volume 4, 1–456. Geneva: WHO.
Anonymous. 2010a. Draft OECD guideline for the testing of chemical—Test guideline 453: Combined chronic toxicity/carcinogenicity studies. http://www.oecd.org/dataoecd/30/42/41753375.pdf. Accessed August 5, 2010.
Anonymous. 2010b. Guidelines for drug master file, version 1.0 (August 2009). Saudi Food & Drug Authority. http://www.sfda.gov.sa. Accessed August 6, 2010.
Anonymous. 2010c. Herbaria and herbarium specimens. University of Florida Herbarium. http://www.flmnh.ufl.edu/herbarium/herbariaandspecimens.html. Accessed May 21, 2010.
Anonymous. 2010d. *ICH Topic Q3C (R4)—Impurities: Guideline for Residual Solvents Step 5*, 1–22. London: European Medicines Agency.
Anonymous. 2010e. *International Conference on Harmonisation of Technical Requirements for Registration of Pharmaceuticals for Human Use, M4—Common Technical Document*. http://www.fda.gov. Accessed August 6, 2010.
Anonymous. 2010f. *OECD Guideline or Testing of Chemicals—Acute Oral Toxicity Acute Toxic Class Method*. http://iccvam.niehs.nih.gov/SuppDocs/FedDocs/OECD/OECD_GL423.pdf. Accessed August 5, 2010.
Anonymous. 2010g. Preparation of plant specimens for deposit as herbarium vouchers. University of Florida Herbarium. http://www.flmnh.ufl.edu/herbarium/voucher.htm. Accessed May 21, 2010.
Anonymous. 2010h. *The Unani Pharmacopoeia of India, Part II*, Volume II: Formulations, 1st edn, 1–299. New Delhi: Department of Ayurveda, Yoga & Naturopathy, Unani, Siddha and Homoeopathy (AYUSH), Government of India.
Anonymous. 2011a. *Clean, Safe Spices—Guidance from the American Spice Trade Association*. Washington, DC: The American Spice Trade Association.
Anonymous. 2011b. *National Formulary of Unani Medicine, Part VI*, 1–189. New Delhi: Central Council for Research in Unani Medicine, Government of India.
Anonymous. 2015. Sustainability issues of traditional Chinese herbal medicine—Part 1: Restricted herb and resource list. https://www.mayway.com/pdfs/maywaymailers/Skye-Sturgeon-QM-Restricted-herbs-P1-10-2011.pdf. Accessed January 28, 2015.
Aslan, N. 2007. Application of response surface methodology and central composite rotatable design for modeling the influence of some operating variables of a multi-gravity separator for coal cleaning. *Fuel* 86: 769–776.
Aslan, N., Cifci, F., Yan, D. 2008. Optimization of process parameters for producing graphite concentrate using response surface methodology. *Separation and Purification Technology* 59: 9–16.

Aslan, N., Ünal, I. 2010. Multi-response optimization of oil agglomeration with multiple performance characteristics. *Fuel Processing Technology* 92: 1157–1163.

Aubry, A.F., Tatersall, P., Ruan, J. 2008. Development of stability indicating methods. In *Handbook of Stability Testing in Pharmaceutical Development*, ed. K. Huynh-Ba, 139–161. Heidelberg: Springer.

Aziz, R.A., Sarmidi, M.R., Kumaresan, S., Taher, Z.M., Foo, D.C.Y. 2003. Phytochemical processing—The next emerging field in chemical engineering: Aspects and opportunities. *Jurnal Kejuruteraan Kimia Malaysia* 3: 45–60.

Banik, R.M., Pandey, D.K. 2008. Optimizing conditions for oleanoic acid extraction from *Lantana camara* roots using response surface methodology. *Industrial Crops and Products* 27: 241–248.

Baser, K.H.Y.C. 2003. Industrial plants as sources of dietary supplements. In *Dietary Supplements of Plant Origin*, ed. M. Maffei, 31–42. London: Taylor & Francis.

Bedi, Y.S. 2008. *Presentation, Chemistry and Manufacturing Control (CMC) for Botanical Products, Herbal International Summit and Expo on Medicinal, Aromatic Products and Spices (Hi-Maps)*, New Delhi, April 4.

Bermejo, D.V., Luna, P., Manic, M.S., Najdanovic-Visak, V., Regleroa, G., Fornari, T. 2012. Extraction of caffeine from natural matter using a bio-renewable agrochemical solvent. *Food and Bioproducts Processing* 91: 303–309.

Bhishagratna, K.K.L. 1916. *An English Translation of The Sushruta Samhita*, 3 Vols. Calcutta: S.L. Bhaduri.

Biran, M., Baykut, S. 1975. Physicochemical properties of gypsophia saponin. *Chimica Acta Turcica* 3: 63–88.

Bombardelli, E. 1991. Technologies for the processing of medicinal plants. In *The Medicinal Plant Industry*, ed. R.O.B. Wijesekera, 85–98. Boca Raton, FL: CRC Press.

Boutekedjiret, C., Bertouche, S., Hellal, A. 2013. Alpha-pinene as a green solvent—Alternative to petrochemical solvents. In *International Congress on Green Extraction of Natural Products—GENP Avignon*, 87, France, April 16–17.

Box, G.E.P., Wilson, K.B. 1951. On the experimental attainment of optimum conditions. *Journal of the Royal Statistical Society* B13: 1–45.

Bradley, N. 2007. The response surface methodology. Master of Science Thesis, Indiana University of South Bend.

Bruhn, J.G., Sandberg, F. 1991. Screening and processing of plant materials for potential pharmaceutical needs—Experience and applications in three continents. In *The Medicinal Plant Industry*, ed. R.O.B. Wijesekera, 223–236. Boca Raton, FL: CRC Press.

Burdick, D.S., Naylor, T.H. 1969. Response surface methods in economics. *Revue de l'Institut International de Statistique* 37: 18–35.

Cacace, J.E., Mazza, G. 2003. Optimization of extraction on anthocyanins from black currants with aqueous ethanol. *Journal of Food Science* 68: 240–248.

Cambrai, A., Marcic, C., Morville, S., Houer, P.S., Bindler, F., Marchioni, E. 2010. Differentiation of chocolates according to the cocoa's geographical origin using chemometrics. *Journal of Agricultural and Food Chemistry* 58: 1478–1483.

Carter, R., Bryson, C.T., Darbyshire, S.J. 2007. Preparation and use of voucher specimens for documenting research in weed science. *Weed Technology* 21: 1101–1108.

Chaloner-Larsson, G., Anderson, R., Egan, A. 1997. *A WHO Guide to Good Manufacturing Practice (GMP) Requirements. Part 1: Standard Operating Procedures and Master Formulae*, 1–111. Geneva: WHO.

Chan, S.Y., Choo, W.S. 2013. Effect of extraction conditions on the yield and chemical properties of pectin from cocoa husks. *Food Chemistry* 141: 3752–3758.

Chemat S., Tomao V., Chemat F. 2012a. Limonene as green solvent for extraction of natural products. In *Green Solvents I—Properties and Applications in Chemistry*, ed. A. Mohammad, and Inamuddin, 175–186. Dordrecht: Springer.

Chemat, F., Vian, M.A., Cravotto, G. 2012b. Green extraction of natural products—Concept and principles. *International Journal of Molecular Sciences* 13: 8615–8627.
Chemat-Djenni, Z., Ferhat, M.A., Tomao, V., Chemat, F. 2010. Carotenoid extraction from tomato using a green solvent resulting from orange processing waste. *Journal of Essential Oil-Bearing Plants* 2: 139–147.
Cohodas, A.M. 1969. Spice grinding process. Canadian Patent No. 808,644.
Cordell, G.A.1981. *Introduction to the Alkaloids—A Biogenetic Approach*. New York: Wiley-Interscience.
Cremer, C.J. 1996. The evolution of a QC guy. *Quality Progress* 29: 42–43.
Cuniff, P. 1995a. *Official Methods of Analysis of AOAC International*, 16th edn, Vol. 1. Arlington, VA: AOAC International.
Cuniff, P. 1995b. *Official Methods of Analysis of AOAC International*, 16th edn, Vol. 2. Arlington, VA: AOAC International.
D'Amelio, Sr., F.S. 1999. *Botanicals—A Phytocosmetic Desk Reference*, 13–38. Boca Raton, FL: CRC Press.
Das, R. 2008. Successful sterilization—Ensuring safe, microbial-free organic ingredients. *Organic Processing* September–October: 22–25.
Dashiell, G.L. 1990. Sources, processing methods and commercial uses of lecithin. In *Edible Fats and Oils Processing—Basic Principles and Modern Practices*, ed. D. R. Erickson, 397–400. Urbana, IL: American Oil Chemists' Society.
de Azeredo, H.M.C., Pereira, A.C., de Souza, A.C.R., Gouveia, S.T., Mendes, K.C.B. 2009. Study on efficiency of betacyanin extraction from red beetroots. *International Journal of Food Science and Technology* 44: 2464–2469.
Du, X. W., Wills, R.B.H., Stuart, D.L. 2004. Changes in neutral and malonyl ginsenosides in American ginseng (*Panax quinquefolium*) during drying, storage and ethanolic extraction. *Food Chemistry* 86: 155–159.
Duan,R., Wang, B., Shi, H.X. 2001. A study on the flavonoid-extracting technology by microwave from *Ginkgo biloba* leaves. *Journal of Huaihai Institute of Technology* 10: 46–48.
Edgar, T.F., Himmelblau, D.M., Lasdon, L.S. 2001. *Optimization of Chemical Processes*, 3–36. London: McGraw-Hill.
Endale, A., Schmidt, P.C., Gebe-Mariam, T. 2004. Standardisation and physicochemical characterization of the extracts of seeds of *Glinus lotoides*. *Pharmazie* 59: 34–38.
Ettre, L.S. 2002. *Milestones in the Evolution of Chromatography*, 129–137. Portland, OR: ChromSource.
Fan, G., Han, Y., Gu, Z., Chen, D. 2008. Optimizing conditions for anthocyanins from purple sweet potato using response surface methodology (RSM). *LWT-Food Science and Technology* 41: 155–160.
Farkas, J. 2000. Spices and herbs. In *The Microbiological Safety and Quality of Food*, Vol. 1, ed. B.M. Lund, T.C. Baird-Parker, and G.W. Gould, 897–918. Gaithersburg, MD: Aspen Publishers.
Finar, I.L. 2001. *Organic Chemistry*, Vol. 2, 696–697. Delhi: Pearson Education Asia.
Fine, F., Pages, X., Carre, P., Fabiano-Tixier, A.S., Abert Vian, M., Chemat, F. 2013. Improvement in oilseed extraction—Evaluation of several bio-solvents. In *International Congress on Green Extraction of Natural Products—GENP*, 112, Avignon, France, April 16–17, 2013.
Freund, R., Littell, R., Creighton, L. 2003. *Regression using JMP*, 156. Cary, NC: S.A.S. Institute.
Gecan, J.S., Bandler, R., Glaze, L.E., Atkinson, J.C. 1986. Microanalytical quality of ground and unground marjoram, sage and thyme, ground allspice, black pepper and paprika. *Journal of Food Protection* 49: 216–221.
Glasgow, S.M. 1997. Crystallization. In *Fermentation and Biochemical Engineering Handbook*, 2nd edn, ed. H.C.Vogel, and C.L. Todaro, 546–547. Park Ridge, NJ: Noyes Publications.
Golmakani, M.T., Mendiola, J.A., Rezaeic, K., Ibáñez, E. 2012. Expanded ethanol with CO_2 and pressurized ethyl lactate to obtain fractions enriched in γ-linolenic acid from *Arthrospira platensis* (Spirulina). *Journal of Supercritical Fluids* 62: 109–115.

Grier, J.B., Bailey, T.G., Jackson, J.A. 1997. Using response surface methodology to link force structure budgets to campaign objectives. In *Proceedings of the 1997 Winter Simulation Conference*, ed. A. Andradóttir, K.J. Healy, D.H. Withers, and B.L. Nelson, 986–973. New York: Association for Computing Machinery.

Güçlü-Üstündağ, Ö., Mazza, G. 2007. Saponins—Properties, applications and processing. *Critical Reviews in Food Science and Nutrition* 47: 231–258.

Hadjmohammadi, M., Sharifi, V. 2009. Investigation of optimum extraction conditions for determination of quercetin and kaempferol in coriander (*Coriandrum sativum* L.) by using experimental design and HPLC. *Journal of Food and Drug Analysis* 17: 293–299.

Harborne, J. B. 1998. *Phytochemical Methods—A Guide to Modern Techniques of Plant Analysis*, 3rd edn. New York: Chapman & Hall.

Hernández, E.J., Luna, P., Stateva, R.P., Najdanovic-Visak, V.,Reglero, G., Fornari, T. 2011. Liquid–liquid phase transition of mixtures comprising squalene, olive oil, and ethyl lactate—Application to recover squalene from oil deodorizer distillates. *Journal of Chemical & Engineering Data* 56, 2148–2152.

Hildreth, J., Hrabata-Robinson, E., Applequist, W., Betz, J., Miller, J. 2007. Standard operating procedure for the collection and preparation of voucher plant specimens for use in the nutraceutical industry. *Analytical and Bioanalytical Chemistry* 389: 13–17.

Hoffmann, D. 2003. *Medical Herbalism—The Science and Practice of Herbal Medicine*, 211. Rochester, VT: Healing Arts Press.

Horn, H. J. 1956. Simplified LD_{50} (or ED_{50}) calculations. *Biometrics* 12: 312–322.

Hostettmann, K., Hostettmann, M., Marston, A. 1991. Saponins. In *Methods in Plant Biochemistry*, ed. P.M. Dey, and J.B. Harbone, Vol. 7, 435–471. Berlin: Springer.

Hurst, W.J. 2008. *Methods of Analysis for Functional Foods and Nutraceuticals*, 2nd edn, 1–532. Boca Raton, FL: CRC Press.

Ishida, B.K., Chapman, M.H. 2009. Carotenoid extraction from plants using a novel, environmentally friendly solvent. *Journal of Agricultural and Food Chemistry* 57: 1051–1059.

Iyer, R., Tomar, S.K., Singh, A.K. 2010. Response surface optimization of the cultivation conditions and medium components for the production of folate by *Streptococcusthermophilus*. *Journal of Dairy Research* 77: 350–356.

Izmailov, N.A., Schreiber, M.S. 1938. Farmatsiya No. 3, p. 1, cited by Wall P.E. (2005) Introduction and history. In *Thin-Layer Chromatography—A Modern Practical Approach*, 1–5. Cambridge: The Royal Society of Chemistry.

Jaijoy, K., Soonthornchareonnon, N., Lertprasertsuke, N., Panthong, A., Sireeratawong, S. 2010. Acute and chronic oral toxicity of standardized water extract from the fruit of *Phyllanthus emblica* Linn. *International Journal of Applied Research in Natural Products* 3: 48–58.

Jambhelkar, S.S. 2005. Micrometrics and rheology. In *Theory and Practice of Contemporary Pharmaceutics*, ed. T.K. Ghosh, and B.R. Jasti, 137–162. Boca Raton, FL: CRC Press.

Jaya, S., Das, H. 2004. Effect of maltodextrin, glycerol monostearate and tricalcium phosphate on vacuum dried mango powder properties. *Journal of Food Engineering* 63: 125–134.

Jones, W.P., Kinghorn, A.D. 2006. Extraction of plant secondary metabolites.In *Natural Products Isolation*, 2nd edn, ed. S.D. Sarker, Z. Latif, and A.I. Gray, 323–351. Totowa, NJ: Humana Press.

Juntachote, T., Berghofer, E., Bauer, F., Siebenhandl, S. 2006. The application of response surface methodology to the production of phenolic extracts of lemon grass, galangal, holy basil and rosemary. *International Journal of Food Science and Technology* 41: 121–133.

Karacabey, E., Mazza, G. 2008. Optimization of solid-liquid extraction of resveratrol and other phenolic compounds from milled grape canes (*Vitis vinifera*). *Journal of Agricultural and Food Chemistry* 56: 6318–6325.

Karimi, P., Abdollahi, H., Aslan, N., Noaparast, M., Shafaei, S.Z. 2011. Application of response surface method and central composite design for modeling and optimization of gold and silver recovery in cyanidation process. *Mineral Processing & Extractive Metallurgy Review* 32: 1–16.

Katušin-Ražem, B., Novak, B., Ražem, D. 2001. Microbial decontamination of botanical raw materials and corresponding pharmaceutical products by irradiation. *Radiation Physics and Chemistry* 62: 261–275.

Kawaguti, H.Y., Manrich, E., Sato, H.H. 2006. Application of response surface methodology for glucosyltransferase production and conversion of sucrose into isomaltulose using free *Erwinia* sp. cells. *Electronic Journal of Biotechnology* 9: DOI: 102225/vol9-issue5. fulltext.6

Kerem, Z., Herman-Shashoua, H., Yarden, O. 2005. Microwave-assisted extraction of bioactive saponins from chickpea (*Cicer arietinum* L.). *Journal of the Science of Food and Agriculture* 85: 406–412.

Kocturk, T.O. 2002. Food rules in the Koran. *Scandinavian Journal of Nutrition* 46: 137–139.

Kong, K.W., Ismail, A.R., Tan, S.T., Prasad, K.M.N., Ismail, A. 2010. Response surface optimization for the extraction of phenolics and flavonoids from a pink guava industrial by-product. *International Journal of Food Science and Technology* 45: 1739–1745.

Koseki, P.M., Villavicencio, A.L.C.H., Brito, M.S., et al. 2002. Effects of irradiation in medicinal and eatable herbs. *Radiation Physics and Chemistry* 63: 681–684.

Kraybill, H.F. 1969. Significance of pesticide residues in foods in relation to total environmental stress. *Canadian Medical Association Journal* 100: 204–215.

Kuk, M.S., Hron, R.J. 1998. Cottonseed extraction with a new solvent system—Isohexane and alcohol mixtures. *Journal of the American Oil Chemists' Society* 75: 927–930.

Kwiatkowski, J.R., McAloon, A.J., Taylor, F., Johnston, D.B. 2006. Modeling the process and costs of fuel ethanol production by the corn dry-grind process. *Industrial Crops and Products* 23: 288–296.

Lachance, P.A. 2008. Nutraceutical/drug/anti-terrorism safety assurance through traceability. In *Nutraceutical and Functional Food Regulations in the United States of America and around the World*, ed. D. Bagchi, 109–112. New York: Academic.

Lee, C.C., Chen, H.C., Ju, H.Y., et al. 2010. Optimization of ultrasound-accelerated synthesis of enzymatic octyl hydroxyphenylpropionate by response surface methodology. *Biotechnology Process* 26:1629–1634.

Lohar, D.R. 2010. *Protocol for Testing Ayurveda, Siddha & Unani Medicines*, 1–200. Ghaziabad: Pharmacopoeial Laboratory for Indian Medicines.

Luthria, D.L. 2008. Influence of experimental conditions on the extraction of phenolic compounds from parsley (*Petroselinum crispum*) flakes using a pressurized liquid extractor. *Food Chemistry* 107: 745–752.

Luthria, D.L. 2009. Phenolic compounds analysis in foods and dietary supplements is not the same using different sample preparation procedures. In *Proceedings of the 2nd International Symposium on Human Health Effects of Fruits and Vegetables*, ed. B. Patil, 381–388, October 9–13, 2007, Houston, TX, *Acta Horticulturae* 841.

Luthria, D.L., Biswas, R., Natarajan, S. 2007. Comparison of extraction solvents and techniques used for the assay of isoflavones from soybean. *Food Chemistry* 105: 325–333.

Ma, H.L., Wang, C., Liu, W.M. 2005. Microwave-assisted extraction of flavonoids from *Radix puerariae*. *Journal of Jiangsu University* 26: 98–100.

Mahady, G.B. 2003. Official and scientific information resources for botanical dietary supplements. In *Dietary Supplements of Plant Origin—A Nutrition and Health Approach*, ed. M. Maffei, 187–202. New York: Taylor & Francis.

Maldonado, L.M., Hernández, V.E., Rivero, E.M., et al. 2007. Optimization of culture conditions for a synthetic gene expression in *Escherichia coli* using response surface methodology—The case of human interferon beta. *Biomolecular Engineering* 24: 217–222.

Mani, S., Jaya, S., Vadivambal, R. 2007. Optimization of solvent extraction of Moringa (*Moringa oleifera*) seed kernel oil using response surface methodology. *Food and Bioproducts Processing* 85 (C4): 328–335.

Martin, S.M., Kau, D.A., Wrigley, S.K. 2006. Scale-up of natural product isolation. In *Natural Products Isolation*, ed. S.D. Sarker, Z. Latif, and A.I. Gray, 439–461. Totowa, NJ: Humana Press.

Meyer, D.L. 1963. Response surface methodology in education and psychology. *The Journal of Experimental Education* 31: 329–336.

Migdal, W., Chmielewski, A.G. 2005. Radiation decontamination of herbs and spices. *Nukleonika* 50: 179–184.

Miller, E.H., Nagorsen, D.W. 1992. Voucher specimens—An essential component of biological surveys. In *Methodology for Monitoring Wildlife Diversity in B.C. Forests*, ed. L. Ramsay, 11–15. Victoria, BC: Ministry of Environment, Lands and Parks.

Mitra, J., Sharma, S. 2006. *Aṣṭāṅgasamgraha of Vāhaṭa or Vṛddha Vāgbhaṭa*. New Delhi: Motilal Banarsidass.

Montgomery, D.C. 2009. *Design and Analysis of Experiments*, 7th edn, 1–22. Hoboken, NJ: Wiley.

Muggeridge, M. 2001. Quality specifications of herbs and spices. In *Handbook of Herbs and Spices*, Vol. 1, ed. K.V. Peter, 13–21. Boca Raton, FL: CRC Press.

Muir, A.D., Paton, D., Ballantyne, K., Aubin, A.A. 2002. Process for recovery and purification of saponins sand sapogenins from quinoa (*Chenopodium quinoa*). U.S. Patent 6,355,249.

Mukhopadhyay, S., Luthria, D.L., Robbins, R.J. 2006. Optimization of extraction process for phenolic acids from black cohosh (*Cimicifuga racemosa*) by pressurized liquid extraction. *Journal of the Science of Food and Agriculture* 86: 156–162.

Murty, K.R.S. 2007. Vāgbhata's *Aṣṭāṅgahṛdayam*, Vols. 1–3. Varanasi: Chowkhamba Krishnadas Academy.

Mutai, M. 2000. National and international guidelines for the conduct of chemical safety studies—Choice of strains. In *The Laboratory Rat*, ed. G. J. Krinke, 17–27. London: Academic.

Myers, R.H., Montgomery, D.C., Anderson-Cook, C.M. 2009. *Response Surface Methodology—Process and Product Optimization Using Designed Experiments*, 3rd edn, 1–12. Hoboken, NJ: Wiley.

Palamakula, A., Nutan, M.T.H., Khan, M.A. 2004. Response surface methodology for optimization and characterization of limonene-based coenzyme Q10 self-nanoemulsified capsule dosage form. *AAPS PharmSci Tech* 5: 114–121.

Pan, X., Liu, H., Jia, G., Shu, Y.Y. 2000. Microwave-assisted extraction of glycyrrhizic acid from licorice root. *Biochemical Engineering Journal* 5: 173–177.

Papamichali, I., Louli, V., Magoulas, K. 2000. Supercritical fluid extraction of celery seed oil. *Journal of Supercritical Fluids* 18: 213–226.

Patel, R.N., Bandyopadhyay, S., Ganesh, A. 2006. Economic appraisal of supercritical fluid extraction of refined cashew nut shell liquid. *Journal of Chromatography A* 1124: 130–138.

Pedec, P. and Klodziejczyk, P.P. 2000. Quality assurance and control for the herbal and tea industry. In *Herbs, Botanicals and Teas*, ed. G. Mazza, and B.D. Oomah, 377–398. Boca Raton, FL: CRC Press.

Pereira, C.G., Rosa, P.T.V., Meireles, M.A.A. 2007. Extraction and isolation of indole alkaloids from *Tabernaemontana cathariensis*—Technical and economical analysis. *The Journal of Supercritical Fluids* 40: 232–238.

Pillai, T.V.S. 1931. *Tamil-English Dictionary of Medicine, Chemistry, Botany and Allied. Sciences*, Vols. 1–5. Madras: The Research Institute of Siddhar's Science.

Rajpal, V. 2002. *Standardization of Botanicals*, Vol. 1, 1–286. Delhi: Eastern Publishers.

Ravindran, P.N., Kallupurackal, J.A. 2001. Black pepper. In *Handbook of Herbs and Spices*, Vol. 1, ed. K.V. Peter, 81. Boca Raton, FL: CRC Press/Woodhead Publishing.

Regenstein, J.M., Chaudry, M.M., Regenstein, C.E. 2003. The kosher and halal food laws. *Comprehensive Reviews in Food Science and Food Safety* 2: 111–127.

RenJie, L. 2008. Optimization of extraction process of *Glycyrrhiza glabra* polysaccharides by response surface methodology. *Carbohydrate Polymers* 74: 858–861.

Rodrigues, S., Pinto, G.A.S. 2007. Ultrasound extraction of phenolic compounds from coconut (*Cocos nucifera*) shell powder. *Journal of Food Engineering* 80: 869–872.

Rosa, P.T.V., Meireles, M.A.A. 2009. Cost of manufacturing of supercritical fluid extracts from condimentary plants. In *Extracting Bioactive Compounds for Food Products—Theory and Applications*, ed. M.A. A. Meireles, 388–401. Boca Raton, FL: CRC Press.

Rouf, S.A., Douglas, P.L., Moo-Young, M., Scharer, J.M. 2001. Computer simulation for large scale bioprocess design. *Biochemical Engineering Journal* 8: 229–234.

Ryan T.P. 2007. *Modern Experimental Design*, 1–601. Hoboken, NJ: Wiley.

Sadeghi-aliabadi, H., Asghari, G., Mostafavi, S.A., Esmaeli, A. 2009. Solvent optimization on taxol extraction from *Taxus baccata* L., using HPLC and LC-MS. *Daru* 17, 192–198.

Said, M. 1997. *Hamdard Pharmacopoeia of Eastern Medicine*, 1–544. Delhi: Sri Sadguru Publications.

Saksena, A.K., Green, M.J., Shue, H.J.,et al. 1985. Identity of colenol with forskolin—Structure revision of a base-catalysed rearrangement product. *Tetrahedron Letters* 26: 551–554.

Sathe, A.Y. 1999. *First Course in Food Analysis*, 5–9. New Delhi: New Age International.

Sharma, R.K., Dash, B. 2004. *Agnivesa's Caraka Samhita*, Vols. 1–6. Varanasi: Chaukhamba Sanskrit Series Office.

Sharma, H.K., Kaur, J., Sarkar, B.C., Singh, C., Singh, B., Shitandi, A.A. 2006. Optimization of pretreatment conditions of carrots to maximize juice recovery by response surface methodology. *Journal of Engineering Science and Technology* 1: 158–165.

Shimoyada, M., Osugi, Y., Shiraiwa, M., Okubo, K., Watanabe, K. 1993. Solubilities of soybean saponins and their solubilization with a bidesmoside saponin. *Journal of the Japanese Society of Food Science and Technology* 40: 210–213.

Shorab. 2008. Traceability—A major challenge for food industry. *Quality World* 5: 11–26.

Sin, H.N., Yusof, S., Hamid, N.S.A., Rahman, R.A. 2006. Optimization of hot water extraction for sapodilla juice using response surface methodology. *Journal of Food Engineering* 74: 352–358.

Siró, I., Kápolna, E., Kápolna, B., Lugasi, A. 2008. Functional food. Product development, marketing and consumer acceptance—A review. *Appetite* 51: 456–467.

Song, J., Yao, H., Li, Y., et al. 2009. Authentication of the family polygonaceae in Chinese pharmacopoeia by DNA barcoding technique. *Journal of Ethnopharmacology* 124: 434–439.

Souza, C.R.F., Schiavetto, I.A., Thomazini, F.C.F., Oliveira, W.P. 2008. Processing of *Rosmarinus officinalis* Linne extract on spray and spouted bed dryers. *Brazilian Journal of Chemical Engineering* 25: 59–69.

Srirama, R., Senthilkumar, U., Sreejayan, N.,et al. 2010. Assessing species admixtures in raw drug trade of *Phyllanthus*, a hepato-protective plant using molecular tools. *Journal of Ethnopharmacology* 130: 208–215.

Stahl, E. 2005. *Thin-Layer Chromatography—A Laboratory Handbook*, 2nd edn. Heidelberg: Springer.

Strati, I.F., Oreopoulou, V. 2011. Effect of extraction parameters on the carotenoid recovery from tomato waste. *International Journal of Food Science & Technology* 46: 23–29.

Sun, J. 2007. D-Limonene—Safety and clinical applications. *Alternative Medicine Review* 12: 259–264.

Sung, H.S., Yang, C.B. 1985. Effect of ethanol concentration on saponin composition of red ginseng extract. *Korean Journal of Food Science and Technology* 17: 227–231.

Sung, H.S., Yang, C.B., Kim, W.J. 1985. Effect of extraction time on saponin composition of red ginseng extract. *Korean Journal of Food Science and Technology* 17: 265–270.

Tainter, D.R., Grenis, A.T. 2001. *Spices and Seasonings—A Food Technology Handbook*, 2nd edn, 9–32. New York: Wiley-VCH.

Takeuchi, T.M., Leal, P.F., Favereto, R., et al. 2008. Study of the phase equilibrium formed inside the flash tank used at the separation step of a supercritical fluid extraction unit. *The Journal of Supercritical Fluids* 43: 447–459.

Takeuchi, T.M., Pereira, C.G., Braga, M.E.M., Marostica Jr., M.R., Leal, P.F., Meireles, M.A.M. 2009. Low pressure solvent extraction (solid-liquid extraction, microwave assisted, and ultrasound assisted) from condimentary plants. In *Extracting Bioactive Compounds for Food Products*, ed. M.A.M. Meireles, 137–218. Boca Raton, FL: CRC Press.

Tombokan, X.C. 2008. Ternary phase equilibria of the sclareol-ethyl lactate-CO_2 system and its application in the extraction and isolation of sclareol from Clary Sage. Ph.D. thesis, North Carolina State University, Raleigh.

Trease, G.E., Evans, W.C. 1983. *Pharmacognosy*, 12th edn, 136. London: Baillière Tindall.
Treybal, R.E. 1980. *Mass Transfer Operations*. New York: McGraw Hill.
Turton,R., Bailie, R.C., Whiting, W.B., Shaeiwitz, J.A. 1998. *Analysis, Synthesis and Design of Chemical Process*. Upper Saddle River, NJ: Prentice Hall.
Tzia, C. 2003. Optimization. In *Extraction Optimization in Food Engineering*, ed. C. Tzia and G. Liadakis, 152–195. New York: Marcel Dekker.
Uddin, M.B., Ainsworth, P., Ibanoğlu, S. 2004. Evaluation of mass exchange during osmotic dehydration of carrots using response surface methodology. *Journal of Food Engineering* 65: 473–477.
Uma, D.B., Ho, C.W., Wan Aida W.M. 2010. Optimization of extraction parameters of total phenolic compounds from henna (*Lawsonia inermis*) leaves. *Sains Malaysiana* 39: 119–128.
Vadas, E.B. 2001. Stability of pharmaceutical products. In *Remington—The Science and Practice of Pharmacy*, Vol. 1, ed. A. R. Gennaro, 986–994. Philadelphia: Lippincott Williams & Wilkins.
van Kleef, E., Van Tripj, H.C.M., Luning, P. 2005. Functional foods—Health claim-food product compatibility and the impact of health claim framing on consumer evaluation. *Appetite* 44: 299–308.
van Kleef, E., Van Trijp, H.C.M., Luning, P., Jongen, W.M.F. 2002. Consumer-oriented functional food development—How well do functional disciplines reflect the 'voice of the consumer'? *Trends in Food Science & Technology* 13: 93–101.
Vasisht, K. 2008. Quality control of medicinal and aromatic plants and their extracted products by HPLC and high performance thin layer chromatography. In *Extraction Technology for Medicinal and Aromatic Plants*, ed. S.S. Handa, S.P.S. Khanuja, G. Lango, and D.D. Rakesh, 239–260. Trieste: International Centre for Science and High Technology, UNIDO.
Venskutonis, P.R. 2002. Harvesting and post-harvest handling in the genus *Thymus*. In *Thyme—The Genus* Thymus, ed. E. Stahl-Biskup and F. Saéz, 197–223. Boca Raton, FL: CRC Press.
Verloop, Q., Marais, A.F., de Villiers, M.M., Liebenberg, W. 2004. Compatibility of sennoside A and B with pharmaceutical excipients. *Pharmazie* 59: 728–730.
Vicente, G., Paiva, A., Fornari, T., Najdanovic-Visak, V. 2011. Liquid–liquid equilibria for separation of tocopherol from olive oil using ethyl lactate. *Chemical Engineering Journal* 172: 879–884.
Vora, P.S., Testa, L.C.A. 1997. Phytochemistry of licorice horticultural and processing procedures. In *Nutraceuticals—Designer Foods III: Garlic, Soy and Licorice*, ed. P.A. Lachance, 243–257. Trumbull, CT: Food & Nutrition Press.
Wade, A., Weller, P.J. 1994. *Handbook of Pharmaceutical Excipients*, 2nd edn. London: American Pharmaceutical Association and The Pharmaceutical Press.
Wagner, H., Bladt, S. 1996. *Plant Drug Analysis—A Thin Layer Chromatography Atlas*, 2nd edn. Heidelberg: Springer.
Walum, E. 1998. Acute oral toxicity. *Environmental Health Perspectives* 106: 497–503.
Wang, Q.E., Ma, S., Fu, B., Lee, F.S.C., Wang, X. 2004. Development of multistage countercurrent extraction technology for the extraction of glycyrrhizic acid (GA) from licorice (*Glycyrrhiza uralensis* Fisch.) *Biochemical Engineering Journal* 21: 285–292.
Wang, Y., Xi, G., Zheng, Y., Miao, F. 2010. Microwave-assisted extraction of flavonoids from Chinese herb *Radix puerariae* (*Ge Gen*). *Journal of Medicinal Plants Research* 4: 304–308.
Wieniawski, W. 2001. Risk assessment as an element of drug control. *WHO Drug Information* 15: 7–11.
Willis, R.A. 2003. Process for removing residual solvent from solids. U.S. Patent No. 6,509,051.
Ye, Z.L., Chen, S.H., Wang, S.M., Lin, L.F., Yan, Y.J., Zhang, Z.J., Chen, J.S. 2010. Phosphorus recovery from synthetic swine wastewater by chemical precipitation using response surface methodology. *Journal of Hazardous Materials* 176: 1083–1088.

4

Supercritical Fluid Extraction of Bioactives

We as a society need to do a better job of environmental performance in today's technological world.

Thomas E. Graedel and Jennifer A. Howard-Grenville
Greening the Industrial Facility (2005)

4.1 Introduction

Many new consumer trends have emerged in recent times. Accordingly, there is a growing concern for the quality, safety, and toxicity of foods and medicines accompanied by a greater preference for "natural" substances. A major technology that has emerged for the production of "natural" and solvent-free substances is supercritical fluid (SCF) extraction technology (Mukhopadhyay, 2000). In 1978, the first industrial-scale SCF extraction plant was commissioned in Germany by Hag AG. The plant was for the decaffeination of green coffee beans. In 1980, Carlton United Breweries of Australia developed a process for the extraction of flavors from hops (Palmer and Ting, 1995).

The concept of critical point or *l'état particulier* was first conceived by the French engineer and physicist, Charles Cagniard de la Tour (1777–1859). The critical point of a pure substance is the highest temperature and pressure at which the substance exists in vapor–liquid equilibrium (Cagniard de la Tour, 1822; Andrews, 1869). In a continuation of the work of Cagniard de la Tour and in an attempt to examine matter up to the limit of the liquid state, Hannay and Hogarth (1879) carried out some interesting experiments, the results of which were reported before the Royal Society in 1879. They described how a solid could dissolve in gas and then precipitate like "snow" or "frost" when the pressure is reduced. A few years later, Eduard Buchner (1906) measured the solubility of naphthalene in supercritical CO_2. Naphthalene achieves a solubility of 7% at 200 bar and 45°C (Val Krukonis, 1998). This solubility behavior forms the basis of all supercritical extraction and separation processes. A single homogeneous fluid is formed at temperatures and pressures above the critical point. This homogeneous fluid is known as SCF. The SCF is neither a gas nor a liquid. It retains the solvent power of liquids and the transport properties of gases. SCF is formed when a gas is heated above its critical temperature or a liquid is compressed above its critical pressure (**Figure 4.1**) (Nahar and Sarker, 2006).

Great variations in the solvent properties of an SCF can be achieved by changes in temperature and pressure. Therefore, it is possible to obtain different products from a single botanical material by using just one solvent, that is, SCF. A number of SCFs such as ethane, butane, pentane, nitrous oxide, ammonia, trifluoromethane, water, and CO_2 are used. However, because of its low critical parameters (31.1°C and 73.8 bar), low cost, and nontoxicity, supercritical CO_2 is the most widely used solvent for SCF extraction. The SCF extraction process makes use of high-purity CO_2, free from water, hydrocarbons, and halocarbons (Nahar and Sarker, 2006).

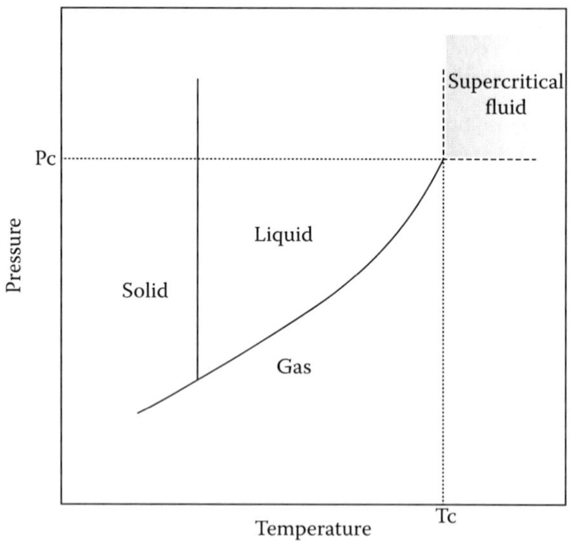

***Figure 4.1** Phase diagram showing the conditions at which thermodynamically distinct phases occur at equilibrium. Pc = critical pressure, Tc = critical temperature. (Reproduced from* Food Chemistry, *98, Herrero, M., Cifuentes, A., Ibañez, E. Sub- and supercritical fluid extraction of functional ingredients from different natural sources: Plants, food- by-products, algae and microalgae—A review, 136–148, Copyright (2006), with permission from Elsevier.)*

4.2 Solubility of compounds in supercritical CO_2

Supercritical CO_2 is nonpolar in nature; therefore, the extractability of compounds depends on the functional groups in these compounds and their molecular weights and polarity. Hydrocarbons and other organic compounds with relatively low polarity can be extracted with supercritical CO_2 at lower pressure in the range of 75–100 bar. Examples are esters, ethers, aldehydes, ketones, lactones, and epoxides. Moderately polar substances are moderately soluble and highly polar compounds such as sugars, polysaccharides, proteins, glycosides, and inorganic salts are generally not soluble. A cosolvent or an entrainer is often injected into supercritical CO_2 to increase its polarity and for the extraction of a certain class of products. Cosolvents facilitate the extraction of compounds in egg yolk, fish oils, carrot, tomato, annatto, and tamarind. Ethanol, methanol, ethyl acetate, and water are used as entrainers (Mukhopadhyay, 2000; Raventós et al., 2002).

The use of an entrainer improves the solvating power of the SCF. While a neat compound may be soluble in supercritical CO_2, it may not be extractable from the matrix without the presence of an entrainer. This is demonstrated in the decaffeination of coffee. Neat caffeine is soluble in dry supercritical CO_2. Nevertheless, moist supercritical CO_2 or moist coffee is essential for the extraction of caffeine from coffee beans. The same phenomenon is reported to occur with decaffeination by traditional organic solvents. It is argued that water frees the "chemically bound" caffeine in the coffee matrix (Calvey and Block, 1997).

4.3 The SCF extraction process

SCF extraction is usually carried out on solid materials (Reverchon, 1997). The material that is to be extracted is first comminuted to the desired particle size and charged into the extractor. CO_2 is then fed into the extractor through a high-pressure pump. The SCF leaches the solute in the material and the extract-laden CO_2 is sent to a separator through a pressure reduction valve. The extract precipitates in the separator at a reduced temperature and pressure. The CO_2 devoid of any extract is returned to the extractor (Mukhopadhyay, 2000). A flow sheet of the process is given in **Figure 4.2**.

SCF extraction can be carried out on systems of different scales. They can be analytical instruments or preparative systems (pilot or industrial scale). Analytical systems are used in the preparation of small samples, ranging from a gram to a few grams. Pilot-scale plants can handle a sample size of several hundred grams to kilograms. Industrial plants can process tons of material.

A preparative system on a pilot scale essentially consists of a solvent pump that delivers the SCF to the system, a modifier pump if necessary, an extraction cell, and one or more separators or fractionation cells, in which the

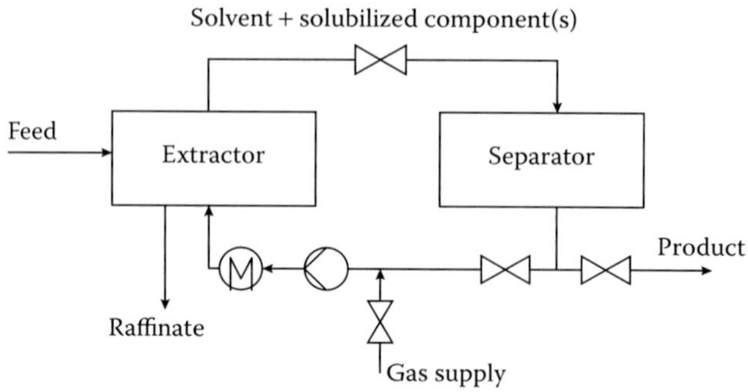

__Figure 4.2__ Flow sheet of SCF extraction. (Reproduced from Danielski, L. Extraction and fractionation of natural organic compounds from plant materials with supercritical carbon dioxide. Doktor-Ingenieur genehmigte Dissertation, Technische Universität Hamburg-Harburg, 2007. With permission.)

solvent-free extract is collected (**Figure 4.3**). The extraction cell and separators are usually equipped with systems for independent control of the temperature and pressure, so that fractionation of the extracted compounds can be carried out by depressurization in a stepwise manner. Different compounds can be collected within each separator, based on their differential solubility in the SCF (Herrero et al., 2006).

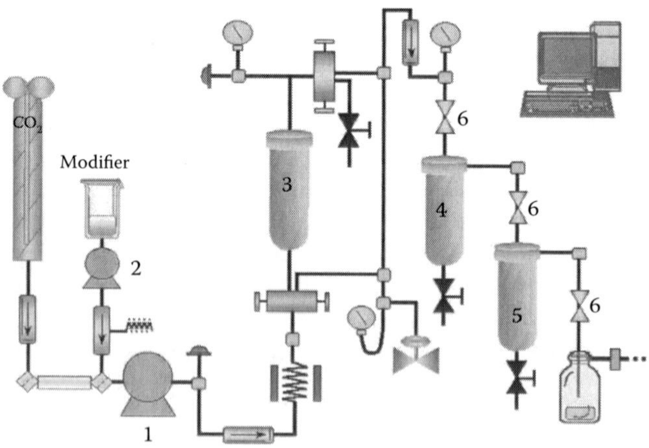

__Figure 4.3__ Diagram of an SCF extraction pilot plant with two fractionation cells. (1) CO_2 pump, (2) modifier pump, (3) sample extraction cell, (4) fractionation cell 1, (5) fractionation cell 2, and (6) valve. (Reproduced from Food Chemistry, *98, Herrero, M., Cifuentes, A., Ibañez, E. Sub- and supercritical fluid extraction of functional ingredients from different natural sources: Plants, food- by-products, algae and microalgae—A review, 136–148, Copyright (2006), with permission from Elsevier.)*

4.3.1 Optimization and scale-up of extraction process

SCF extraction processes are usually optimized using a factorial design. Extraction is carried out at different pressures and temperatures. The results of these extractions are plotted to obtain overall extraction curves. A typical overall extraction curve for SCF extraction from a solid matrix has three distinctive steps: (i) the constant extraction rate (CER) period, characterized by the predominance of convective effects; (ii) the decreasing (falling) extraction rate (FER) period, for which both convective as well as diffusional effects are important; and (iii) the diffusion-controlled rate period that is characterized by the diffusion of the solvent and the solute/solvent mixture in the solid matrix (**Figure 4.4**).

The experimental data on the mass of the extract as a function of the duration of extraction are used to derive the kinetic parameters for the CER period, namely, the mass transfer during the CER period (M_{CER}), the duration of the CER period (t_{CER}), the duration of the FER (t_{FER}) period, the mass ratio of solute in the fluid phase at the bed outlet (Y_{CER}), the yield achieved during the CER period (R_{CER}), and the total yield (R_{total}) (Pasquel et al., 2000; Prado et al., 2011; Prado, 2014).

Mathematical models are used for the development of scale-up procedures from laboratory to pilot to industrial scale. Such models are used for the prediction of large-scale extraction curves because large-scale real experiments are expensive (Nei et al., 2007; Martinez and Martinez, 2008). Nonetheless, very simple models have proven to be effective for scale-up studies as demonstrated by Prado et al. (2011).

The extraction curve of a material is completely determined by parameters such as the solvent mass flow rate (Q_{CO2}), the extract solubility in the solvent

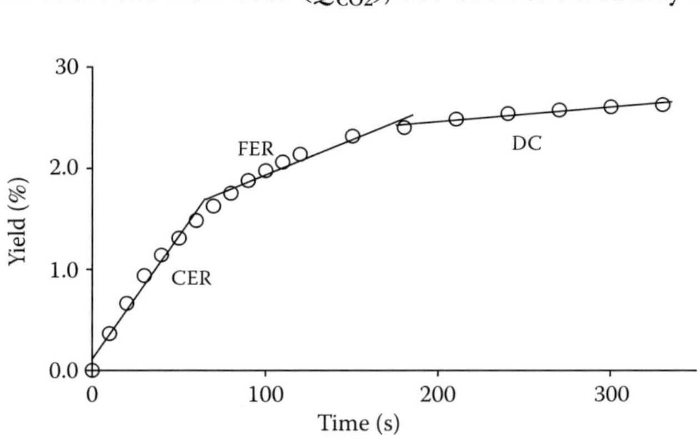

***Figure 4.4** Overall extraction curve of the SCFE process for a hypothetical vegetable matrix+CO_2 system, where CER is the constant extraction rate, FER is the falling extraction rate, and DC is the diffusion-controlled rate. (Reproduced from Prado, I.M.D. Supercritical and subcritical fluid technology utilization for the processing of flaxseed hull. MSc thesis, University of Alberta, Edmonton, 2014.)*

at a given temperature and pressure ($Y*$), the mass of the nonextractable material (m_{SI}), the initial mass of the extractable material relative to the mass of the nonextractable material (X_0), the extraction bed diameter (d_b), the extraction bed height (H), and the extraction bed porosity (ϵ). These parameters are called *contextual parameters*. Extraction curves are ideally predicted as functions of these contextual parameters. However, for such a prediction, a small numbers of model parameters are adopted in addition to the contextual parameters. Thus, the model parameters are functions of the contextual parameters and help in the prediction of extraction curves. Examples are the model parameter related to diffusion in the solid phase of the extraction bed (W) and the model parameter related to convection in the fluid phase of the extraction bed (Z) (Martinez and Martinez, 2008).

Many small-scale experiments are carried out to discover the functional relationship between contextual parameters and model parameters. Contextual parameters are measured and model parameters are estimated from individual extraction curves (Martinez and Martinez, 2008). An appropriate mathematical model derived with algorithms is fitted to the experimental data using regression techniques (Martinez and Martinez, 2008; Rosa and Meireles, 2009; Hrnčič et al., 2010). Lack's (1985) model extended by Sovová (1994) is one such simple solution to explain the mass balance and to give a good physical description of the process (**Figure 4.5**).

A comparison of the overall extraction curve of ginger oleoresin extraction with various mathematical models is provided in **Figure 4.6**. Data on the mathematical modeling of the processes of SCF extraction of many products

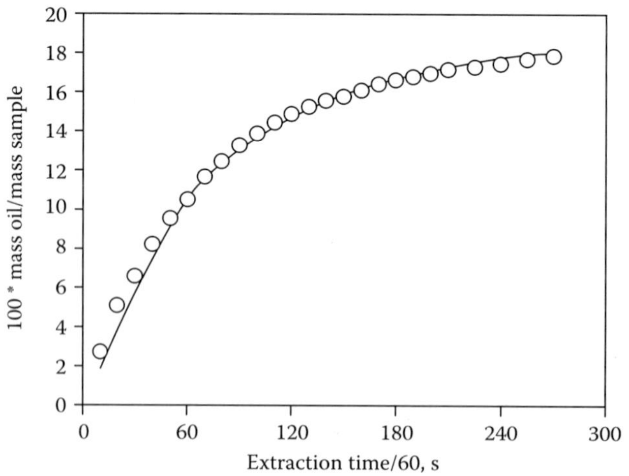

Figure 4.5 *Comparison of the experimental overall extraction curve (O) with the calculated overall extraction curve (——) at 288.15 K, 66.7 bar, and 3.16×10⁻⁵ kg/s. (Reproduced from Sousa, E.M.B.D. et al.,* Braz. J. Chem. Eng., *19: 229–241, 2002. With permission.)*

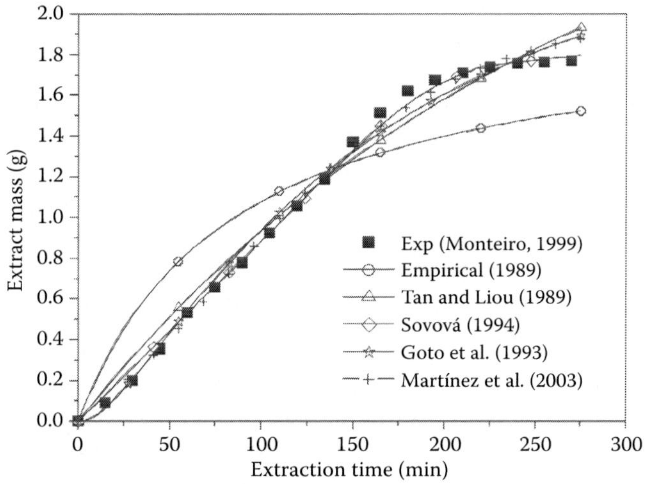

Figure 4.6 *Overall extraction curve of ginger oleoresin extraction with various mathematical models. Extraction conditions: 15 MPa, 313.2 K, and CO_2 mass flow rate of 3.5 kg/min. (Reproduced from Rosa, P.T.V., Meireles, M.A.A. Fundamentals of supercritical extraction from solid matrices. In* Extracting Bioactive Compounds for Food Products: Theory and Applications, *ed. M.A.A. Meireles, 272–288. CRC Press, Boca Raton, 2009. With permission.)*

are available (Reverchon, 2000; Reverchon and Marrone, 2001; Del Valle et al., 2004; Carvalho et al., 2005; Moura et al., 2005; Özkal et al., 2005a; Zizovic et al., 2005; Kotnik et al., 2007; Martinez et al., 2007; Takeuchi and Meireles, 2007; Han et al., 2009; Mezzomo et al., 2009; Prado et al., 2011; Shokri et al., 2011).

Increasingly, response surface methodology is being used for the identification of conditions to obtain the highest yield of the desired compounds in the extract. The particle size, pressure, temperature, the flow rate of CO_2, the extraction time (static and dynamic), and the volume of the entrainer are selected as independent variables. The yield of the extract or its content of actives serves as the dependent variable (Özkal, 2009; Li et al., 2010; Nie et al., 2010; Ghoreishi and Bataghva, 2011; Ghasem et al., 2011).

4.3.2 Some scale-up problems

One major problem encountered in scale-up studies is the use of small extraction vessels to obtain overall extraction curves (Prado et al., 2011). This influences the results as some quantity of extract is lost in the walls of the tubes of the equipment. Meireles (2008a, 2008b) remarks that vessels larger than 50 mL should be used for determining the overall extraction curves. Del Valle et al. (2004) observe that because many parameters influence the process of SCF extraction, the scale-up criteria should be more complex. However,

Prado et al. (2011) successfully scaled-up an SCF extraction process for clove and sugarcane residue using a simple scale-up criterion. Laboratory-scale SCF extraction equipment with a 290 mL extraction vessel and pilot-scale equipment with a 5.15 L extraction vessel were used in the study. The scale-up criterion consisted of maintaining the solvent mass to the feed mass ratio constant. This criterion was used successfully for a 15-fold scale-up of the overall extraction curves for clove and sugarcane residue from 290 mL to 5.15 L. The overall extraction curves presented similar shapes. However, the pilot-scale overall extraction curves showed higher yield than the laboratory-scale extraction curves. Clove yielded 20% more extractive at 130 min and sugarcane residue yielded 15% more extractive at 180 min.

The SCF extraction process of a herb at the industrial scale is designed after successful scale-up to the pilot-plant scale. Mass transfer mechanisms differ among species and plant parts used for extraction (Reverchon and Marrone, 1997). Therefore, before validating the scale-up criteria, it is essential to assess their suitability to different types of raw materials. The principles of scale-up are described by Clavier et al. (1996).

On account of the important hazards posed by the handling of liquefied gases, safety must be taken into account in the design of the equipment, building, installation, operation, and maintenance. A detailed analysis of potential hazards should be carried out. Mechanical hazards such as plugging, rupture of tubing connections, metal fatigue, fragilization, thermodynamic hazards such as dry ice, boiling liquid expanding vapor explosion (BLEVE), chemical hazards such as flammable fluids, corrosion, biological hazards such as asphyxia, and chemical toxicity are likely to be encountered in SCF extraction. Clavier and Perrut (2005) detail the scale-up issues encountered within SCF processing.

The recovery of the extract, fractionation, energy management, and improvement of the procedures of extractor loading and emptying are to be considered in the design of the industrial process. A picture of a commercial SCF extraction plant is provided in **Figure 4.7**.

4.4 Applications of SCF technology

SCF extraction technology has been successfully used for the decaffeination of tea and coffee (Zosel, 1980; Díaz et al., 1997; Saldaña et al., 1999; Saldaña and Mohamed, 2003; Kopcak and Mohamed, 2004; Sun et al., 2010), the extraction of cholesterol and lipids from egg yolk, fish, and meat (Froning et al., 1990, 1994, 1998; Merkle and Larick, 1993; Cooke, 1995; Riha and Brunner, 2000; Mohamed et al., 2003), the fractionation of milk (Rizvi et al. 1995), the extraction of colorants (França et al., 1999; Baysal et al., 2000; Nobre et al., 2006), the extraction, refining, and fractionation of vegetable oils (Robin et al., 1996), the extraction of natural flavorings (Chouchi et al., 1996), and the dealcoholization of drinks (Raventós et al., 2002).

Figure 4.7 *An industrial SCF extraction facility. (Reproduced with permission from Dr. Swapneshu A. Baser, Deven Supercriticals Pvt Ltd., Mumbai; www.scfe.in.)*

Increasingly, extraction with supercritical CO_2 is being used for the isolation of medicinal substances from herbs (Chassagnez-Méndez et al., 2000; Meireles, 2008a, 2008b). Flavors, edible oils, essential oils, antioxidants, and bioactive compounds have been extracted from whole herbs, roots, stems, barks, leaves, flowers, fruits, peels, and seeds. Some recent applications of this technology and their extraction conditions are available in Reverchon (1997), Raventós et al. (2002), Temelli and Güçlü-Üstündağ (2005) Cheah et al. (2006), Nahar and Sarker (2006), Herrero et al. (2009), and Rosa et al. (2009).

4.5 Disadvantages of SCF technology

4.5.1 High capital investments

An SCF extraction unit is relatively more expensive than conventional extraction equipment to invest in and maintain, on account of the need to maintain a high-pressure system. The extracts are also relatively more expensive (Perrut, 2000; Rosa and Meireles, 2005).

4.5.2 More variables to optimize

SCF extraction is affected by a large number of variables related to sample preparation, extraction parameters, and trapping of the extract (Cheah et al., 2006). Sample preparation involves sample size, particle size, comminution procedure, homogeneity of sample, moisture content, drying agent, vessel volume, and packing density. The extraction parameters to be optimized are the SCF used, pressure, temperature, flow rate, entrainer (type and volume), static/action mode, extraction time, and restrictor temperature.

Collecting or trapping the SCF extract is also an important step. During the collection step, the SCF extract decompresses from a pressure of 75–600 atm to atmospheric pressure, with a volume increase from 1 mL to approximately 500 mL of gas. During trapping, the extract may be lost through volatilization or aerosol formation (McDaniel and Taylor, 1999). Several parameters need to be optimized for efficient trapping: collection technique, sorbent type or solvent used, trap size or solvent volume, collection temperature, trap elution solvent used, trap elution volume, temperature, and flow rate (Cheah et al., 2006).

The altering of parameters such as pressure, temperature, entrainer volume, and extraction time affects the quality and quantity of the extract (Khajeh et al., 2004; Pourmortazavi et al., 2003). Using six different extraction conditions, da Costa et al. (1999) observed that varying amounts of flavones and xanthones were extracted from the root bark of the Osage orange tree. Similarly, an ethanol-modified supercritical CO_2 extract of the *Tamarindus indica* seed coat obtained at higher pressure and temperature contained more antioxidant compounds and possessed more antioxidant activity (Tsuda et al., 1995).

4.5.3 Strong dependence on matrix–analyte interactions

It is observed that the efficiency of SCF extraction depends heavily on matrix–analyte interactions. The compound that is to be extracted may be deposited, adsorbed, or chemically bonded to the surface of the matrix or distributed within it (Cheah et al., 2006). A unique feature of all SCF extractions is that most of the extraction takes place early in the process and the rate is considerably reduced thereafter (Sovová et al., 1994; Pan et al., 1995; Damanjović et al., 2005). Free analytes adsorbed on the matrix surface are extracted rapidly at first followed by slower extraction of analytes held inside the matrix (Özkal et al., 2005b; Modey et al., 1996a) or trapped within internal structures such as a gland (Damanjović et al., 2005). Due to the strong matrix–analyte interaction, poor extractability may occur, despite the favorable solubility of the analyte in the SCF (Benner, 1998). Notwithstanding the good solubility of caffeine and coumarin in supercritical CO_2, these compounds were poorly extracted from coffee beans, kola nuts, and guaco leaves (Ndiomu and Simpson, 1988; Vilegas et al., 1997).

4.5.4 Difficulties in scale-up

The development of SCF extraction methods has been hampered by a general lack of experience, especially with reference to matrix–analyte interactions. Difficulty is also experienced in transferring an optimized method from one extractor to another (Cheah et al., 2006).

4.6 Concluding remarks

Despite some minor difficulties, extraction technology with SCF is gaining wider acceptance. The efficiency of extraction is increased in this system, as SCF has a gas-like flow behavior and a liquid-like solvating power (Lang and Wai, 2001). In comparison with conventional extraction techniques, SCF extraction is more precise and selective (Jeong and Chesney, 1999; Ziemons et al., 2005; Chun et al., 1996; Macias-Sanchez et al., 2005; Khajeh et al., 2004; Pourmortazavi et al., 2003). SCF extraction also offers greater time efficiency, as extraction can be completed within 30 min (Ziemons et al., 2005; Ozcan and Ozcan, 2004; Pourmortazavi et al., 2005; Smith and Burford, 1992). The extraction of chamomile flowers by SCF extraction for 30 min yielded 4.4 times more essential oil than conventional extraction by steam distillation for 4 h (Scalia et al., 1999).

As the SCF is easily removed from the extract by depressurization, there is no residual solvent left in it. Even when an organic solvent is used as an entrainer, the volume needed is much less than that used in conventional solvent extraction (Hsu et al., 2001; Ozcan and Ozcan, 2004; Luque de Castro and Jimenez-Carmona, 2000; Barth et al., 1995). In addition to minimizing the disposal of organic solvent wastes, the safety of the operators is also ensured due to reduced exposure to hazardous vapors.

On account of the mild extraction conditions adopted in SCF extraction, this process is suitable for volatile and thermolabile compounds (Song et al., 1992; Smith and Burford, 1992; Scalia et al., 1999; Bartley and Foley, 1994). The extraction system offers a light- and oxygen-free environment, thus minimizing the degradation of compounds and preserving the bioactive properties of the extracts (Murga et al., 2000; Yepez et al., 2002; Dean and Khundker, 1997; Wang et al., 2004; Nguyen et al., 1994). These aspects are of value in the extraction of natural flavors, as steam distillation is likely to cause hydrolysis of the molecules (Modey et al., 1996b).

The milder conditions prevailing in an SCF extraction system also ensure that chemical reactions do not take place between the compounds and the extractive is truly representative of the original product (Bartley and Foley, 1994; Dean and Khundker, 1997; Doneanu and Anitescu, 1998).

SCF extraction may also possess sterilizing properties. Hong and Pyun (2001) reported that *Lactobacillus plantarum* cells underwent irreversible cellular membrane damage when treated with CO_2 at 7 MPa and 30°C for 10 min. Bacteriostatic effects are also reported to take place during SCF processing (Ambrosino et al., 2004; Spilimbergo et al., 2002, 2003).

Functional food products manufactured by SCF extraction technology have gained significant consumer acceptance. There are about 110 SCF extraction plants distributed in Germany, the United States, South Korea, France, and Japan. There has been a significant increase in the number of SCF extraction plants in India. In the beginning, this technology was confined to pilot-scale plants at educational institutions such as the Indian Institute of Technology, Mumbai. However, several companies are now manufacturing production-scale plants in India. Despite the high investment costs, SCF extraction technology is enjoying a sustained growth in interest. Though initially restricted only to high-cost, low-volume products, SCF technology is currently employed in the extraction of a wide variety of products (King, 2004; Srinivas and King, 2010).

References

Ambrosino, P., Galvano, F., Fogliano, V., Logrieco, A., Fresa, R., Riteni, A. 2004. Supercritical fluid extraction of Beauvericin from maize. *Talanta* 62: 523–530.

Andrews, T. 1869. The Bakerian Lecture: On the continuity of the gaseous and liquid states of matter. *Philosophical Transactions of the Royal Society of London* 159: 575–590.

Barth, M.M., Zhou, C., Kute, K.M., Rosenthal, G.A. 1995. Determination of optimum conditions for supercritical fluid extraction of carotenoids from carrot (*Daucus carota* L.) tissue. *Journal of Agricultural and Food Chemistry* 43: 2876–2878.

Bartley, J.P., Foley, P. 1994. Supercritical fluid extraction of Australian-grown ginger (*Zingiber officinale*). *Journal of the Science of Food and Agriculture* 66: 365–371.

Baysal, T., Ersus, S., Starmans, D.S.J. 2000. Supercritical CO_2 extraction of β-carotene and lycopene from tomato waste. *Journal of Agricultural and Food Chemistry* 48: 5507–5511.

Benner, Jr., B.A. 1998. Summarizing the effectiveness of supercritical fluid extraction of polycyclic aromatic hydrocarbons from natural matrix environmental samples. *Analytical Chemistry* 70: 4594–4601.

Buchner, E.G. 1906. Die beschrankte Mischbarkeit von Flussigkeiten das System Diphenylamin und Kohlensaure. *Zeitschrift für Physikalische Chemie* 56: 257.

Cagniard de la Tour, C. 1822. Exposé de quelques résultats obtenu par l'action combinée de la chaleur et de la compression sur certains liquides, tels que l'eau, l'alcool, l'ether sulfurique et l'essence de pétrole rectifiée. *Annales de Chimie et de Physique* 21: 127–132.

Calvey, E.M., Block, E. 1997. Supercritical fluid extraction of *Allium* species. In *Spices: Flavor Chemistry and Antioxidant Properties*, ed. S.J. Risch and C.T. Ho, 113–124. Washington, DC: American Chemical Society.

Carvalho, Jr., R.N., Moura, L.S., Rosa, P.T.V., Meireles, M.A.A. 2005. Supercritical fluid extraction from rosemary (*Rosmarinus officinalis*): Kinetic data, extract's global yield, composition, and antioxidant activity. *The Journal of Supercritical Fluids* 35: 197–204.

Chassagnez-Méndez, A.L., Machado, N.T., Araujo, M.E., Maia, J.G., Meireles, M.A.A. 2000. Supercritical CO_2 extraction of curcumins and essential oil from the rhizomes of turmeric (*Curcuma longa* L.). *Industrial & Engineering Chemistry Research* 39: 4729–4733.

Cheah, E.L.C., Chan, L.W., Heng, P.W.S. 2006. Supercritical carbon dioxide and its application in the extraction of active principles from plant materials. *Asian Journal of Pharmaceutical Sciences* 1: 59–71.

Chouchi, D., Barth, D., Reverchon, E., Della Pota, G. 1996. Bigarade peel oil fractionation by supercritical carbon dioxide desorption. *Journal of Agricultural and Food Chemistry* 44: 1100–1104.

Chun, M.K., Shin, H.W., Lee, H. 1996. Supercritical fluid extraction of paclitaxel and baccatin III from needles of *Taxus cuspidata*. *The Journal of Supercritical Fluids* 9: 192–198.

Clavier, J.Y., Majewski, W., Perrut, M. 1996. Extrapolation for pilot plant to industrial scale SFE. In *High Pressure Chemical Engineering Process Technology Proceedings*, Vol. 12, ed. R. von Rohr and C. Trepp, 639–644. Amsterdam: Elsevier.

Clavier, J.Y., Perrut, M. 2005. Scale-up issues for supercritical fluid processing in compliance with GMP. In *Supercritical Fluid Technology for Drug Product Development*, ed. P. York, U.B. Kompella, and B.Y. Shekunov, 565–597. New York: Marcel Dekker.

Cooke, L. 1995. Supercritical fluid fat extraction. *Agricultural Research* 43: 18.

da Costa, C.T., Margolis, S.A., Benner, B.A., Horton, D. 1999. Comparison of methods for extraction of flavones and xanthones from the root bark of the osage orange tree using liquid chromatography. *Journal of Chromatography A* 831: 167–178.

Damanjović, B., Lepojević, Ž., Živković, V., Tolić, A. 2005. Extraction of fennel (*Foeniculum vulgare* Mill.) seeds with supercritical CO_2: Comparison with hydrodistillation. *Food Chemistry* 92: 143–149.

Danielski, L. 2007. Extraction and fractionation of natural organic compounds from plant materials with supercritical carbon dioxide. Doktor-Ingenieur genehmigte dissertation, Technische Universität Hamburg-Harburg, 6.

Dean, J.R., Khundker, S. 1997. Extraction of pharmaceuticals using pressurized carbon dioxide. *Journal of Pharmaceutical and Biomedical Analysis* 15: 875–886.

Del Valle, J.M., Rivera, O., Mattea, M., Ruetsch, L., Daghero, J., Flores, A. 2004. Supercritical CO_2 processing of pretreated rosehip seeds: Effect of process scale on oil extraction kinetics. *The Journal of Supercritical Fluids* 31: 159–174.

Díaz, O., Cobos, A., de la Hoz, L., Ordóñez, J.A. 1997. El dióxidio de carbono supercrítico en la elaboración de alimentos de origen vegetal. Otras aplicaciones. *Alimentación, Equipos y Tecnología* 8: 55–63.

Doneanu, C., Anitescu, G. 1998. Supercritical carbon dioxide extraction of *Angelica archangelica* root oil. *The Journal of Supercritical Fluids* 12: 59–67.

França, L.F., Reber, G., Meireles, M.A.A., Machado, N.T., Brunner, G. 1999. Supercritical extraction of carotenoids and lipids from Buriti (*Mauritia flexuosa*), a fruit from the Amazon Region. *The Journal of Supercritical Fluids* 14: 247–256.

Froning, G.W., Fieman, F., Wehling, R.L., Cuppett, S.L., Niemann, L. 1994. Supercritical carbon dioxide extraction of lipids and cholesterol from dehydrated chicken meat. *Poultry Science* 73: 571–575.

Froning, G.W., Wehling, R.L., Cuppett, S.L., Niemann, L. 1998. Moisture content and particle size of dehydrated egg yolk affect lipid and cholesterol extraction using supercritical carbon dioxide. *Poultry Science* 77: 1718–1722.

Froning, G.W., Wehling, R.L., Cuppett, S.L., Pierce, M.M., Niemann, L., Siekman, D.K. 1990. Extraction of cholesterol and other lipids from dried egg yolk using supercritical carbon dioxide. *Journal of Food Science* 55: 95–98.

Ghasem, E., Raofie, F., Najafi, N.M. 2011. Application of response surface methodology and central composite design for the optimisation of supercritical fluid extraction of essential oils from *Myrtus communis* L. leaves. *Food Chemistry* 126: 1449–1453.

Ghoreishi, S.M., Bataghva, E. 2011. Supercritical extraction of evening primrose oil: Experimental optimization via response surface methodology. *AIChE Journal* 57: 3378–3384.

Han, X., Cheng, L., Zhang, R., Bi, J. 2009. Extraction of safflower seed oil by supercritical CO_2. *Journal of Food Engineering* 92: 370–376.

Hannay, J.B., Hogarth, J. 1879. On the solubility of solids in gases. *Proceedings of the Royal Society* (London) 29: 324–326.

Herrero, M., Cifuentes, A., Ibañez, E. 2006. Sub- and supercritical fluid extraction of functional ingredients from different natural sources: Plants, food- by-products, algae and microalgae—A review. *Food Chemistry* 98: 136–148.

Herrero, M., Mendiola, J.A., Cifuentes, A., Ibañez, E. 2009. Supercritical fluid extraction: Recent advances and applications. *Journal of Chromatography* 1217: 2495–2511.

Hong, S.I., Pyun, Y.R. 2001. Membrane damage and enzyme inactivation of *Lactobacillus plantarum* by high pressure CO_2 treatment. *International Journal of Food Microbiology* 63: 19–28.

Hrnčič, D., Mernik, M., Hrnčič, M.K., Knez, Ž. 2010. Fitting Sovova's mass transfer model using an evolutionary algorithm and differential evolution. *International Journal of Innovative Computing and Applications* 2: 237–243.

Hsu, R.C., Lin, B.H., Chen, C.W. 2001. The study of supercritical carbon dioxide extraction for *Ganoderma lucidum*. *Industrial and Chemical Engineering Research* 40: 4478–4481.

Jeong, M.L., Chesney, D.J. 1999. Investigation of modifier effects in supercritical CO_2 extraction from various solid matrices. *The Journal of Supercritical Fluids* 16: 33–42.

Khajeh, M., Yamini, Y., Bahramifar, N., Sefidkon, F., Pirmoradei, M.R. 2004. Comparison of essential oils compositions of *Ferula assa-foetida* obtained by supercritical carbon dioxide extraction and hydrodistillation methods. *Food Chemistry* 90: 636–644.

King, J.W. 2004. Development and potential of critical fluid technology in the nutraceutical industry. In *Supercritical Fluid Technology for Drug Product Development*, ed. P. York, U.B. Kompella, and B.Y. Shekunov, 579–614. New York: Marcel Dekker.

Kopcak, U., Mohamed, R.S. 2004. Caffeine solubility in supercritical carbon dioxide/co-solvent mixtures. *The Journal of Supercritical Fluids* 34: 209–214.

Kotnik, P., Škerget, M., Knez, Ž. 2007. Supercritical fluid extraction of chamomile flower-heads: Comparison with conventional extraction, kinetics and scale-up. *The Journal of Supercritical Fluids* 43:192–198.

Krukonis, V. 1998. Supercritical fluids: Their proliferation in the pharma industry. *European Pharmaceutical Contractor*, May. http://www.phasex4scf.com/supercritical-fluids-applications/. Accessed September 12, 2010.

Lack, E.A. 1985. *Kriterien zur Auslegung von Anlagen für die Hochdruckextraktion von Naturstoffen*. PhD thesis, Technische Universität Graz, Austria.

Lang, Q., Wai, C.M. 2001. Supercritical fluid extraction in herbal and natural product studies—A practical view. *Talanta* 53: 771–782.

Li, B., Xu, Y., Jin, Y.X., Wu, Y.Y., Tu, Y.Y. 2010. Response surface optimization of supercritical fluid extraction of kaempferol glycosides from tea seed cake. *Industrial Crops and Products* 32: 123–158.

Luque de Castro, M.D., Jimenez-Carmona, M.M. 2000. Where is supercritical fluid extraction going? *Transactions of Analytical Chemistry* 19: 223–228.

Macias-Sanchez, M.D., Mantell, C., Rodriguez, M., de la Ossa, E.M., Lubian, L.M., Montero, O. 2005. Supercritical fluid extraction of carotenoids and chlorophyll A from *Nannochloropsis gaditana*. *Journal of Food Engineering* 66: 245–251.

Martinez, J., Martinez, J.M. 2008. Fitting the Sovova's supercritical fluid extraction model by means of a global optimization tool. *Computers & Chemical Engineering* 32: 1735–1745.

Martinez, J., Rosa, P.T.V., Meireles, M.A.A. 2007. Extraction of clove and vetiver oils with supercritical carbon dioxide: Modeling and simulation. *The Open Chemical Engineering Journal* 1: 1–7.

McDaniel, L.H., Taylor, L.T. 1999. Modification of the collection solvent to enhance liquid trapping efficiencies after supercritical fluid extraction. *Journal of Chromatographic Science* 37: 203–209.

Meireles, M.A.A. 2008a. Extraction of bioactive compounds from Latin American plants. In *Supercritical Fluid Extraction of Nutraceuticals and Bioactive Compounds*, ed. J. Martinez, 243–274. Boca Raton: CRC Press.

Meireles, M.A.A. 2008b. Supercritical fluid extraction of medicinal plants. *Electronic Journal of Environmental, Agricultural and Food Chemistry* 7: 3254–3258.

Merkle, J.A., Larick, D.K. 1993. Triglyceride content of supercritical carbon dioxide extracted fractions of beef fat. *Journal of Food Science* 58: 1237–1240.

Mezzomo, N., Martinez, J., Ferreira, S.R.S. 2009. Supercritical fluid extraction of peach (*Prunus persica*) almond oil: Kinetics, mathematical modeling and scale-up. *The Journal of Supercritical Fluids* 51: 10–16.

Modey, W.K., Mulholland, D.A., Raynor, M.W. 1996a. Application of a dynamic extraction model to the supercritical fluid extraction of the liminoid cedrelone from *Cedrela toona*. *The Journal of Chromatographic Science* 34: 320–325.

Modey, W.K., Mulholland, D.A., Raynor, M.W. 1996b. Analytical supercritical fluid extraction of natural products. *Phytochemical Analysis* 7: 1–15.

Mohamed, R.S., Saldaña, M.D.A., de Azevedo, A.B.A., Kopcak, U. 2003. Removal of cholesterol from food products using supercritical fluids. In *Extraction Optimization in Food Engineering*, ed. C. Tzia and G. Liadakis, 380–401. New York: Marcel Dekker.

Moura, L.S., Carvalho, Jr., R.N., Stefanini, M.B., Ming, L.C., Meireles, M.A.A. 2005. Supercritical fluid extraction from fennel (*Foeniculum vulgare*): Global yield, composition and kinetic data. *The Journal of Supercritical Fluids* 35: 212–219.

Mukhopadhyay, M. 2000. Introduction. In *Natural Extracts Using Supercritical Carbon Dioxide*, 1–9. Boca Raton: CRC Press.

Murga, R., Ruiz, R., Beltran, S., Cabezas, J.L., 2000. Extraction of natural complex phenols and tannins from grape seed by using supercritical mixture of carbon dioxide and alcohol. *Journal of Agricultural and Food Chemistry* 48: 3408–3412.

Nahar, L., Sarker, S.D. 2006. Supercritical fluid extraction. In *Natural Products Isolation*, 2nd edn, ed. S.D. Sarker, Z. Latif, and A.I. Gray, 47–76. Totowa, NJ: Humana Press.

Ndiomu, D.P., Simpson, C.F. 1988. Some applications of supercritical fluid extraction. *Analytica Chimica Acta* 213: 237–243.

Nei, H.Z.N., Fatemi, S., Salimi, A., Mehrnia, M.R. 2007. Mathematical modeling of supercritical fluid extraction of fatty acids from trout powder: Correlation of mass transfer parameters. *Chemical Product and Process Modeling* 2: DOI: 10.2202/1934-2659.1094.

Nguyen, U., Evans, D.A., Frakman, G. 1994. Natural antioxidants produced by supercritical extraction. In *Supercritical Fluid Processing of Food and Biomaterials*, ed. S.S.H. Rizvi, 103–113. Glasgow: Blackie Academic and Professional.

Nie, S.P., Li, J.E., Yang, C., Qiu, Z.H., Xie, M.Y. 2010. Optimization of supercritical fluid extraction of essential oil from *Herba Moslae* by response surface methodology and its chemical composition analysis. *Food Science and Technology Research* 16: 185–190.

Nobre, B.P., Mendes, R.L., Queiroz, E.M., Pessoa, F.L.P., Coelho, J.P., Palavra, A.F. 2006. Supercritical carbon dioxide extraction of pigments from *Bixa orellana* seeds (experiments and modeling). *Brazilian Journal of Chemical Engineering* 23: 251–258.

Ozcan, A., Ozcan, A.S. 2004. Comparison of supercritical fluid and Soxhlet extractions for the quantification of hydrocarbons from *Euphorbia macroclada*. *Talanta* 64: 491–495.

Özkal, S.G. 2009. Response surface analysis and modeling of flaxseed oil yield in supercritical carbon dioxide. *Journal of the American Oil Chemists' Society* 86: 1129–1135.

Özkal, S.G., Salgin, U., Yener, M.E. 2005a. Supercritical carbon dioxide extraction of hazelnut oil. *Journal of Food Engineering* 69: 217–223.

Özkal, S.G., Yener, M.E., Bayyndyrly, L. 2005b. Mass transfer modeling of apricot kernel oil extraction with supercritical carbon dioxide. *The Journal of Supercritical Fluids* 35: 119–127.

Palmer, M.V., Ting, S.S.T. 1995. Applications for supercritical fluid technology in food processing. *Food Chemistry* 52: 345–352.

Pan, W.H.T., Chang, C.C., Su, T.T., Lee, F., Fuh, M.R.S. 1995. Preparative supercritical fluid extraction of pyrethrin I and II from pyrethrum flower. *Talanta* 42: 1745–1749.

Pasquel, A., Meireles, M.A.A., Marques, M.O.M., Petenate, A.J. 2000. Extraction of Stevia glycosides with CO_2+water, CO_2+ethanol, and CO_2+water+ethanol. *Brazilian Journal of Chemical Engineering* 17: 261–270.

Perrut, M. 2000. Supercritical fluid applications: Industrial developments and economic issues. *Industrial and Engineering Chemistry Research* 39: 4531–4535.

Pourmortazavi, S.M., Ghadiri, M., Hajimirsadeghi, S.S. 2005. Supercritical fluid extraction of volatile components from *Bunium persicum* Boiss. (black cumin) and *Mespilus germanica* L. (medlar) seeds. *Journal of Food Composition and Analysis* 18: 439–446.

Pourmortazavi, S.M., Sefidkon, F., Hosseini, S.G. 2003. Supercritical carbon dioxide extraction of essential oils from *Perovskia atriplicifolia* Benth. *Journal of Agricultural and Food Chemistry* 51: 5414–5419.

Prado, I.M.D. 2014. Supercritical and subcritical fluid technology utilization for the processing of flaxseed hull, 22. MSc thesis, University of Alberta, Edmonton.

Prado, J.M., Prado, G.H.C., Meireles, M.A.A. 2011. Scale-up study of supercritical fluid extraction process for clove and sugarcane residue. *The Journal of Supercritical Fluids* 56: 231–237.

Raventós, M., Duarte, S., Alarcón, R. 2002. Application and possibilities of supercritical CO_2 extraction in food processing industry: An overview. *Food Science and Technology International* 8: 269–284.

Reverchon, E. 1997. Supercritical fluid extraction and fractionation of essential oils and related products. *The Journal of Supercritical Fluids* 10: 1–37.

Reverchon, E. 2000. Mathematical modeling and simulation of pennyroyal essential oil supercritical extraction. *Chemical Engineering Science* 55: 2917–2922.

Reverchon, E., Marrone, C. 1997. Supercritical extraction of clove bud essential oil: Isolation and mathematical modeling. *Chemical Engineering Science* 52: 3421–3428.

Reverchon, E., Marrone, C. 2001. Modeling and simulation of the supercritical CO_2 extraction of vegetable oils. *The Journal of Supercritical Fluids* 19: 161–175.

Riha, V., Brunner, G. 2000. Separation of fish oil ethyl esters with supercritical carbon dioxide. *The Journal of Supercritical Fluids* 17: 55–64.

Rizvi, S.S.H., Mulvaney, S.J., Sokhey, A.S. 1995. The combined application of supercritical fluid and extrusion technology. *Trends in Food Science and Technology* 6: 232–240.

Robin, Y.Y., Chiou, Y.Z., Wu, P., Chen, W., Weng, Y. 1996. Partial defatting of roasted peanut meals and kernels by supercritical CO_2 using semicontinuous and intermittently depressurized processes. *Journal of Agricultural and Food Chemistry* 44: 574–578.

Rosa, P.T.V., Meireles, M.A.A. 2005. Rapid estimation of the manufacturing cost of extracts obtained by supercritical fluid extraction. *Journal of Food Engineering* 67: 235–240.

Rosa, P.T.V., Meireles, M.A.A. 2009. Fundamentals of supercritical extraction from solid matrices. In *Extracting Bioactive Compounds for Food Products: Theory and Applications*, ed. M.A.A. Meireles, 272–288. Boca Raton: CRC Press.

Rosa, P.T.V., Parajó, J.C., Domínguez, H., et al. 2009. Supercritical and pressurized fluid extraction applied to the food industry. In *Extracting Bioactive Compounds for Food Products: Theory and Applications*, ed. M.A.A. Meireles, 269–401. Boca Raton: CRC Press.

Saldaña, M.D.A., Mohamed, R.S. 2003. Extraction of alkaloids from natural plants using supercritical fluids. In *Extraction Optmization in Food Engineering*, ed. C. Tzia and G. Liadakis, 358–379. New York: Marcel Dekker.

Saldaña, M.D.A., Mohamed, R.S., Baer, M.G.G., Mazzafera, P. 1999. Extraction of purine alkaloids from mate (*Ilex paraguariensis*) using supercritical CO_2. *Journal of Agricultural and Food Chemistry* 47: 3804–3808.

Scalia, S., Giuffreda, L., Pallado, P. 1999. Analytical and preparative supercritical fluid extraction of Chamomile flowers and its comparison with conventional methods. *Journal of Pharmaceutical and Biomedical Analysis* 21: 549–558.

Shokri, A., Hatami, T., Khamforoush, M. 2011. Near critical carbon dioxide extraction of anise (*Pimpinella anisum* L.) seed: Mathematical and artificial neural network modeling. *The Journal of Supercritical Fluids* 58: 49–57.

Smith, R.M., Burford, M.D. 1992. Supercritical fluid extraction and gas chromatographic determination of sesquiterpene lactone parthenoline in the medicinal herb feverfew (*Tanacetum parthenium*). *Journal of Chromatography* 627: 255–261.

Song, K.M., Park, S.W., Hong, W.H., Lee, H., Kwak, S.S., Liu, J.R. 1992. Isolation of vindoline from *Catharanthus roseus* by supercritical fluid extraction. *Biotechnology Progress* 8: 583–586.

Sousa, E.M.B.D., Chiavone-Filho, O., Moreno, M.T., Silva, D.N., Marques, M.O.M., Meireles, M.A.A. 2002. Experimental results for the extraction of essential oil from *Lippia sidoides* Cham. using pressurized carbon dioxide. *Brazilian Journal of Chemical Engineering* 19: 229–241.

Sovová, H. 1994. Rate of the vegetable oil extraction with supercritical CO_2. I. Modelling of extraction curves. *Chemical Engineering Science* 49: 409–414.

Sovová, H., Kučera, J., Jež, J. 1994. Rate of the vegetable oil extraction with supercritical CO_2. II. Extraction of grape oil. *Chemical Engineering Science* 49: 415–420.

Spilimbergo, S., Bertucco, A., Lauro, F.M., Bertolini, G. 2003. Inactivation of *Bacillus subtilis* spores by supercritical CO_2 treatment. *Innovative Food Science and Emerging Technologies* 4: 161–165.

Spilimbergo, S., Elvassore, N., Bertucco, A. 2002. Microbial inactivation by high pressure. *Journal of Supercritical Fluids* 22: 55–63.

Srinivas, K., King, J.W. 2010. Supercritical carbon dioxide and sub critical water: Complementary agents in the processing of functional foods. In *Functional Food Product Development*, ed. J. Smith and E. Charter, 39–78. Chichester: Blackwell.

Sun, Q.L., Hua, S., Ye, J.H., Lu, J.L., Zheng, X.Q., Liang, Y.R. 2010. Decaffeination of green tea by supercritical carbon dioxide. *Journal of Medicinal Plants Research* 4: 1161–1168.

Takeuchi, T.M., Meireles, M.A.A. 2007. Study of bed geometry influence on the kinetics of supercritical fluid extraction of *Achyrocline satureioides*. In *1st Iberoamerican Conference on Supercritical Fluids, Foz do Iguaçu, Brazil, April 10–13*. Full paper available in PROSCIBA 2007 CD-ROM (SC-027).

Temelli, F., Güçlü-Üstündağ, Ö. 2005. Supercritical technologies for further processing of edible oils. In *Bailey's Industrial Oil and Fat Products*, ed. F. Shahidi, 397–432. Hoboken, NJ: Wiley.

Tsuda, T., Mizuno, K., Oshima, K., Kawaishi, S., Osawa, T. 1995. Supercritical carbon dioxide extraction of antioxidative components from tamarind (*Tamarindus indica* L.) seed coat. *Journal of Agricultural and Food Chemistry* 43: 2803–2806.

Vilegas, J.H.Y., de Marchi, E., Lancas, F.M. 1997. Extraction of low-polarity compounds with emphasis on coumarin and kaurenoic acid from *Mikania glomerata* (Guaco) leaves. *Phytochemical Analysis* 8: 266–270.

Wang, Z., Ashraf-Khorassani, M., Taylor, L.T. 2004. On-line coupling of supercritical CO_2 evaluation with reversed-phase liquid chromatography for the quantitative analysis of analytes in aqueous matrices. *Journal of Chromatography A* 1033: 221–227.

Yepez, B., Espinosa, M., López, S., Bolaños, G. 2002. Producing antioxidant fractions from herbaceous matrices by supercritical fluid extraction. *Fluid Phase Equilibria* 194–197: 879–884.

Ziemons, E., Goffin, E., Lejeune, R., Proenca da Cunha, A., Angenot, L., Thunus, L. 2005. Supercritical carbon dioxide extraction of tagitinin C from *Tithonia diversifolia*. *The Journal of Supercritical Fluids* 33: 53–60.

Zizovic, I., Stamenic, M., Orlovic, A., Skala, D. 2005. Supercritical carbon dioxide essential oil extraction of Lamiaceae family species—Mathematical modeling on the micro-scale and process optimization. *Chemical Engineering Science* 60: 6747–6756.

Zosel, I., 1980. Separation with supercritical gases: Practical applications. In *Extraction with Supercritical Gases*, ed. G.M. Schneider, E. Stahl, and G. Wilke, 1–23. Weinheim: Verlag Chemie.

5

Formulation of Food Supplements

One should consume regularly such substances that help maintain good health and prevent the appearance of diseases.

Caraka Samhita
300 BC

5.1 Introduction

Extracts derived from herbs can be used in the manufacture of food supplements and fortified food. Food supplements are also known as dietary supplements or nutritional supplements. Some countries define them as drugs, whereas others treat them as foods. According to the laws in force in the United States, Europe, Australia, and Japan, a food supplement cannot claim to prevent, treat, or cure any disease. Violation of these laws implies that the product in question is an unapproved or illegal drug. Despite this apparent constraint, wise physicians and the lay public in those countries make good use of these innocuous products. Food supplements are generally manufactured according to the principles of modern pharmaceutics, which uses various dosage forms to deliver the medicinal substances.

5.2 Food supplements

The development of dosage forms from herbal extracts is a challenging task. Pure active compounds derived from natural products or synthetic reactions contain only negligible amounts of extraneous substances. Nevertheless, extracts of herbs contain small quantities of the active substance and large amounts of other materials such as organic and inorganic salts, resins, waxes, tannins, sugars, saponins, polyphenols, polysaccharides, and so on. These secondary metabolites can affect the pharmaceutical technology and stability of the finished products. On account of these peculiarities, the development of pharmaceutical products from herbal extracts calls for close interaction between the formulator and the person in charge of the production of the extracts (Bonati, 1991).

Herbal extracts can be either total extracts or purified extracts. Total extracts are usually prepared using water or a water–alcohol mixture as the extraction solvent. These extracts contain all the extractive matter. Purified extracts, on the other hand, are devoid of those inert substances that are either not necessary for the desired biological effect or are considered to be detrimental to that effect. Purified extracts are prepared by extracting the herb with a selective solvent or by removing the undesirable substances by defatting, passing though absorption resins, back extraction (partition chromatography), or precipitation in aprotic solvents such as acetone, dichloromethane, ethyl acetate, dimethylformamide, or acetonitrile. Defatting is usually done by treating the extract with a nonpolar solvent such as hexane, which washes away the oils and waxes (Bonati, 1991).

5.3 Hygroscopicity of extracts

The inert substances are hygroscopic in nature and render the extracts sticky soon after extraction. They are a major impediment to the development of solid dosage forms from such extracts. The most important among such substances are the carbohydrates that are invariably present in large amounts in most plants. They are soluble in the water or water–alcohol mixtures that are commonly used in the extraction of bioactives from herbs. Foremost among these carbohydrates are the simple sugars such as fructose and glucose, which, in the case of Radix Ophiopogonis (*Ophiopogon japonicus*) and Rhizoma Polygonati (*Polygonatum* species), account for 50% of the extract by weight (Tong et al., 2008). The complex composition of these extracts hinders the crystallization of such carbohydrates during the drying process, turning the dried extract into an amorphous form. On account of their high thermodynamic activity, these amorphous solids interact with ambient water vapor and become sticky (Chiu and Chow, 2000).

The removal of hygroscopic substances is a prelude to the successful development of dosage forms. Nevertheless, it needs to be established that the biological activity in question is not lost during the process of purification. Saponins can be removed by partitioning the extract with butanol. Driver and Francis (1979) have reported a similar method for the removal of a toxic saponin from pokeberries (*Phytolacca americana*) using the butanol-acetone-diethyl ether system for the solvent extraction of the berries.

5.4 Removal of interfering substances

Polyphenolics are brown or red-colored substances that oxidize rapidly when exposed to air, forming insoluble precipitates. Gel filtration with polyamide,

Sephadex LH-20, solvent partition, or adsorption on Whatman DE52 resin are recommended for the removal of polyphenolics in extracts (Houghton and Raman, 1998).

To overcome the problems posed by the hygroscopicity of extracts, the manufacturing, packaging, and storage operations are stringently controlled. These activities are carried out in dehumidified atmospheres, and waterproof polymeric coatings and moisture-resistant packaging are used (Tong et al., 2008). Traditional extracts can be divided into fluid, soft, and dry extracts. Purified extracts are always in solid form (Bonati, 1991).

An extract is primarily a finished product and it can also be the constituent of another product such as a nutraceutical or a fortified food. Therefore, its quality needs to be controlled. Quality tests have to focus on the physical characteristics (appearance, pH, solubility, ash content, and total solids), solubility in solvents commonly used in formulations (sugar syrup, sorbitol), particle size of dry extracts, tolerance limit for fine powders, content of actives, and total anaerobic microbial count (Bonati, 1991).

5.5 Dosage forms

A dosage form is commonly defined as the physical form of a dose of the medicine intended for administration or consumption. It is thus a sophisticated delivery system. Some of the common dosage forms recommended by modern pharmaceutics are tablets, capsules, syrups, aerosols, injections, and powders. All dosage forms can be prepared from herbal extracts. However, from the nutraceutical point of view, the important forms are solid dosage forms, such as tablets and gelatin capsules, and liquid dosage forms. Tablets can be classified into several major varieties such as sugar-coated tablets, film-coated tablets, enteric-coated tablets, multiple compressed tablets, controlled-release tablets, effervescent tablets, and chewable tablets. Capsules can be either hard gelatin capsules or soft gelatin capsules. Liquid dosage forms can come in the form of syrups, drops, solutions, and suspensions for soft gelatin capsules (Bonati, 1991; Rudnic and Schwartz, 2001).

5.5.1 Solid dosage forms

5.5.1.1 Compressed tablets

To qualify to be made into tablets, a medicinal substance should have the ability to flow freely, cohesiveness and lubrication. The other ingredients used in the production of tablets (excipients) should not interfere with these qualities of the drug. Many extracts do not have all these qualities. Therefore, suitable excipients need to be added for the successful preparation of tablets from these extracts (Rudnic and Schwartz, 2001).

Suitably formulated extract–excipient powder is fed into a tablet punching machine. The powder enters the cavity of the die of appropriate size and pressure is applied by the punches. A tablet is formed as a result of the compaction of the powder and it is ejected from the die. Tablets are manufactured using one of these three methods: wet granulation, dry granulation, or direct compression.

5.5.1.2 Ingredients of tablets

Tablets are made of the active substance (herbal extract) and additives or excipients. Excipients are of two kinds. One group imparts compression characteristics to the formulation and helps in its processing. Diluents, binders, glidants, and lubricants fall into this variety. The second group gives the tablet its desirable physical characteristics. It includes disintegrants, colors, flavors, sweetening agents, polymers, and waxes (Wade and Weller, 1994a; Rudnic and Schwartz, 2001).

5.5.1.2.1 Diluents Sometimes, a single dose of the extract may be too small, calling for an inert substance to increase the bulk of the formulation and make a tablet of a practical size. Such diluents include microcrystalline cellulose, dicalcium phosphate, calcium sulfate, lactose, cellulose, kaolin, dry starch, and powdered sugar (Rudnic and Schwartz, 2001).

5.5.1.2.2 Binders These are agents that improve the cohesive property of the powdered active substance. They are used in the granulation of extracts. Binders improve the free-flowing nature of the formulation and ensure that the tablet remains intact after compression. Examples are starch, gelatin, sugars such as glucose and dextrose, gums such as acacia and sodium alginate, mucilage of isaphgol husks, carboxymethylcellulose, methylcellulose, polyvinylpyrrolidone (PVP), and so on. Binders are used as a solution or in dry form (Rudnic and Schwartz, 2001).

5.5.1.2.3 Lubricants Lubricants prevent the adhesion of the formulation powder to the surface of the dies and punches, reduce friction between the particles, cause the ejection of tablets from the die cavity, and improve the flowability of the powder. Examples are magnesium stearate, calcium stearate, stearic acid, hydrogenated vegetable oils, and polyethylene glycol (Rudnic and Schwartz, 2001).

5.5.1.2.4 Glidants Glidants are agents that improve the free-flowing nature of the formulation. They are added in the dry state, prior to the compression of the tablets. Colloidal silicon dioxide and asbestos-free talc are the commonly used glidants (Rudnic and Schwartz, 2001).

5.5.1.2.5 Disintegrants Some substances or a mixture of substances are added to the drug material to enhance disintegration in the stomach after ingestion of the tablet. These substances are called *disintegrants* and they cause rapid dissolution of the active ingredient(s). Common disintegrants include varieties of starches, clays, celluloses, algins, gums, and polymers. Starch is the oldest and most popular disintegrant. It swells on contact with moisture and causes the tablet matrix to rupture. To obtain the disintegration effect of starches, they are added in the dry state to the mixture of powders (Rudnic and Schwartz, 2001).

5.5.1.2.6 Coloring agents Colors are used in tablets for aesthetic reasons and for their usefulness in the identification of the product. Such colorants include approved food colors, their mixtures, or corresponding lakes. Colors are usually dissolved in the binding solution prior to granulation (Rudnic and Schwartz, 2001).

5.5.1.2.7 Flavoring agents Sweeteners are added to tablet powder blends to enhance the flavor of the product (Rudnic and Schwartz, 2001).

5.5.1.3 Tablet compression

Tablets are made by compressing the active substance–excipients mixture on tablet punching machines. Prior to tabletting, the ingredients of the tablet need to be blended into a homogeneous mixture. If this cannot be achieved with simple blending, the constituents need to be granulated prior to compression. Granulation is a technique that is used to improve the pharmacotechnological properties of extracts. Bonds are created between the particles during granulation, improving the flowability and compactibility of the powder. There are two methods of granulation, namely, wet granulation and dry granulation (Marshall, 1991).

5.5.1.3.1 Wet granulation The wet granulation method makes use of a solution, suspension, or slurry containing a binder, which is added to the active substance–excipients mixture. Wet granulation is carried out in several steps. To begin with, the liquid binder is added to carefully weighed quantities of the active substance and excipients and stirred. The commonly used binders are aqueous suspensions of corn starch, methyl cellulose, gelatin, or natural gums such as acacia. The binding agent is added until the mass of material acquires the consistency of soft dough. The addition of the liquid has to be controlled, as over-wetting will cause the granules to be very hard and insufficient wetting can cause them to be too soft, causing difficulty in compression. Thereafter, the damp mass is passed through a screen to form granules. The granules are then dried using tray dryers or fluid-bed dryers. The dried granules are finally passed through a smaller size screen than the one used earlier so that granules of uniform size are obtained (Gokhale and Trivedi, 2010).

5.5.1.3.2 Dry granulation This method is adopted when the tablet ingredients are sensitive to moisture, unable to withstand high temperatures during drying, or have insufficient binding properties. Dry granulation is also called *precompression* or *double compression*. The powder of the blended ingredients is subjected to slugging using slugging tooling or a roller compactor. The considerable amount of air contained in the powder is expelled during slugging and granules are formed. These granules are compressed into tablets after the addition of suitable lubricants (Peck et al., 2008).

5.5.1.3.3 Direct compression This approach is adopted when the medicinal substance has optimum physical and chemical properties such as low stickiness and high compactibility. Thus, it does not undergo processes such as granulation but is made into tablets directly. On account of the emphasis in the pharmaceutical industry on simplifying manufacturing processes to reduce the necessary floor space and labor force, increasing attention is now being paid to the process of direct compression of tablets. There is growing interest in research for improving the flowability of powders with and without additives (Bolhuis and Armstrong, 2006; Saha and Shahiwala, 2009).

5.5.1.3.4 Production of tablets Tablets have been popular ever since William Brockedon of England invented the tablet press, for which he received British Patent No. 9977, 1843 (Pietsch, 2005). The blended tablet powder is placed in two steel punches inside the cavity of a steel die. When pressure is applied on the powder by the punches, the powder gets compacted and acquires the shape and size of the punches and dies used. The tablets are ejected from the machine as soon as compression is over. Rotary tablet machines carrying a series of punches and dies are used today for the industrial production of tablets.

5.5.1.4 Characteristics of tablets

Compressed tablets have several characteristics such as diameter, size, shape, thickness, weight, hardness, disintegration time, and dissolution. These features form the specifications of the tablet in question. Strict adherence to these specifications ensures that the tablets have the same physical, chemical, and pharmacological properties (Rudnic and Schwartz, 2001).

5.5.1.4.1 Hardness On their journey from the point of manufacture to the user, tablets undergo storage, transportation, and handling. Tablets need to have sufficient hardness to withstand abrasion or breakage. The hardness of tablets (crushing strength) is nowadays determined objectively using electronically operated tablet hardness testers (Rudnic and Schwartz, 2006).

5.5.1.4.2 Friability Friability is a measure of the force required to crush a tablet. This factor decides the tablets' ability to withstand abrasion during packaging and transportation. A fixed number of tablets (at least 6 g of the product, equivalent to approximately 20 tablets, depending on weight) is weighed and placed in a Roche friabilator (tumbling apparatus) in which they are made to roll and undergo free fall. After a specific number of rotations, the tablets are weighed and their friability is calculated from the loss of material. Normally, less than 1% of loss is acceptable (Rhines, 2009).

5.5.1.4.3 Thickness The thickness of tablets can vary without a change in weight, as it is dependent on the bulk density of the granules and the pressure applied on the tablet dies. If tablets vary in thickness, they cannot be packed in a bottle of a given size. Uniform thickness also makes all tablets look identical in appearance. Calipers are used in the determination of tablet thickness, which is invariably expressed in millimeters (Rudnic and Schwartz, 2006).

5.5.1.4.4 Tablet weight The weight of tablets is determined by the weight of the granulated powder that fills the die. The volumetric fill of the die cavity is adjusted during the manufacturing operation to control the filling of the cavity. The weight of tablets is also checked frequently (Rudnic and Schwartz, 2006).

5.5.1.4.5 Content uniformity All tablets produced from a given quantity of tablet powder should have very little variation in their content of the active compound. This has become important on account of the increased awareness of the bioavailability of medicinal substances (Rudnic and Schwartz, 2006).

5.5.1.4.6 Disintegration of tablets The medicinal substance contained in a tablet should dissolve in the gastrointestinal fluid before it is absorbed into the system. The disintegration test measures the time taken by the tablet to disintegrate into particles, under a set of conditions, and serves as a quality control parameter for tablets. It is determined using the Vanderkamp tablet disintegration tester (Rudnic and Schwartz, 2006).

5.5.1.4.7 Dissolution The dissolution test measures the span of time required for the active substance in a tablet to dissolve under a given set of conditions. It indicates the bioavailability of the medicinal substance. This test serves as a quality control parameter. The granulation process is known to improve the dissolution of poorly soluble drugs. Starches, lactose, and microcrystalline cellulose, which are used in granulation, improve the dissolution characteristics of many drugs (Rudnic and Schwartz, 2001; Abdou et al., 2001).

5.5.1.5 Coating of tablets

Most modern tablets are given a coating of polymers or polysaccharides to protect them from environmental factors, to control the release of active substances, to enhance their appearance, or to improve their mechanical integrity. Three major coating methods are in vogue.

5.5.1.5.1 Sugar coating This is the oldest method of tablet coating. Sucrose is the sugar employed in most cases and is coated on the tablets in a multistep process. The aesthetics of coated tablets is very important and many companies employ groups of skilled workers to work on this (Allen et al., 2005a).

5.5.1.5.2 Film coating Film coating was introduced in the early 1950s to overcome the drawbacks of sugar coating, which proved to be tedious, time consuming, and specialized. Moreover, film-coated tablets are more resistant to abrasion. Cellulose ethers or acrylics are used in film coating. A plasticizer such as glycerin or propylene glycol is incorporated into the formulation to improve the adhesion of the film to the tablet surface. Film-coating solutions may be aqueous or nonaqueous. Tablets are film-coated by spraying the coating solution on the tablets in coating pans (Allen et al., 2005a).

5.5.1.5.3 Modified-release film coatings The release of the drug from tablets can be modified by coating them appropriately. Delayed-release tablets resist drug release in the stomach and disintegrate in the intestine. Such tablets have an enteric coating. The widely used enteric coating is cellulose acetate phthalate, followed by polyvinyl acetate phthalate, which is preferred for drug release in the duodenum (Porter, 2001).

Reasonably constant blood levels of actives over a long period can be achieved with sustained-release coatings. Such coatings are made of mixtures of waxes, shellac, ethyl cellulose, acrylic resins, cellulose acetate, and silicone elastomers (Porter, 2001).

5.5.1.6 Effervescent tablets

Effervescence is the evolution of carbon dioxide when acids and bases react in water. The reaction often employed in effervescent tablets is that between a soluble acid and an alkali metal carbonate to produce carbon dioxide. They are a good dosage form to deliver drugs that are pH sensitive, require a large dose, or are susceptible to light, oxygen, and moisture. Effervescent formulations have less than 0.5% of moisture. To protect the formulation from light, oxygen, and moisture, the tablets should be packed in 0.001 in. thick aluminum (Mohrle, 1989).

The solubility of the raw materials is very important in the formulation of effervescent tablets. The effervescent reaction will not occur if the tablet components are not soluble in water and tablet disintegration will be unsatisfactory. The most commonly used acid is citric acid. Its citrus-like taste enhances the acceptability of the product. The common bases used are sodium bicarbonate, potassium bicarbonate, sodium carbonate, and potassium carbonate. Sodium bicarbonate is preferred to the others on account of the clear solution it produces on the disintegration of the tablet (Mohrle, 1989).

Water-soluble binders having very little moisture are necessary to impart hardness to the tablet. Dextrose, sorbitol, xylitol, and lactose are commonly used. PVP is also used as a binder. There is little need for diluents as the effervescent materials themselves are present in large amounts. The lubrication of effervescent tablets is a difficult task, as the conventional lubricant, magnesium stearate, is water insoluble. In its place, many formulators use sodium benzoate, polyethylene glycol, and adipic acid. Colors, sweeteners, and flavors are included in the formulation to improve the appeal or to mask off-notes of ingredients (Mohrle, 1989).

The formula of a representative example of effervescent tablets is available in Mohrle (1989). Light mineral oil (15 g) is mixed with 200 g of sodium bicarbonate (granular). Color (5 g) is dispersed on 35 g of sodium bicarbonate. Anhydrous granular citric acid (1300 g) is placed in the bowl of a planetary mixer. The mixer is started and 4 g of water is slowly added with thorough mixing. While mixing, the remainder of the sodium bicarbonate, anhydrous sodium carbonate (80 g), amino acid (50 g), spray-dried flavor (50 g), color dispersion, and mineral oil dispersion are added to the mixer in sequence. Mixing is continued till the mass is homogeneously mixed. The mass is compressed into ¾ in., flat-faced, beveled edge tablets weighing 2.23 g each. The tablets are passed through a curing oven, cooled, and packaged in aluminum foil.

5.5.1.6.1 Production Effervescent tablets are produced more or less in the same way as compressed tablets. However, production must be carried out in a strictly low humid atmosphere, as trace amounts of moisture can activate the effervescent system. Both wet and dry granulation methods are employed. The granules are fed into tablet presses that can deliver high compression forces (Mohrle, 1989).

Effervescent tablets are invariably packed in aluminum foil. They can degrade when the packaging material does not have a moisture vapor transmission rate of 0, the seal of the foil pouch is imperfect due to machine malfunctions, or the ingredients are not compatible with each other or the effervescent components (Mohrle, 1989).

When formulated and produced in the desired way, effervescent tablets can serve as effective drug delivery systems. They can incorporate a large amount of active substances and are self-mixing and flavored. The carbon dioxide produced by the effervescent reaction can induce enhanced permeability of the active ingredients (Mohrle, 1989; Eichman and Robinson, 1998).

5.5.1.6.2 Stability of effervescent tablets The stability of effervescent tablets depends on the stability of the functional ingredients and the effervescent system itself. Effervescent tablets are hygroscopic and to obtain a reasonable shelf life they must be protected from atmospheric moisture. The tablets should be hermetically sealed irrespective of the container used. They are usually packaged in glass, plastic, or metal tubes and individual foil pouches.

5.5.1.7 Rapidly dissolving tablets

Rapidly dissolving tablets (RDT) are a new generation of formulations, combining the advantages of liquid and tablet formulations. They are also known as mouth dissolving, fast-melting, fast-dissolving, oral disintegrating, and orodisperse tablets. They are defined as tablets that can be placed in the mouth where they disperse rapidly before swallowing (Shukla et al., 2009a, 2009b). They are rapidly absorbed from the pregastric area and are of advantage to patients who cannot swallow.

RDT disintegrate in the mouth within a few seconds due to the quick entry of water into the tablet. The formulation technology therefore maximizes the porous structure of the tablet matrix, incorporates appropriate disintegrating agents, and uses highly water-soluble excipients (Shukla et al., 2009a).

Various technologies such as lyophilization (Jaccard and Leyder, 1985), molding (Pebley et al., 1994), direct compression (Bi et al., 1996), cotton candy process (Myers et al., 1995), spray drying (Allen et al., 1998), sublimation (Koizumi et al., 1997), mass extrusion (Bhaskaran and Narmada, 2002), and fast-dissolving films (Bess et al., 2006) are used in the manufacture of RDTs.

RDTs are evaluated on the basis of the tablets' tensile strength, friability, moisture uptake, porosity, wetting time and water absorption ratio, fineness of dispersion, disintegration time, and dissolution (Shukla et al., 2009b).

5.5.2 Development of tablets from herbal extracts

Several reports are available on the development of tablet formulations with herbal extracts. The reports of Plazier-Vercamen and Bruwier (1986), Diaz et al. (1996), and Renoux et al. (1996) indicate that obtaining the good flowability and compressibility of extracts is a formidable task. Nevertheless, Palma et al. (2002) succeeded in producing tablets through direct compression of the dry extract of the Chilean herb, *Peumus boldus*. The rheological properties of the dry extract were evaluated on the basis of its density, angle of repose, and compressibility. The direct compression at 1000 mPa of a formulation containing dry plant extract (170 mg), microcrystalline cellulose (Avicel PH101, 112 mg), lactose (112 mg), and magnesium stearate (6 mg) yielded the best results.

Similar results were obtained with a spray-dried extract of *Tanacetum parthenium* (Feverfew), an Argentine herb reputed in the treatment of migraine. Chaves et al. (2009) developed enteric-coated tablets from the spray-dried extract. **Table 5.1** shows the composition of the extract–excipient mixture. Tablets of 200 mg were obtained by the direct compression method. The tablets were film-coated (**Table 5.2**) and exhibited good pharmacotechnical properties.

However, wet granulation cannot be adopted in the case of dry extracts on account of their high hygroscopicity. Dry granulation is the ideal alternative (Rocksloh et al., 1999; von Eggelkraut-Gottanka et al., 2002). The use of lubricants during direct compression is known to increase the disintegration time, and the incorporation of high amounts of magnesium stearate into tablet granules shortens the disintegration time (Rocksloh et al., 1999; von Eggelkraut-Gottanka et al., 2002). Therefore, Soares et al. (2005) evaluated the effects of concentrations of sodium carboxymethylcellulose and colloidal

Table 5.1 Composition of Direct Compressed Tablets from Spray-Dried Extract of *Tanacetum parthenium*

Ingredients	%
Polyvinylpyrrolidone (PVP) K30	5
Aerosil	1
Microcrystalline cellulose PH 102	44.03
Sodium starch glycolate	3
Magnesium stearate	2
Lactose	14.68
Talc	3
Spray-dried extract of *Tanacetum parthenium*	27.29

Source: Reproduced from Chaves et al., *Braz. J. Pharm. Sci.*, 45, 573–584, 2009. With permission.

Table 5.2 Composition of Coating Suspension for Tablets of *Tanacetum parthenium*

Ingredients	%
EUDRAGIT L30 D55	15
Simethicone	0.5
Polysorbate 80	0.6
Talc	3
Titanium dioxide	1.5
Triethyl citrate	2.1
Water	100 q.s. ad

Source: Reproduced from Chaves et al., 2009, *Braz. J. Pharm. Sci.*, 45, 573–584, 2009. With permission.

silicon dioxide on the crushing strength and friability of tablets formulated with spray-dried granulations of the Brazilian plant *Maytenus ilicifolia*. The optimum formula for minimum disintegration time and friability and maximum crushing strength was recommended to contain 1.2% (w/w) of colloidal silicon dioxide and 5% (w/w) of sodium carboxymethylcellulose. The tablets that were prepared showed a crushing strength of 107.9 N, a friability of 0.56% (w/w), and a maximum disintegration time of 6.8 min.

Morazzoni and Bombardelli (2002) formulated an extract of *Coleus forskohlii* into a beverage powder and a compressed tablet for use in the treatment of alcohol addiction. Each sachet of the beverage powder contained 2300 mg of the formulation, which had the following composition: forskolin (100 mg), saccharose (2000 mg), maltodextrin (110 mg), citric acid (30 mg), orange flavor (40 mg), and hydrogenated vegetable oil (20 mg).

Each 100 mg tablet contained forskolin (25 mg), microcrystalline cellulose (25 mg), lactose (37 mg), colloidal silica (1 mg), cross-linked sodium carboxymethylcellulose (6 mg), PVP (5 mg), and magnesium stearate (1 mg) (Morazzoni and Bombardelli, 2002).

A novel method to prepare tablets containing essential oils was disclosed by Ninkov (2005). Cinnamon leaf oil, ginger oil, and turmeric oil were mixed with olive oil, incorporated into untreated fumed silica, and powdered well. The powder was blended with cocoa extract, citric acid, and excipients, and was compressed into tablets. The tablets had the following composition: cinnamon leaf oil (10–20 mg), ginger oil (1–7 mg), turmeric oil (0.5–3 mg), olive oil (3–10 mg), cocoa extract (50–100 mg), citric acid or citrus oil (1–5 mg), protein from kidney bean or wheat (00–1000 mg), fumed silica (25–50 mg), crystalline cellulose (Avicel PH 102) (100–200 mg), crystalline cellulose (Avicel PH 101) (150–350 mg), and methyl cellulose (100–200 mg).

Xiong et al. (2001) disclosed the process of preparing effervescent tablets containing green tea extracts. The tablets had the following composition: green tea extract (10%–50%), sodium bicarbonate (5%–30%), anhydrous citric acid (10%–45%), polyethylene glycol 6000 (1%–10%), PVP (1%–10%), and flavoring agent (0.1%–3%).

Ayurveda employs many herbs as single drug remedies. Considering the value of the tablet dosage form for popularizing the use of such single herb remedies, Patra et al. (2008) attempted to prepare compressed tablets from the root powder of *Asparagus racemosus*. Tablets were prepared by wet granulation and direct compression methods. For wet granulation, root powder (80% w/w) and Avicel PH 101 (15% w/w) were mixed and moistened with appropriate amounts of 10% w/w starch paste. The wet mass was granulated and the granules were dried in a hot air oven for 4 h at 60°C. The granules were resieved and mixed with talc (1%) and magnesium stearate (1%). For direct compression, root powder (80% w/w), Avicel PH102 (10% w/w), silicified microcrystalline cellulose (8%), talc (1%), and magnesium stearate (1%) were mixed well. Evaluation tests revealed that both wet granulation and direct compression methods can be applied to the development of tablets from *Asparagus racemosus* root powder. Nevertheless, granules prepared according to the wet granulation method had better flowability and compressibility.

5.5.3 Capsules

The hard gelatin capsule is another convenient solid dosage form. It can be used as a container for powdered drugs, multiparticulate systems, a liquid-fill matrix, or an oily vehicle (Washington et al., 1989). A travelogue of 1730 mentions the Viennese pharmacist de Pauli, who made oval-shaped capsules to hide the taste of turpentine, which he prescribed to his patients who were suffering from gout (Feldhaus, 1954). Modern gelatin capsules were invented in 1834 by Joseph Gérard Auguste Dublanc and his student François Achille Barnabé Mothes (French Patent, 1834). Like tablets, they can be used conveniently by adults. Although gelatin is insoluble in cold water, it can often absorb up to 10 times its weight of water. Two types of capsules, namely, hard and soft gelatin capsules, are in use. Hard gelatin capsules are made of pure gelatin, with a sorption water content of 13%–16%. Soft gelatin capsules have thicker shells that contain 20%–30% of plasticizers such as glycerol or sorbitol. They have a water content of about 30%. Soft gelatin capsules are intended to contain only liquids and pastes. Hard gelatin capsules can contain powders and liquids. Empty gelatin capsules come in a variety of lengths, diameters, and capacities. The size of the capsule used depends on the amount of material to be encapsulated (Allen et al., 2005b). Information on capsule sizes, fill volume, and fill weight is available in Augsburger (2002).

Compared with tablets, the powders that are encapsulated in hard gelatin capsules can be formulated with minimum effort. Such powders contain diluents, such as lactose, mannitol, calcium carbonate, or magnesium carbonate, and lubricants to enhance flowability. These powders are filled into capsules by large-scale capsule-filling machines. Modern capsule-filling machines can produce around 200,000 capsules per hour and can fill a number of different substances in a single process run (Rudnic and Schwartz, 2001; Allen et al., 2005b; Stegemann and Bornem, 2002).

5.5.3.1 Powder filling of capsules

Immediate release capsules are filled with a simple powder and the entire process consists of four or five steps: weighing and preparation of ingredients, mixing, filling into capsules, and packing. In contrast, other types of dosage forms require more steps. For example, tablets require around nine steps: weighing and preparation of ingredients, mixing, granulation, drying, sieving, the addition of lubricants (mixing/sieving), compression, and packing (Stegemann and Bornem, 2002).

The formulation of hard gelatin capsules involves diluents (mannitol, lactose, corn starch, microcrystalline cellulose, starch 1500), lubricants (magnesium stearate, stearic acid, glyceryl monostearate), glidants (aerosil, talc), disintegrants (croscarmellose, crospovidone, sodium glycyl starch, corn starch, starch 1500, alginic acid), and wetting agents (sodium lauryl sulfate, Tween 80). The medicinally active substance and the excipients are mixed homogeneously and filled into the capsules. At times, a light precompression is necessary to form a plug of powder. The force used for precompression is around 20–30 N, whereas the force required for the compression of tablets is 30,000 N (Stegemann and Bornem, 2002).

Carr's index is used to predict the required capsule size and to estimate or adjust the powder flow. Carr's index of <15% indicates very good flow, 16%–26% indicates good, 27%–35% indicates fairly good, and >35% indicates poor flow (Carr, 1965).

$$C_i = \frac{T_d - B_d}{B_d} \times 100 \tag{5.1}$$

where:

C_i = Carr's index
T_d = tapped density
B_d = bulk density

5.5.3.2 Factors to be considered during formulation

5.5.3.2.1 Compatibility with gelatin The first step in the formulation of a herbal extract in a hard gelatin capsule is determining its compatibility with the gelatin shell. It is known that some substances rich in reactive aldehydes can react with gelatin by forming cross-links (Stegemann and Bornem, 2002).

The water content of the gelatin shell itself can cause incompatibility. A hygroscopic extract filled in a hard gelatin capsule can absorb moisture from the shell, rendering it brittle and liable to break. If an extract is sensitive to humidity, the water content of the capsule shell (13%–16%) can cause degradation of the compounds in the extract (Stegemann and Bornem, 2002).

5.5.3.2.2 Dose The dose of the medicinal substance to be formulated is a major concern. Milligram amounts of the drug substance have to be mixed homogeneously in the powder. For doses above 100 mg, the characteristics of the active are more important than those of the excipients, which in this case will be smaller in volume (Stegemann and Bornem, 2002).

High concentrations of the extract can cause problems during the filling of the capsules. These problems can be prevented by the addition of appropriate quantities of diluents and lubricants (Hogan, 1995).

It is not possible to put more than 600 mg of powder into the largest capsule available. However, such volumes can be filled after granulating the powder (Stegemann and Bornem, 2002).

5.5.3.2.3 Shape of particles To achieve perfect content uniformity in capsules, the powder should have good flowability, which in turn is dependent on the shape of the particles, intraparticulate cohesion, and surface films. Isometric (round) particles form a dense shape and are ideal for capsule filling. The flowability of powders containing anisometric particles (needle shaped) can be improved by grinding or granulation (Stegemann and Bornem, 2002).

5.5.3.2.4 Solubility To achieve satisfactory disintegration and dissolution of the powder, the active substance and the excipients need to be water soluble. Poorly soluble actives can be improved by formulating them with appropriate disintegrants and diluents (Stegemann and Bornem, 2002).

5.5.3.2.5 Particle size The fluidity of the powder is dependent on the particle size of the active substance. Too fine or too coarse powders are unsuitable, as they hinder the free flow of the powder. The ideal particle size is between 10 and 150 μm (Hogan et al., 1996).

5.5.3.2.6 Hygroscopicity Hygroscopic extracts can absorb moisture from the capsule shell, which causes brittleness of the shell. Additionally, moisture can increase during the manufacturing process and cause the formation of sorption films, which directly affect the powder flow and the filling of the capsules. Hygroscopic extracts should be formulated with mannitol, which is an inert excipient as far as water absorption is concerned (Wade and Weller, 1994b).

5.5.3.2.7 Adhesion Powders that exhibit an adhesive nature create problems during filling, as the particles can stick to the surfaces of the capsule-filling machine. Consequently, the filled substance breaks up, leading to a variation in the quantities filled in the capsules. The stickiness of the actives and excipients can be controlled by the addition of a glidant or a combination of a glidant and a lubricant, such as aerosil/magnesium stearate or talc/stearic acid (Stegemann and Bornem, 2002).

5.5.3.2.8 Wetting nature The ability of the material filled in the capsule to mix with water is vital to the release of the active compounds. The release of hydrophobic substances can be improved by including lactose as a diluent or a wetting agent such as sodium lauryl sulfate. It is to be noted that magnesium stearate can cause the opposite effects, such as a reduction in the wetting action and a slowing down of disintegration and dissolution (Samyn and Jung, 1970; Frömming and Gröbler, 1983).

A suitable disintegrant should be included in the formulation when hydrophobic substances are used in high doses or when they form a major part of the formulation. Sodium croscarmellose and crospovidone are strong disintegrants, while sodium glycol starch and corn starch are moderate disintegrants (Stegemann and Bornem, 2002).

5.5.3.2.9 Moisture sensitivity of actives The moisture sensitivity of the medicinal substance can interfere with the stability of the formulation. The moisture of the capsule shell can damage such substances. The addition of mannitol to the formulation prevents such damage (Wade and Weller, 1994b).

5.5.3.2.10 Lubrication Lubricants have different functions in a formulation. They prevent powder sticking to the metal surfaces of the capsule-filling machine and optimize the powder flow and compressibility characteristics. Over lubrication, on the other hand, significantly reduces dissolution and interferes with content uniformity, powder density, and plug formation (Jones, 1998).

The time required to mix the lubricant with the formulation is also known to affect the release profile of some synthetic drugs, even if their concentration remains unchanged (Murthy and Samyn, 1977). Therefore, lubricants should be added in low quantities.

5.5.3.2.11 Dissolution and disintegration The dissolution of hard gelatin capsules is studied by immersing them in a dissolution fluid at 37°C. On immersion in the fluid, the capsules initially rupture at the shoulders of the cap. As the dissolution fluid enters the contents of the capsule, the powder disintegrates and exposes the drug particles for dissolution. Disintegration and dissolution are dependent on the wettability and on the overall composition of the formulation (Augsburger, 2002: 359). Hard gelatin capsules disintegrate fully within 10 min (Ludwig et al., 1979, 1980). Disintegration and dissolution tests on hard and soft capsules are carried out in the same way as for uncoated tablets (Allen et al., 2005b).

5.5.3.2.12 Weight variation Capsules filled with the medicinal substance should have uniform weight. This is ensured by determining the weight of the shells of 10 capsules and their contents. Similar to tablets, the contents of capsules also need to be dissolved in a short span of time (Allen et al., 2005b).

5.5.3.2.13 Capsules for multiple units Single dosage forms that disintegrate into several units after ingestion in the gastrointestinal tract are called *multiple units*. Today, machines are available that allow hard gelatin capsules to be filled with different types of pellets and/or formulations via several filling stations within a single process. Multiple units are developed as enteric-coated and controlled-release pellets using the appropriate film coating. The derivatives of acrylic acid and cellulose are commonly used for the coating (Hogan, 1995).

5.5.4 Computer-aided formulation development

In the olden days, pharmaceutists used to adopt a trial-and-error approach to formulation development, based on the experiences of individual formulators. Their modern counterparts encounter numerous and complex drug substances that need to be formulated efficiently in a short span of time. Hard gelatin capsules are generally perceived to be a simple dosage form. Nevertheless, their design presents many challenges to the formulator who has to control factors such as ingredient compatibility and stability, powder blending and homogeneity, and powder fluidity and lubrication. To help the formulators of hard gelatin capsules, Guo et al. (2002) created a prototype intelligent hybrid system, by linking a decision module with a prediction module. Validation experiments proved that the hybrid system is capable of yielding formulations of a model Biopharmaceutics Classification System Class II drug, piroxicam. The framework of this hybrid system can be extended to the development of other dosage forms as well.

5.5.5 Syrups

Syrups are concentrated solutions of sugars such as glucose in water or aqueous liquids. When pure water alone is used for dissolving sucrose, the product is called *syrup* or *simple syrup*. Polyols such as sorbitol, or glycerin may also be added at times to retard the crystallization of sucrose or to improve the solubility of the other ingredients. Alcohol and preservatives are added to retard microbial growth. Syrups containing medicinal substances are called *medicated syrups*. Syrups also serve as ideal vehicles for bitter drugs (Nairn, 2001).

Syrups are prepared by four methods, namely, with the aid of heat, by agitation, by adding sugar to a medicated liquid, and by percolation of the medicinal substance or sugar. Syrups offer the advantages of precision of dosage, ease of dispensing, and enhanced stability (Kolling and Ghosh, 2005).

Medicated syrups are generally prepared by dissolving the medicinal substance and other ingredients, such as flavoring agents, colorants, citric acid, and preservatives, in sucrose syrup. The commonly known vasaka syrup is prepared by mixing 500 mL of vasaka liquid extract with 100 mL of glycerin and making up the volume to 1000 mL with a sufficient quantity of syrup (Gokhale et al., 2007). Commercially prepared syrups contain special solvents, solubilizing agents, thickeners, or stabilizers. The common preservatives are benzoic acid (0.1%–0.2%), sodium benzoate (0.1%–0.2%), and various combinations of methylparabens, propylparabens, and butylparabens totaling about 0.1% (Allen et al., 2005c).

Syrups can be transformed into dry syrup formulations to improve their stability and to minimize microbial attack. The production of dry syrups does not involve the use of water, other solvents, or heat. The medicinal substance can be added to the dry product before granulation or it can be dissolved or suspended in the granulating liquid. The granules are dried and screened to obtain granules of uniform particle size. Granulated dry syrup powders have better appearance and flowability and less aggregation of particles (Nairn, 2001).

5.5.6 Common problems and their solutions

Liquid dosage forms such as syrups can be prepared from fluid, soft, or dry extracts of herbs. These extracts need to be diluted to make liquid preparations. It is at this point that the formulator is faced with the problem of solubility. If the extract is not dissolved completely, the product will develop turbidity or precipitate immediately or on storage. This problem can be solved in the following ways (Bonati, 1991).

1. While redissolving an extract or when diluting a fluid extract, the formulator should use the exact same strength of solvent as that used for preparing the extract. For example, if a herb has been extracted with 50% aqueous alcohol, the extract should be dissolved in 50% aqueous alcohol only (Bonati, 1991).
2. High-strength alcohol or other organic solvents are sometimes used in the extraction of certain herbs. The preparation of the liquid dosage form many not require such a strong solvent. Such instances are common with extracts of essential oil-bearing herbs. In situations like

this, the formulator should use cosolvents or surfactants. The commonly used cosolvents are glycols, polyglycols, glycerin, and sorbitol. At times, their use may be restricted on account of the large quantities that are used in the formulation. Such situations call for the use of nonionic surfactants such as polysorbates and polyoxyethylene derivatives of oleic alcohol. An example from Bonati (1991) clearly illustrates this point. When the soft extract of *Ruscus aculeatus* (2 g) is formulated into a syrup using sucrose (20 g), alcohol (10 g), polyethylene glycol 400 (5 g), a mixture of *p*-hydroxybenzoates (0.1 g), and purified water (100.0 q.s. ad.), the extract is insoluble in the formulation. Nevertheless, a clear syrup is formed when 2.5% polyoxyethylen-20-sorbitan monostearate is added to the formulation (Bonati, 1991).

3. Alterations in the pH can also enhance the solubility of extracts. This technique works more in the case of substances such as alkaloids whose solubility increases through the formation of salts. Organic acids such as lactic acid, citric acid, and tartaric acid are often used for dissolving extracts containing alkaloids. The chemical stability of the alkaloids is also ensured in this way (Bonati, 1991).
4. Cloudy solutions of extracts containing easily soluble constituents can be clarified by filtration. This approach is recommended only when there is no decrease in the content of the active substances following filtration (Bonati, 1991).

The stability of the solution of an extract can be jeopardized by microbial growth. This can be circumvented by the use of appropriate preservatives. Sometimes, constituents of the liquid formulation can cross-react producing precipitates, as is the case with tannins and alkaloids. Cosolvents, surfactants, and the removal of the interfering substances can solve the problem (Bonati, 1991).

5.5.6.1 Masking of tastes

Herbal extracts are of varying tastes and masking their tastes is important to the pharmaceutical and food industries. Flavors are commonly used for masking and complimenting tastes. The extent to which taste masking is feasible cannot be predicted, as taste perception is a complex phenomenon. Cocoa syrup is quite effective in masking a bitter taste. It is followed by raspberry syrup, cherry syrup, cinnamon syrup, citric acid, licorice syrup, and orange syrup (Reiland and Lipari, 2007). Many bitter substances are hydrophobic and therefore hydrophobic compounds can be used to mask the target sites of bitter substances. The water-dispersible lipoproteins PA-LG, made of phosphatidic acid and β-lactoglobulin, suppress the taste response to bitter substances (Katsuragi and Kurihara, 1993). Polymers, complexing agents, and low molecular weight substances are very effective in masking a bitter taste. Readers are directed to an excellent review on the topic by Ley (2008).

A sour taste can be best masked by raspberry syrup or other fruit syrups. Acacia syrup and similar mucilaginous substances disguise the pungent feeling imparted by capsicum and pepper, by forming a protective colloidal coating over the buccal cavity and taste buds. Tragacanth gum can be used in the medium of alcohol (Reilly, 2001).

5.5.7 Stability of food supplements

The testing of stability is an essential part of the formulation of food supplements. The stability of all dosage forms is studied as a function of time against numerous environmental factors such as temperature, humidity, and light and their combinations. The data accrued from stability studies are pivotal to the establishment of recommended storage conditions, expiry dates, and shelf lives. The stability of a food supplement is an index of its strength, purity, identity, safety, degradation, and physical or biological changes and their effects on the biological actions of the product (Valvani, 2000).

The International Conference on Harmonization of Technical Requirements for Registration of Pharmaceuticals for Human Use (ICH) guidelines Q1E and Q1A (R2) provide guidance for the design methodology and analysis of stability studies (Anonymous, 2003a, 2003b). Huynh-Ba (2009) exhaustively treats the various aspects of stability testing, such as stability regulations, stability methodologies, and best practices. The basic method of stability testing is briefly described next.

Samples of the dosage form manufactured to a pilot scale are packed in containers that are identical to the one proposed for distribution. Thereafter, the samples are stored in chambers maintained at a specific temperature and humidity. For long-term studies, samples are stored for 12 months at 25°C ± 2°C/60 ± 5% RH (relative humidity). Accelerated stability is tested by keeping the samples at 40°C ± 2°C/75 ± 5% RH for 6 months. For intermediate stability studies, samples are stored at 30°C ± 2°C/65 ± 5% RH for 6 months. These samples are evaluated on predecided parameters at 0, 1, 2, 3, 6, 9, and 12 months stations. Real-time studies can be conducted by storing the packets of products at 30°C ± 2°C/65 ± 5% RH for 12, 24, 36, and 48 months. In this case, the samples are to be evaluated at 0, 1, 3, 6, 12, 18, 24, 36, and 48 months.

The organoleptic, chemical, and physical characteristics of the dosage forms are monitored during the course of the stability study and recorded. The stability of the product is determined on the basis of the recorded data. The types of packing and storage conditions of the products are decided on the basis of the stability data (Vadas, 2001). A 6 month accelerated stability study will provide data to support a 2 year shelf life.

5.5.8 Pharmaceutical excipients of herbal origin

5.5.8.1 In modified-release tablets

Substances extracted from herbs have recently been proved to be of use in pharmaceutical technology. For example, the tubers of *Amorphophallus konjac* are rich in the hydrocolloidal polysaccharide, glucomannan.

Alvarez-Manceñido et al. (2008) observed that konjac glucomannan gel systems could maintain the integrity and control the release of theophylline for 8 hours. Matrix tablets prepared from konjac glucomannans alone sustained the release of cimetidine in the fluids of the stomach and the intestine. β-Mannanase present in the colon accelerated the release of the drug. Konjac glucomannan cross-linked with trisodium trimetaphosphate formed hydrogels that could sustain the release of hydrocortisone (Liu et al., 2007).

Pectin is a complex polysaccharide predominantly found in fruits. It has high potential in controlled-release matrix delivery systems. However, its water solubility is an impediment (Beneke et al., 2009). This problem can be overcome by salt formation. Matrix tablets prepared with calcium pectinate were found to be suitable for colon-targeted drug delivery systems. There were additional effects such as the suppression of dissolution and swelling of the systems (Sinha et al., 2001; Chourasia et al., 2004; Bhatia et al., 2008).

Natural gums such as guar gum, locust bean gum, and aloe gel possess properties that qualify them for use in controlled-release tablets (Varshosaz et al., 2006; Vendruscolo et al., 2005; Jani et al., 2007). Singhal et al. (2010) recently tested the feasibility of using guar gum as an excipient in the formulation of colon-targeted curcumin tablets. Tablets formulated with 40% guar gum showed satisfactory drug release. The other components of the formulation were curcumin (10%), lactose (45%), 8% starch paste (q.s.), magnesium stearate (1%), talc (2%), and sodium lauryl sulfate (2%). Guar gum is a highly branched galactomannan that hydrates quickly to produce a viscous pseudoplastic solution, having greater low-shear viscosity than ordinary hydrocolloids. This gelling property does not permit the release of the drug from the solid dosage form. However, on account of its susceptibility to degradation by colonic bacteria, the formulated drug readily disperses in the colon. Guar gum can thus be useful in the formulation of colon-targeted curcumin delivery systems to treat diseases such as irritable bowel syndrome, colon cancer, colitis, and ulcerative colon (Yang et al., 2002).

5.5.8.2 In film coatings

Inulin is a mixture of oligomers and polymers belonging to the group of glucofructans. Colon-specific films prepared in combination with EUDRAGIT RS and inulin could resist breakdown by the gastric and intestinal fluids (Vervoort et al., 1996; Akhgari et al., 2006).

Another natural polymer rosin has been studied for its utility as a film-coating agent and matrix material in modified-release tablets (Mandaogade et al., 2002; Fulzele et al., 2003; Satturwar et al., 2003).

5.5.8.3 Direct compression excipients

Starches from various species of *Dioscorea* can be used as excipients in the direct compression of tablets. Tablets made with acid-modified starch had a higher crushing strength and an acceptable disintegration time (Odeku et al., 2009).

Alginates or alginic acid extracted from seaweeds and marine algae are used as stabilizers in emulsions, suspending agents, binding agents, and disintegrants (Shirwaikar et al., 2008; Beneke et al., 2009).

5.6 Concluding remarks

While working with the traditional methods of mixing, granulation, and drying, binders such as cellulose derivatives and PVP should be used preferably in a solution with organic solvents such as chloroform, acetone, or alcohol. When solutions of cellulose derivatives, gum arabic, gelatin, or aqueous gels are used, problems may occur in the course of their processing. Sugars and saponins present in the extracts dissolve in the water used for mixing. They form granules that are difficult to dry or they may form hard granules that are difficult to compress.

Tablets compressed after granulation with aqueous solutions may show poor disintegration. After coming into contact with water, such extracts form a gel on the surface of the tablet, preventing further penetration of the water into the tablet. However, tablets made from granulated extracts disintegrate quickly (Kopelman et al., 2008). The best remedy for these problems is wet granulation with organic solvents (Bonati, 1991).

While formulating more than one extract together, interactions may occur between the chemical constituents of the extracts, forming products with new solubility characteristics. Such cross-reaction can occur while formulating an alkaloid-containing extract with one having tannins or having organic acids that can form salts with the alkaloids. The organic acid–alkaloid interaction on the other hand, leads to the formation of salts and consequent rapid bioavailability (Bonati, 1991).

Free-flowing and nonhygroscopic granules are essential for preparing hard gelatin capsules. As the hard gelatin capsule shell does not offer protection against humidity, the granules should be protected by coating them with resins or methylcellulose. This coating prevents degradation of the extracts due to moisture slowly entering the capsule, resulting in a sold lump that is poorly soluble (Bonati, 1991).

Some valuable publications have appeared in recent times aimed at improving the quality of food supplements. One such publication is the *Quality Guide for Botanical Food Supplements* (Anonymous, 2011). Developed by the European Federation of Associations of Health Product Manufacturers, this document provides guidelines for best practice in the manufacture, quality control, packaging, storage, and distribution of food supplements containing extracts of herbs. It includes mandatory requirements from legislations enacted by the European Union. It deals with topics such as quality management, food premises and equipment, personnel and training, hazard analysis critical control point (HACCP), safety, product and process development, manufacture, recovery of reworking material, warehousing, transport and distribution, documentation, complaints procedure, self-inspections, and laboratory testing. It is an excellent manual on the subject.

References

Abdou, H.M., Hanna, S., Muhammed, N. 2001. Dissolution. In *Remington: The Science and Practice of Pharmacy*, Vol. 1, ed. A.R. Gennaro, 654–666. Baltimore, MD: Lippincott Williams & Wilkins.

Akhgari, A., Farahmand, F., Garekani, H., Sadeghi, F., Vandamme, T.F. 2006. Permeability and swelling studies on free films containing inulin in combination with different polymethacrylates aimed for colonic drug delivery. *European Journal of Pharmacology* 28: 307–314.

Allen, Jr., L.V., Popovich, N.G., Ansel, H.C. 2005a. Tablets. In *Ansel's Pharmaceutical Dosage Forms and Drug Delivery Systems*, 8th edn, 227–256. Baltimore, MD: Lippincott Williams & Wilkins.

Allen, Jr., L.V., Popovich, N.G., Ansel, H.C. 2005b. Capsules. In *Ansel's Pharmaceutical Dosage Forms and Drug Delivery Systems*, 8th edn, 204–226. Baltimore, MD: Lippincott Williams & Wilkins.

Allen, Jr., L.V., Popovich, N.G., Ansel, H.C. 2005c. Solutions. In *Ansel's Pharmaceutical Dosage Forms and Drug Delivery Systems*, 8th edn, 336–384. Baltimore, MD: Lippincott Williams & Wilkins.

Allen, L.V., Wang, B., Davis, J.D. 1998. Rapidly dissolving tablet. US Patent No. 5,807, 576.

Alwarez-Manceñido, F., Landin, M., Lacik, I., Martinez-Pacheco, R. 2008. Konjac glucomannan and konjac glucomannan/xanthan gum mixture as excipients for controlled drug delivery systems. Diffusion of small drugs. *International Journal of Pharmaceutics* 349: 11–18.

Anonymous. 2003a. ICH Harmonised Tripartite Guideline, Stability Testing of New Drug Substances and Products Q1A(R2), Current Step 4 Version, 1-18. http://www.ich.org/fileadmin/Public_Web_Site/ICH_Products/Guidelines/Quality/Q1A_R2/Step4/Q1A_R2__Guideline.pdf. Accessed May 27, 2015.

Anonymous. 2003b. ICH Harmonised Tripartite Guideline, Evaluation for Stability Data Q1E, Current Step 4 Version, 1–15. http://wenku.baidu.com/view/6ab8891bff00bed5b9f31d61.html?re=view. Accessed February 5, 2015.

Anonymous. 2011. *Quality Guide for Botanical Food Supplements*. Brussels: European Botanical Forum.

Augsburger, L.L. 2002. Hard and soft shell capsules. In *Modern Pharmaceutics*, 4th edn, ed. G.S. Banker and C.T. Rhodes, 335–380. New York: Marcel Dekker.

Beneke, C.E., Viljoen, A.M., Hamman, J.H. 2009. Polymeric plant-derived excipients in drug delivery. *Molecules* 14: 2602–2620.

Bess, W.S., Kulkarni, N., Ambike, S.H., Ramsay, M.P. 2006. Fast dissolving orally consumable solid film containing a taste masking agent and pharmacologically active agent at weight ratio of 1:3 to 3:1. US Patent No. 7,067,116.

Bhaskaran, S., Narmada, G.V. 2002. Rapidly dissolving tablet—A novel dosage form. *Indian Pharmacist* 1: 9–12.

Bhatia, M.S., Deshmukh, R., Choudhari, P., Bhatia, N.M. 2008. Chemical modifications of pectins, characterization and evaluation for drug delivery. *Scientia Pharmaceutica* 76: 775–784.

Bi, Y., Sunada, K., Yanezawa, Y., Danjo, K., Otsuka, A., Iida, K. 1996. Preparation and evaluation of a compressed tablet rapidly disintegrating in the oral cavity. *Chemical and Pharmaceutical Bulletin* 44: 2121–2127.

Bolhuis, G.K., Armstrong, N.A. 2006. Excipients for direct compaction—An update. *Pharmaceutical Development and Technology* 11: 111–114.

Bonati, A. 1991. Formulation of plant extracts into dosage forms. In *The Medicinal Plant Industry*, ed. R.O.B. Wijesekera, 107–113. Boca Raton, FL: CRC Press.

Carr, R.L. 1965. Evaluating flow properties of solids. *Chemical Engineering* 72: 163–168.

Chaves, J.S., Da Costa, F.B., Pedro de Freitas, L.A. 2009. Development of enteric coated tablets from spray dried extracts of feverfew (*Tanacetum parthenium* L.). *Brazilian Journal of Pharmaceutical Sciences* 45: 573–584.

Chiu, K.K.W., Chow, A.H.L. 2000. Impact of carbohydrate constituents on moisture sorption of herbal extracts. *Pharmaceutical Research* 17: 1133–1137.

Chourasia, M.K., Jain, S.K. 2004. Polysaccharides for colon-targeted drug delivery. *Drug Delivery* 11: 129–148.

Diaz, L., Souto, C., Concheiro, A., Gomez-Amoza, L.M., Martinez-Pacheco, R. 1996. Evaluation of Eudragit E as excipient in tablets of dry plant extracts. *S.T.P. Pharma* 6: 105–109.

Driver, M.G., Francis, F.J. 1979. Purification of phytolaccin (betanin) by removal of phytolaccatoxin from *Phytolacca americana*. *Journal of Food Science* 44: 521–523.

Eichmann, J.D., Robinson, J.R. 1998. Mechanistic studies on effervescent-induced permeability enhancement. *Pharmaceutical Research* 15: 925–930.

Feldhaus, F.M. 1954. Zur Geschichte der Arzneikapsel. *Deutsche Apotheke Zeitung* 94: 321.

French Patent, 1834. Patent No. 5648, dated March 25, 1834.

Frömming, K.H., Gröbler, S. 1983. Einfluβ von Füllmitteln und von Magnesiumstearat auf die Wikstofffreisetzung aus Hartgelatinekapseln. *Pharmazeutische Zeitung* 128: 786–793.

Fulzele, S.V., Satturwar, P.M., Dorle, A.K. 2003. Study of the biodegradation and *in vivo* biocompatibility of novel biomaterials. *European Journal of Pharmaceutical Sciences* 2: 53–61.

Gokhale, S.B., Gattani, S.G., Bakliwal, S.R. 2007. Syrups. In *Practical Pharmaceutics*, 1–2. Pune: Nirali Prakashan.

Gokhale, R., Trivedi, N.R. 2010. Wet granulation in low- and high shear mixers. In *Handbook of Pharmaceutical Granulation Technology*, ed. D.M. Parikh, 183–203. New York: Informa Healthcare.

Guo, M., Kalra, G., Wilson, W., Peng, Y., Augsburger, L. 2002. A prototype intelligent hybrid system for hard gelatin capsule formulation development. *Pharmaceutical Technology* 26: 44–60.

Hogan, J.E. 1995. Film coating materials and their properties. In *Pharmaceutical Coating Technology*, ed. G. Cole, J. Hogan, and M. Aulton, 6–52. London: Taylor & Francis.

Hogan, J., Shue, P.I., Podczeck, F., Newton, J.M. 1996. Investigations into the relationship between drug properties, filling and the release of drug from hard gelatin capsules using multivariate statistical analysis. *Pharmaceutical Research* 13: 944–949.

Houghton, P.J., Raman, A. 1998. Methods for extraction and sample clean-up. In *Laboratory Handbook for the Fractionation of Natural Extracts*, 45–53. London: Chapman and Hall.

Huynh-Ba, K. 2009. *Handbook of Stability Testing in Pharmaceutical Development*. New York: Springer.

Jaccard, T.T., Leyder, S. 1985. Une nouvelle forme galenique: le lyoc. *Annales Pharmaceutiques Françaises* 43: 123–131.

Jani, G.K., Shah, D.P., Jain, V.C., Patel, M.J., Vithalan, D.A. 2007. Evaluating mucilage from *Aloe barbadensis* Miller as a pharmaceutical excipient for sustained-release matrix tablets. *The Journal of Pharmacy Technology* 31: 90–98.

Jones, B.E. 1998. New thoughts on capsule filling. *STPP* 8: 277–283.

Katsuragi, Y., Kurihara, K. 1997. Specific inhibitor for bitter taste. In *Modifying Bitterness: Mechanisms, Ingredients and Applications*, ed. G. Roy, 255–284. Boca Raton, FL: CRC Press.

Koizumi, K.I., Watanabe, Y., Morita, K., Utoguchi, N., Matsumoto, M. 1997. New method for preparing high porosity rapidly saliva soluble compressed tablets using mannitol with camphor, a subliming material. *International Journal of Pharmaceutics* 152: 127–131.

Kolling, W.M., Ghosh, T.K. 2005. Oral liquid dosage forms: Solutions, elixirs, syrups, suspensions and emulsions. In *Theory and Practice of Contemporary Pharmaceutics*, ed. T.K. Ghosh and B.R. Jasti, 367–386. Boca Raton, FL: CRC Press.

Kopelman, S.H., Jin, P., Augsburger, L.L. 2008. Botanicals and their formulation into oral solid dosage forms. In *Pharmaceutical Dosage Forms: Tablets*, 3rd edn, Vol. 2: Rational Design and Formulation, ed. L.L. Augsburger and S.W. Hoag, 333–360. New York: Informa Healthcare.

Ley, J.P. 2008. Masking bitter taste by molecules. *Chemosensory Perception* 1: 58–77.

Liu, M., Fan, J., Wang, K., He, Z. 2007. Synthesis, characterization and evaluation of phosphated cross-linked konjac glucomannan for colon-targeted drug delivery. *Drug Delivery* 14: 397–402.

Ludwig, A., van Ooteghem, M. 1980. Disintegration of hard gelatin capsules, Part 2: Disintegration mechanism of hard gelatin capsules investigated with a stereoscopic microscope. *Pharmazeutische Industrie* 42: 405–406.

Ludwig, A., van Ooteghem, M., Delva, A. 1979. Disintegration of hard gelatin capsules, Part 1: Composition and structure of the capsule wall. *Pharmazeutische Industrie* 41: 796–798.

Mandaogade, P.M., Satturwar, P.M., Fulzele, S.V., Gogte, B.B., Dorle, A.K. 2002. Rosin derivatives: Novel film forming materials for controlled drug delivery. *Reactive and Functional Polymers* 50: 233–243.

Marshall, K.M. 1991. Compression and consolidation of powdered solids. In *The Theory and Practice of Industrial Pharmacy*, ed. L. Lachman, H.A. Lieberman, and J.L. King, 66–99. Bombay: Varghese Publishing House.

Marshall, R.T. 1993. *Standard Methods for the Examination of Dairy Products*, 16th edn. Washington, DC: American Public Health Association.

Mohrle, R. 1989. Effervescent tablets. In *Pharmaceutical Dosage Forms: Tablets*, Vol. 1, 2nd edn, ed. H.A. Lieberman, L. Lachman, and J.B. Schwartz, 285–328. New York: Marcel Dekker.

Morazzoni, P., Bombardelli, E. 2002. Use of forskolin or extracts containing it in the manufacture of a medicament for the treatment of alcohol addiction. European Patent No. EP 0825859.

Murthy, K.S., Samyn, W.Y. 1977. Effect of shear mixing on *in vitro* drug release of capsule formulations containing lubricants. *Journal of Pharmaceutical Sciences* 66: 1215–1219.

Myers, G.L., Battist, G.E., Fuisz, R.C. 1995. Process and apparatus for making rapidly dissolving dosage units and product therefrom. PCT Patent WO 95/34293-A1.

Nairn, J.G. 2001. Solutions, emulsions, suspensions and extracts. In *Remington: The Science and Practice of Pharmacy*, Vol. 1, ed. A.R. Gennaro, 721–752. Baltimore, MD: Lippincott Williams & Wilkins.

Ninkov, D. 2005. Method of protection of biologically active essential oils and pharmaceutical composition for weight control and enhance fat burning in adults, adolescents and children. US Patent Application No. 20050277657.

Odeku, O.A., Picker-Freyer, K. 2009. Characterization of acid modified *Dioscorea* starches as direct compression excipient. *Pharmaceutical Development and Technology* 14: 259–270.

Palma, S., Lujan, C., Llabot, J.M., Barboza, G., Manzo, R.H., Allemandi, D.A. 2002. Design of *Peumus boldus* tablets by direct compression using a novel dry plant extract. *International Journal of Pharmaceutics* 233: 191–198.

Patra, N.C., Singh, S.P., Kumar, P.H., Devi, M.V. 2008. A systematic study on flowability and compressibility of *Asparagus racemosus* root powder for tablet preparation. *International Journal of Pharmaceutical Sciences and Nanotechnology* 1: 129–135.

Pebley, W.S., Jager, N.E., Thomson, S.J. 1994. Rapidly disintegrating tablet. US Patent No. 5,298,261.

Peck, G.E., Soh, J.L.P., Morris, K.R. 2008. Dry granulation. In *Pharmaceutical Dosage Forms*, 3rd edn, Vol. 1. ed. L.L. Augsburger and S.W. Hoag, 303–336. New York: Informa Healthcare.

Pietsch, W. 2005. Pressure agglomeration technologies. In *Agglomeration in Industry*, Vol. 1, 114–146. Weinheim: Wiley-VCH.

Plazier-Vercamen, J.A., Bruwier, C. 1986. Evaluation of excipients for direct compression of the spray-dried extract of *Harpagoyphytum procumbens. S.T.P. Pharma* 2: 525–530.

Porter, S.C. 2001. Coating of pharmaceutical dosage forms. In *Remington: The Science and Practice of Pharmacy*, Vol. 1, ed. A.R. Gennaro, 894–902. Baltimore, MD: Lippincott Williams & Wilkins.

Reiland, T.L., Lipari, J.M. 2007. Flavors and flavor modifiers. In *Encyclopedia of Pharmaceutical Technology*, Vol. 3, 3rd edn, ed. J. Swarbrick, 1763–1772. New York: Informa Healthcare.

Reilly, Jr., W.J. 2001. Pharmaceutical necessities. In *Remington: The Science and Practice of Pharmacy*, Vol. 1, ed. A.R. Gennaro, 1015–1050. Baltimore, MD: Lippincott Williams & Wilkins.

Renoux, R., Demazieres, J.A., Cardot, J.M., Aiache, J.M. 1996. Experimentally designed optimisation of direct compression tablets. *Drug Development and Industrial Pharmacy* 22: 103–105.

Rhines, T. 2009. Non-chromatographic methods to support stability program. In *Handbook of Stability Testing in Pharmaceutical Development: Regulations, Methodologies and Best Practices*, ed. K. Huynh-Ba, 201–222. Heidelberg: Springer.

Rocksloh, K., Rapp, F.R., Abu Abed, S., et al. 1999. Optimization of crushing strength and disintegration time of a high-dose plant extract tablet by neural networks. *Drug Development and Industrial Pharmacy* 25: 1015–1025.

Rudnic, E.M., Schwartz, J.D. 2001. Oral solid dosage forms. In *Remington: The Science and Practice of Pharmacy*, Vol. 1, ed. A.R. Gennaro, 858–893. Baltimore, MD: Lippincott Williams & Wilkins.

Rudnic, E.M., Schwartz, J.D. 2006. Oral solid dosage forms. In *Remington: The Science and Practice of Pharmacy*, 21st edn, ed. D.B. Troy, 889–928. Baltimore, MD: Lippincott Williams & Wilkins.

Saha, S., Shahiwala, A.F. 2009. Multifunctional co-processed excipients for improved tabletting performance. *Expert Opinion on Drug Delivery* 6: 197–208.

Samyn, J.C., Jung, W.Y. 1970. In vitro dissolution from several experimental capsule formulations. *Journal of Pharmaceutical Sciences* 59: 169–175.

Satturwar, P.M., Fulzele, S.V., Dorle, A.K. 2003. Biodegradation and *in vivo* biocompatibility of rosin: A natural film forming polymer. *AAPS PharmSciTech* 4: 1–6.

Shirwaikar, A., Shirwaikar, A., Prabhu, S.L., Kumar, G.A. 2008. Herbal excipients in novel drug delivery systems. *Indian Journal of Pharmaceutical Sciences* 70: 415–422.

Shukla, D., Chakraborty, S., Singh, S., Mishra, M. 2009a. Mouth dissolving tablets. I: An overview of formulation technology. *Scientia Pharmaceutica* 76: 309–326.

Shukla, D., Chakraborty, S., Singh, S., Mishra, M. 2009b. Mouth dissolving tablets. II: An overview of evaluation techniques. *Scientia Pharmaceutica* 77: 327–341.

Singhal, A.K., Nalwaya, N., Jarald, E.E., Ahmed, S. 2010. Colon targeted curcumin delivery using guar gum. *Pharmacognosy Research* 2: 82–85.

Sinha, V.R., Kumria, R. 2001. Polysaccharides in colon-specific drug delivery. *International Journal of Pharmaceutics* 224: 19–38.

Soares, L.A.L., Ortega, G.G., Petrovick, P.R., Schmidt, P.C. 2005. Optimization of tablets containing high dose of spray-dried plant extract: A technical note. *AAPS PharmSciTech* 6: E367–E371.

Stegemann, S., Bornem, C. 2002. *Hard Gelatin Capsules Today—And Tomorrow*, 2nd edn. Morristown, NJ: Capsugel Library.

Tong, H.H.Y., Wong, S.Y.S., Law, M.W.L., Chu, K.K.W., Chow, A.H.L. 2008. Anti-hygroscopic effect of dextrans in herbal formulations. *International Journal of Pharmaceutics* 363: 99–105.

Vadas, E.B. 2001. Stability of pharmaceutical products. In *Remington: The Science and Practice of Pharmacy*, Vol. 1, 20th edn, 986–994. Baltimore, MD: Lippincott, Williams & Wilkins.

Valvani, S.C. 2000. Industrial stability testing in the United States and computerization of stability data. In *Drug Stability: Principles and Practices*, 3rd edn, ed. J.T. Carstensen and C.T. Rhodes, 515–552. New York: Marcel Dekker.

Varshosaz, J., Tavakoli, N., Eram, S.A. 2006. Use of natural gums and cellulose derivatives in production of sustained release metoprolol tablets. *Drug Delivery* 13: 113–119.

Vendruscolo, C.W., Andreazza, I.F., Gantor, J.L.M.S., Ferrero, C., Bresolin, T.M.B. 2005. Xanthan and galactomannan (from *M. scabrella*) matrix tablets for oral controlled delivery of theophylline. *International Journal of Pharmaceutics* 296: 1–11.

Vervoort, L., Kinget, R. 1996. *In vitro* degradation by colonic bacteria of inulinHP incorporated in Eudragit RS films. *International Journal of Pharmaceutics* 129: 185–190.

Von Eggelkraut-Gottanka, S.G., Abu Abed, S., Müller, W., Schmidt, P.C. 2002. Roller compaction and tabletting of St. John's wort plant dry extract using a gap width and force controlled roller compactor. I. Granulation and tabletting of eight different extract batches. *Pharmaceutical Development and Technology* 7: 433–445.

Wade, A., Weller, P.J. 1994a. *Handbook of Pharmaceutical Excipients*, 2nd edn. Washington, DC/London: American Pharmaceutical Association/The Pharmaceutical Press.

Wade, A., Weller, P.J. 1994b. Mannitol. In *Handbook of Pharmaceutical Excipients*, 2nd edn, 294–298. Washington, DC/London: American Pharmaceutical Association/The Pharmaceutical Press.

Washington, N., Washington, C., Wilson, C.G. 1989. The stomach. In *Physiological Pharmaceutics*, 2nd edn, 94. London: Taylor & Francis.

Xiong, W., Quan, D., Patel, D.C. 2001. Effervescent green tea extract formulation. US Patent No. 6,299,925 B1.

Yang, L., Chu, J.S., Fix, J.A. 2002. Colon-specific drug delivery: New approaches and *in vitro/in vivo* evaluation. *International Journal of Pharmaceutics* 235: 1–15.

6

Food and Beverages Fortified with Phytonutrients

Let food be thy medicine and medicine thy food!

Hippocrates of Kos
(460 BC)

6.1 Introduction

The lifestyles of people worldwide have changed in the last century due to a rise in income, increased leisure time, and reduced physical activity. These new lifestyles have a considerable impact on health. Consequently, there is a global rise in the incidence of diseases such as obesity, cardiovascular diseases, diabetes mellitus, and rheumatoid arthritis. The medical world is therefore looking for better strategies to contain this trend. As a parallel development, there is a worldwide increase in health awareness and interest in herbal alternatives. Nowadays, elderly people worldwide are more concerned about the quality of the food and beverages that they consume. Many of them periodically monitor their biomarkers, such as low-density lipoprotein (LDL), blood glucose, C-reactive protein, and so on, in an effort to reduce the risk of diseases. Functional foods have therefore emerged as an effective means for the prevention of diseases. This change in consumer outlook has encouraged the food and beverage industry to apply modern manufacturing technology in food fortification (Venugopal, 2009a). A reduction in weight, a reduction of cholesterol, the promotion of bone health, an increase in energy, the enhancement of disease resistance through the immune system, and the improvement of digestive functions are the major health concerns that influence the purchase of functional foods (Arvanitoyannis and van Houwelingen-Koukaliaroglou, 2005).

Up to 1970, food fortification was mostly focused on the identification of vitamin and mineral deficiencies in human health. Between 1970 and 1990, nutritionists became aware of the need to consume nutrients to prevent signs of deficiency. An example is the intake of antioxidants such as vitamin C to lower the risk of cancers. Since 1990, attention has been paid to substances other than vitamins, minerals, or trace elements that can exert beneficial effects on the body. These substances are collectively called *nutraceuticals* (Venugopal, 2009a). An avalanche of scientific reports on natural products and nutraceuticals have in fact done away with the line of demarcation between food and medicine, taking us back to the advice of Hippocrates that food can be the medicine for our ailments. Nutraceuticals generally refer to the chemical constituents of natural products. Functional foods are foods or food ingredients that provide a health benefit beyond their nutritive value.

The concept of functional foods was born in Japan. As a result of the systematic and large-scale research on chemical compounds derived from natural products, the concept of foods for specific health use (FOSHU) was first developed in 1991 (Kubomara, 1998). These foods are intended to improve the lives of people, and the manufacturer is permitted to display the foods' specific health claims on the container. Functional foods are also known as *nutraceuticals*, *designer foods*, *farmafoods*, *pharmafoods*, *medifoods*, *vitafoods*, *dietary supplements*, *fortified foods*, or *foodaceuticals* (Roberfroid, 2008). These are "foods marketed with the message of a benefit to health" (Riemersma, 1996) or "foods or isolated food ingredients that deliver specific non-nutritive physiological effects that may enhance health" (Stephen, 1998).

Fortification is achieved by adding a nutrient (fortificant or additive) to the food in question, which serves as a vehicle for carrying this nutrient. Early examples of food fortification involved iodine, vitamins, and minerals. The fortification of foods is now achieved using numerous beneficial compounds derived from natural products. Examples of fortified foods are wheat flour, corn flour, rice, salt, sugar, biscuits, curry powder, fish sauce, and soy sauce (Venugopal, 2009b).

6.2 Demonstration of functional effects

The functional nature of fortified foods needs to be proved on the basis of objective parameters. This can be achieved by studying the ability of the functional food to modulate target functions in the body, as these target functions are directly related to an improved state of health or a reduced risk of developing a disease. Well-defined biochemical, physiological, or behavioral markers are used in the assessment of the modulatory effect. Markers may represent an event of interest or correlated events (Nowicka and Naruszewicz, 2004). For example, anthropometry, body fat mass, total body water, procollagen propeptide excretion, and urinary creatinine excretion can be the possible markers for assessing the effects on growth and body composition.

Similarly, tests of behavior, cognitive function, and visual acuity can be the markers for psychomotor and cognitive development (Diplock et al., 1999). Components of the fortified food are used as markers for exposure to the food. Functional claims for fortified foods can be made after assessing the improvement of biological processes, the reduction in the risk of pathological processes, and the safety of the product. Human nutrition studies for all members of a population or for a particular group (age, sex, etc.) are also essential in this regard (Nowicka and Naruszewicz, 2004).

Several problems need to be addressed to achieve the successful fortification of foods. Physicochemical properties such as pH, water or oil content, proteins, and fibers can influence the stability of the nutrient added. Fortification can alter the sensory characteristics of the food and some fortificants can change the color and flavor of the product. At times, the ingredients of fortified food can interact with each other. The processes used in the manufacture of fortified food should be designed in such a way that they do not cause denaturation or loss of nutrients. The packaging of the fortified food can affect the stability of the nutrients. The color of the product may change, haziness or sedimentation can develop in liquid preparations, or ingredients such as vitamin C or β-carotene may undergo oxidation (Venugopal, 2009b).

Fortified food can be made more palatable to the consumer by using sweeteners such as aspartame or saccharin to overcome the intensity of off-flavors. Similarly, food acids such as DL-malic acid or citric acid can be used to reduce bitterness and to counter off-flavors. The taste of the food can be further improved by protecting sulfur groups of amino acids such as cysteine and methionine or by employing bitterness inhibitors. The release of the nutrient into the food matrix can be slowed by coating it with an inert layer. Bitter substances can be encapsulated in wax or fat (Venugopal, 2009b).

6.3 Interesting classes of foods that can be fortified

Advances in food technology have made it possible to manufacture many ready-to-eat foods. Even in a country like India, where food habits have been guided by tradition, several ready-to-serve foods have appeared in recent years (Sethi, 2008). Many of them can be fortified with phytonutrients. The following account can serve as a guide to innovative food formulators.

6.3.1 Beverages

Unfermented and fermented beverages have been in use in different cultures since ancient times. Archaeological oncology research has demonstrated that the remnants of alcoholic beverages from ancient Egypt and China contained many plant-derived compounds with lung and colon cancer-fighting activity. Residue from an ancient Egyptian wine was obtained from an urn dating to 3150 BC found in the tomb of Pharaoh Scorpion I of Dynasty 0. Using sophisticated analytical tools, it is now established that the substance was grape wine, with such other ingredients as savory (*Satureja* spp.), wormwood (*Artemisia annua*), tansy (*Tanacetum* spp.), balm (*Melissa* spp.), and so on. It is inferred that this medicinal wine was perhaps intended to be dispensed as a drug. The Chinese beverage tested was a rice-based drink that was present in a jug excavated from a 1050 BC Changzikou tomb in Hunan province. Based on the chemical compounds identified in the sample, it has been established that the beverage also contained Chinese fir (*Cunnighamia lanceolata*), chrysanthemum (*Dendranthema* spp.), and wormwood (*Artemisia* spp.) (McGovern et al., 2010; Anonymous 2011a).

With the progress in food technology, there is an ever-expanding set of choices, tastes, and drinks with diverse nutritional qualities. The category of soft drinks, which made its appearance with John Pemberton's Coca Cola in 1886, has been an immensely successful segment in the United States and elsewhere (Grivetti and Wilson, 2004). Protagonists of the soft-drink industry consistently portray their products as being positively healthful (Ladas et al., 2013; Tóthova et al., 2013). Nevertheless, these products are now being blamed for causing diseases, such as obesity, osteoporosis, dental caries, dental erosion, and heart disease, among their users (Jacobson, 2004). Thus, there is an obvious need for formulating healthy beverages.

Beverages are easily consumed along with meals and so they are good vehicles for the delivery of nutritional supplements into the body (Pszczola, 1998). The first fortified beverage to appear in the world market was Aqua Libra, launched in England in the 1980s. According to the product's label, it contains the juices of fruits such as apple, grape, passion fruit, and melon, carbonated spring water, malic acid, flavors, and extracts of sesame seed, sunflower seed, and tarragon (Anonymous, 2012). Over the years, many such products have appeared in the market, addressing health problems such as aging, stress, and weakness. Otsuka Pharmaceuticals produced Fibe Mini, which contains dietary fibers, minerals, and vitamins, and is perhaps Japan's best-selling soft drink. Recently, the beverage giant Coca Cola introduced another fiber-rich drink named Fibi. The Japanese company Otsuka introduced Pocari Sweet Stevia containing stevioside from *Stevia rebaudiana* as the sweetener (Goldberg, 1999). The world's first stevia-sweetened Coca Cola was launched in Argentina in 2013 (Bouckley, 2013).

Recently, several fruit juice–based products have been developed. Gonzalez-Molina et al. (2009) developed new polyphenol-rich beverages by blending lemon juice and pomegranate juice in different proportions. They estimated the content of bioactives (flavonoids and vitamin C), antioxidant activity, and color stability over a 70-day period of storage. The results suggest that the blend containing 75% pomegranate juice and 25% lemon juice exhibited high antioxidant activity and improved color properties. The antioxidant activity was due to punicalgin isomers, anthocyanins, and vitamin C.

Schubert et al. (1999) studied the antioxidant activity, total phenolic content, and eicosanoid enzyme inhibition properties of fermented pomegranate juice.

The fermented pomegranate juice exhibited strong antioxidant activity very similar to that of butylated hydroxyanisole (BHA) and higher than that of red wine.

The pomegranate, the fruit of the *Punica granatum* tree, is native to the Himalayan region of India. It is now widely cultivated in Mediterranean countries, India, China, Japan, Russia, the United States, and Afghanistan. More than 100 bioactive compounds have been isolated from the pomegranate fruit. The majority of them are polyphenols such as flavonoids (anthocyanins, cyanidin, delphinidin, pelargonidin), flavonols (luteolin, quercetin, kaempferol), and hydrolysable tannins (ellagitannins, punicalgins, gallotannins) (Toniolo, 2012). In an attempt to determine the suitability of pomegranate juice as a nondairy probiotic drink, Mousavi et al. (2011) fermented pomegranate juice using *Lactobacillus plantarum*, *L. delbruekii*, *L. paracasei*, and *L. acidophilus* at 30°C for 72 h under microaerophilic conditions. The results showed better growth for *L. plantarum* and *L. delbruekii* during fermentation, indicating that pomegranate juice could be a suitable medium for the production of a fermented probiotic drink.

The segment of fortified beverages includes four subsegments: sports drinks, enriched beverages, nutraceutical drinks, and energy drinks (Whitehead, 2005). While formulating these products, the food technologist takes into consideration many factors such as the compatibility of the ingredients, the pH of the medium, and the nature of flavorings used and their stability during production and on storage (Mathews, 1999).

The major components in a beverage are sugars, characterizing ingredients such as glucose syrup, acids, flavors, colorings, emulsifiers, stabilizers, antioxidants, preservatives, acidity regulators, and water. Sugars provide sweetness, mouthfeel, and fruitiness. Acids, flavors, colorings, emulsifiers, and stabilizers impart flavor, body, and appearance. Antioxidants and preservatives enhance the stability of the product. Acidity regulators reduce corrosion of the cans and water adds bulk and mass to the beverage. It also helps in thirst quenching (Mathews, 1999). Beverages can be fortified by the addition of herbal extracts. However, this is beset with the following problems.

6.3.1.1 Stability

Some extracts, especially liquid extracts, have a tendency to sediment and it may be necessary to shake the container before use. In such cases, the label should clearly communicate this information to the consumer. Some extracts in the beverage may change color over a period of time. These points need to be taken care of during product development (Whitehead, 2005). Tipvarakarnkoon et al. (2010) have reported that modified Acacia gum (Super Gum) can be recommended as an emulsifier and stabilizer for use in low-viscous emulsions such as coconut milk drinks and other beverages.

6.3.1.2 Haziness

Some liquid extracts dissolve well in the beverage. However, after some time, haziness (cloudiness) appears in the beverage. This is due to differences in the pH of the extract and the beverage, with soft drinks usually having a pH of 3. In such cases, the liquid extract should be preconditioned to the pH of the final product. This will allow some components in the extract to precipitate. They can be removed by filtration before the final product is bottled (Whitehead, 2005).

Haziness can also occur due to the incompatibility of the polarity of the extract and the beverage. Resins and other nonpolar components that are extracted by alcohol remain clear in the liquid extract stage. Nevertheless, on coming into contact with water, they can come out of the solution, remain suspended in the medium, and impart haziness to the beverage (Whitehead, 2005).

Israeli scientists recently developed a way to solve the problem of haziness while enriching transparent beverages with nutraceuticals. They formed Maillard reaction–based conjugates of casein and maltodextrin and coassembled them with hydrophobic nutraceuticals, vitamin D, and epigallocatechin gallate (EGCG). These conjugates conferred better protection on vitamin D and EGCG than the unconjugated casein maltodextrin mixture. The solution was more transparent than the vitamin dispersion. There was greater colloidal stability and better protection against degradation at low pH and during shelf life. Entrapment efficiency measured using Nile red, a fluorescent model for a hydrophobic nutraceutical, was found to be 90%. When subjected to simulated gastric digestion, Nile red was not released from the conjugates, suggesting a potential application in enteric delivery. The conjugates are suitable for sports drinks, iced tea, mineral water, soft drinks, and even beer and hot drinks (Markman and Livney, 2012).

6.3.1.3 Herbal extracts in beverages

Herbal extracts, especially the infusion type, can be incorporated into soft drinks, mineral water–based drinks, and energy drinks. On account of their mind-calming and soporific properties, *Panax ginseng*, *Echinacea purpurea*, *Chamomilla recuttiana*, *Passiflora incarnate*, and *Valeriana officinalis* are the herbs used in European mineral water–based drinks. On the other hand, herbs commonly used in energy drinks are guarana (*Paullinia cupana*), kola nut (*Cola acuminata, Cola itida, Cola vera*), coffee (*Coffea arabica, Coffea canephora*), tea (*Camellia sinensis*), mate (*Ilex paraguariensis*), and cocoa beans (*Theobroma cacao*) (Whitehead, 2005). The formulas for two nutraceutical beverages are given in **Tables 6.1** and **6.2**.

In 2001, a Swedish beverage manufacturer launched Nexcite, claimed to be an all-natural blend of proven herbal aphrodisiacs. The product contains

Table 6.1 Beverage for Cardiovascular Health

Ingredients	Composition (%)
Water extract of *Terminalia arjuna* bark	1.50
Sugar	96.75
Citric acid	0.75
Malic acid	0.65
Trisodium citrate	0.30
Strawberry flavor	0.05
Preparation: Dissolve 20 g of the mixture in 200 mL of cold water and serve.	

Table 6.2 Beverage for Liver Health

Ingredients	Composition (%)
Water extract of *Wedelia calendulacea* herb	1.5
Glucose	75.0
Sucrose	12.0
Citric acid	7.0
Trisodium citrate	1.5
Calcium lactate	1.0
Orange flavor	2.0
Preparation: Dissolve 25 g of the mixture in 200 mL of cold water and serve.	

extracts of damiana (*Turnera diffusa*), guarana (*P. cupana*), mate (*I. paraguariensis*), schizandra (*Schisandra chinensis*), and ginseng (*P. ginseng*). This cobalt blue beverage is claimed to be a powerful aphrodisiac (Anonymous, 2011b).

The fruit of *Euterpe oleracea*, known as the açai berry, is a nutritious food that is consumed by the indigenous people of the Amazon. A functional juice beverage named MonaVie Active, having a good safety profile, was recently launched in the United States (Schauss et al., 2010). It has açai berry as its predominant ingredient, along with lesser amounts of 18 fruits and berries in descending order of dominance (Schauss et al., 2006). MonaVie Active has significant antioxidant and anti-inflammatory activities (Jensen et al., 2008; Schauss et al., 2009). An open-label, clinical pilot study involving 14 subjects demonstrated that consumption of this beverage resulted in significant pain reduction (Jensen et al., 2011).

Dartsch et al. (2010) investigated the health benefits of some functional drinks designed to improve the body's performance and health. Using cell-based assays, they investigated the potential of the products eQ Brain, eQ Beauty, and Let's Get Red. These drinks were prepared from ingredients such as natural green tea flavor, acerola flavor, dragon fruit flavor, rooibos extract, cinnamon extract, grape-seed extract (GSE), and so on. The drinks caused a dose-dependent stimulation of the metabolic activity of cultured connective tissue fibroblasts. They were also able to inactivate exogenous and endogenous free oxygen radicals.

Attempts have been made to develop nonalcoholic beverages using combinations of maize powder, mango fruit powder, and soy flour. A beverage fortified with mango and maize had low titratable acidity and ranked highest in sensory evaluation (Ade-Omowaye et al., 2006). The study also demonstrated the feasibility of developing good-quality beverages from sour water that is generally discarded as waste from the manufacture of the popular Nigerian fermented food product *ogi* (Akingbala et al., 1994).

Evans et al. (2014b) conducted a pilot study on 20 obese adults to investigate the effects of freeze-dried mango supplementation. The subjects consumed 10 g of the mango pulp daily for 12 weeks. Anthropometric measurements, body composition, and biochemical parameters were assessed at baseline and at the end of the study. Blood glucose was significantly reduced in both male and female subjects after 12 weeks of mango supplementation. Additionally, hip circumference was reduced significantly in males but not females. This is the first report on the effects of mango supplementation in obese persons, suggesting that the fortification of beverages with mango pulp can promote wellness.

Pomegranate seeds are by-products from the pomegranate juice industry. The seed oil has nutritional and medicinal properties. Mohagheghi et al. (2011) formulated a stable pomegranate seed oil-in-water emulsion that could be used as the base formula for a new functional beverage.

Dahl et al. (2005) conducted a double-blind, 3-week study testing an inulin-fortified beverage against a starch-thickened one. The test beverage was well accepted by the subjects and it increased their stool output. Temelli et al. (2004) developed an orange-flavored beverage containing β-glucan extracted from barley. Trained panelists observed a fruity orange aroma and a sufficient intensity of sweetness. Its shelf stability was also satisfactory (Temelli et al., 2004).

Nemzer et al. (2011) examined the effect of a single dose of a polyphenol-rich beverage on serum anti-inflammatory and antioxidant markers. Subjects were given a single 3 oz. dose of the beverage following the baseline blood draw. The beverage consisted of two major components—a proprietary, powdered blend of selected freeze-dried whole fruit and vegetable powders and a preservative-free liquid delivery system to dissolve the powder mixture. The liquid consisted of a blend of grape, pomegranate, pear, apple, strawberry, açai, yumberry, cucpuacu, and camu camu juices, as well as a standardized extract of *Withania somnifera*, natural flavors, and colors. One hour after the intake of the beverage, the serum values for 8-iso-PGF2-alpha and advanced oxidation protein products decreased significantly by 40% and 30%, respectively. C-reactive protein decreased and nitric oxide levels increased. There were also statistically significant increases in hydroxyl radical antioxidant capacity (HORAC) values. This study shows that a mixture of polyphenol-rich fruits and vegetables may work acutely on specific oxidative and inflammatory markers in the human blood through rapidly acting and currently unidentified mechanisms (Nemzer et al., 2011). Mullen et al. (2011) conducted a pilot study on the effect of short-term consumption of a polyphenol-rich drink on biomarkers of coronary artery disease (CAD). They observed that the drink may have beneficial effects on urinary markers of CAD. Polyphenolic beverages are also reported to reduce initial bacterial adherence to enamel, thus contributing to the prevention of biofilm-induced diseases in the oral cavity (Hannig et al., 2009).

Recent research shows that whole grape extract may improve antioxidant and cholesterol levels. Evans et al. (2014a) recruited 24 prehypertensive, overweight, and/or prediabetic subjects in their randomized, double-blind, placebo-controlled, pilot study. The subjects were randomly assigned to receive either a placebo or 350 mg of whole grape extract per day, providing 60%–70% proanthocyanidins (catechin, epicatechin, and epicatechin-3-O-gallate). At the conclusion of the study, the participants in the whole grape extract group had significantly lower levels of superoxide dismutase, lower ratios of total:high-density lipoprotein (HDL) cholesterol, and higher levels of HDL cholesterol. The authors also observed in the whole grape extract group, a lower level of 8-isoprostane, a biomarker of coronary heart disease (Vassalle and Andreassi, 2004).

The enrichment of fruit juices with polyphenolic extracts is a strategy to compensate for the possible loss of phenolic compounds through gastrointestinal processing. Therefore, Frontela et al. (2010) studied the stability of Pycnogenol, a standardized extract of the bark of the French maritime pine (*Pinus pinaster* Ait.), as a potential additive to fruit juices by comparing the phenolic contents of commercial and Pycnogenol-enriched juices following *in vitro* gastrointestinal digestion. The addition of Pycnogenol to fruit juices before and after processing resulted in higher levels of phenolic compounds following *in vitro* gastrointestinal digestion, when compared with nonenriched juices. Pycnogenol fortification produced an increase in Folin-reactive substances and an increase in specific monomeric phenolic compounds such as gallic acid, chlorogenic acid, caffeic acid, ferulic acid, and taxifolin. Therefore, Pycnogenol is suggested as a good fortificant of fruit juices.

Seeram et al. (2007) observed that pomegranate juice inhibited the growth of prostate cancer cells. They reported that urolithins (metabolites of ellagic acid) localize in the prostate, inhibiting the growth hormone-sensitive and hormone-refractory cancer cell. Koyama et al. (2010) have recently shown that pomegranate juice induces apoptosis in prostate cancer cells by the inhibition of insulin-like growth factor (IGF). These reports suggest that the consumption of pomegranate juice may retard the growth of prostate cancer. The results of some clinical trials suggest that pomegranate juice is effective in delaying the progression of prostate cancer (Pantuck et al., 2006; Carducci et al., 2011; Toniolo, 2012).

A recent clinical study suggests that the consumption of pomegranate juice can help to control blood glucose levels. Eighty-five subjects with type 2 diabetes were enrolled in the study and their fasting blood glucose levels were determined. Thereafter, they were administered 1.5 mL of pomegranate juice per kilogram of body weight. An analysis of blood samples collected 1 and 3 h after the juice administration revealed that 3 h after consuming pomegranate juice, the blood glucose had decreased, β-cell function had increased, and insulin resistance had decreased. The hypoglycemic response was dependent on the initial fasting serum glucose, as subjects with lower fasting sugar levels experienced a greater hypoglycemic response, in comparison with those who had higher fasting sugar levels. The hypoglycemic effect of pomegranate juice was unaffected by the sex of the subjects and was less forceful in elderly patients (Banihani et al., 2014).

Morand et al. (2010) investigated the effect of orange juice and its major flavonoid constituent hesperidin on microvascular reactivity, blood pressure, and cardiovascular risk biomarkers through postprandial and chronic intervention studies. It was observed that in healthy, middle-aged, overweight men, orange juice decreases the diastolic blood pressure when consumed regularly. When administered postprandially, it increases endothelium-dependent microvascular reactivity. This study lends support to the view that the consumption of citrus fruits is associated with a lower risk of acute coronary events and stroke (Johnsen et al., 2003; Dauchet et al., 2004).

Chudnovskiy et al. (2014) allowed C57Bl/6 mice to drink sweetened and clarified grapefruit juice *ad libitum*. The clarified grapefruit juice decreased weight gain, hepatic triacylglycerol accumulation, and fasting blood glucose, and improved insulin sensitivity in mice fed a high-fat diet. In mice that consumed a low-fat diet, the consumption of clarified grapefruit juice caused a twofold decrease in fasting insulin. The bioactive compound naringin, which is present in grapefruit juice, lowered blood glucose and improved insulin tolerance, but it did not reduce weight gain. This suggests that grapefruit juice contains more than one nutraceutical.

Tomato juice is also now known to be medicinal. Harms-Ringdahl et al. (2012) reported that the level of 8-oxodG in the blood of 15 healthy subjects was increased significantly after 20 min of acute physical activity, caused by an increase in intracellular reactive oxygen species. However, no increase was observed when these individuals had been drinking 150 mL of tomato juice per day over a period of 5 weeks. This intervention study suggests that antioxidants such as lycopene, which is present in tomato juice, may offer protection from oxidative stress induced by intense physical exercise.

A recent study has shown that a tomato-rich diet could help to protect at-risk postmenopausal women from breast cancer. For 10 weeks, 70 postmenopausal women ate tomato products containing at least 25 mg of lycopene daily. At the end of the study, they were found to have higher levels of adiponectin, a hormone involved in the regulation of blood glucose and fat (Llanos et al., 2014).

Recent studies have highlighted the functional nature of watermelon juice. L-Citrulline from watermelon is efficiently converted into L-arginine, which is the substrate for endothelial nitric oxide production concerned with the regulation of vascular tone (Collins et al., 2007). Tarazona-Díaz et al. (2013) reported that the administration of watermelon juice helped to reduce the recovery heart rate and muscle soreness in seven athletes after 24 h of exercise in a cycloergometer.

Carillon et al. (2014) conducted a randomized, double-blind, placebo-controlled clinical trial to investigate the effects of superoxide dismutase-melon concentrate, containing superoxide dismutase, catalase, glutathione peroxidase, coenzyme Q10, lipoic acid, vitamins, and so on. Sixty-one volunteers were advised to take one capsule per day for 12 weeks. Four psychometric scales were employed to evaluate stress and fatigue. At the end of the trial, it was observed that the melon-concentrate supplementation significantly decreased perceived stress in comparison with a placebo.

A low ankle-brachial index (ABI), which is the ratio of ankle to brachial systolic blood pressure (SBP), indicates the presence of lower limb atherosclerosis and increased pressure wave reflection (augmentation index, AIx) from the peripheral to the central arteries (Resnick et al., 2004; Khaleghi and Kullo, 2007). High ankle SBP is a marker of subclinical atherosclerosis and predicts cardiovascular mortality in adults with a normal ABI and a high brachial blood pressure (Hietanen et al., 2008). L-Citrulline and L-arginine from synthetic or watermelon sources are known to decrease brachial blood pressure and aortic blood pressure in adults with high blood pressure through improved endothelial function and AIx (Figueroa et al., 2011; Orea-Tejeda et al., 2010; Ochiai et al., 2010). Figueroa et al. (2012) assigned 14 obese, middle-aged adults with prehypertension or Stage 1 hypertension and a normal ABI to 6 weeks of watermelon extract supplementation or a placebo, followed by a 2-week washout period and then crossover. Watermelon supplementation reduced the ankle blood pressure, brachial blood pressure, and carotid wave reflection in the subjects. The study shows that watermelon extract improves arterial function independent of the reduction in peripheral blood pressure.

The beneficial effects of beetroot juice supplementation have been evaluated during walking, running, and swimming. Lansley et al. (2011) investigated changes in blood pressure, mitochondrial oxidative capacity (Q_{max}), and physiological responses to walking and moderate- and severe-intensity running following supplementation with beetroot juice in nine healthy subjects. The subjects consumed 500 mL of beetroot juice per day for 6 days. They completed treadmill exercise tests on Days 4 and 5 and knee-extension exercise tests for estimating Q_{max} on Day 6 of supplementation. In comparison with a placebo, the beetroot juice elevated plasma NO_2^- concentration and reduced SBP. The O_2 cost of walking, moderate-intensity running, and severe-intensity running was reduced by the supplementation, and time to exhaustion during severe-intensity running was increased by 15%. These positive effects are attributed to the high NO_3^- content of beetroot juice.

Pinna et al. (2014) studied the effects of beetroot supplementation on the performance of 14 moderately trained male swimmers. After 6 days of supplementation with 500 mL of organic beetroot juice per day, the workload at anaerobic threshold was significantly increased by the supplementation. Additionally, aerobic energy cost was reduced. These effects positively affected the swimming performance.

6.3.1.4 Tea

After water, tea is the most widely consumed beverage in the world today. Tea as a beverage is said to have originated in China around 2700 BC. The word *tea* comes from a Chinese ideogram pronounced *tay* in Amoy dialect. The word entered the English language and the pronunciation changed to its present form in the eighteenth century. In Cantonese dialect, it is known as *chah*, and the word was accepted in India and the USSR. Tender leaves of the tea plant *C. sinensis* are plucked and transported to the factory as soon as possible. There, they are subjected to withering and rolling. Thereafter, the leaves are fermented at high humidity and low temperature. Flavanols present in the fresh leaves turn to different types of polymeric compounds during fermentation. The fermented leaves are dried, sorted, and packed. There are three general types of manufactured tea: green (unfermented), oolong (partially fermented), and black (fully fermented) (Anonymous, 1992; Harbowy and Balentine, 1997; Weinberg and Bealer, 2001). Tea contains polyphenols (catechins, gallocatechins, flavonols, simple polyphenols, tannins), caffeine, methyl xanthine, proteins, amino acids, carbohydrates, pectin, fiber, organic acids, vitamin C, lipids, chlorophylls, carotenoids, and minerals (Harbowy and Balentine, 1997).

Unlike oolong and black tea, green tea is subjected to minimal oxidation. In the processing of green tea, the tea leaves are either steamed or pan-fried. These heat treatments inactivate the enzymes in tea leaves. Following heat treatment, tea leaves are subjected to subsequent rolling and drying processes. The composition of green tea is very similar to that of the fresh leaf except for a few enzymatically catalyzed changes, which occur extremely rapidly following plucking. Green tea contains unoxidized green leaf polyphenols (Graham, 1992). Epicatechin (EC), epigallocatechin (EGC), epicatechin gallate (ECG), and EGCG are dominant in green tea (Zhu et al., 2000). Green tea is considered to be medicinal in nature.

Kriangkraiphiphat and Cheowtirakul (2000) attempted to find out the optimum extraction conditions for the production of a green tea beverage. They reported that the best result was obtained with the extraction of green tea at 50°C in water containing ascorbic acid and sodium thiosulfate. On the basis of sensory evaluation, the authors observed that a beverage with 1% Chinese green tea leaf, 1% fresh ginger rhizome, and 8% sweetener (honey and cane sugar in equal parts) was the best accepted formulation.

Peters et al. (2010) tried to arrive at the best formulation of green tea extract by designing several formulations, each containing 50 mg of green tea extract per serving. They reported that, in rats, the formulation with sucrose (1.25 g) and ascorbic acid (10 mg) increased the absorption of EGC and EGCG, the polyphenol species that are responsible for the health benefits of green tea.

Continued heating at higher temperatures causes a reduction in the catechin content of the final product, which was demonstrated by Kim et al. (2006). Using high-performance liquid chromatography (HPLC) and gas chromatography–mass spectrometry, they observed that at higher temperatures of extraction (85–120°C), EGCG, EGC, EC, and ECG epimerize partially, decreasing the concentration of catechins. Twenty volatile compounds were detected and some of the unpleasant ones such as indole (animal like) and α-terpeneol (ammoniacal) were found to be responsible for the off-flavor of these beverages. Kim et al. (2006) recommend 85°C or less as the ideal temperature for the extraction and pasteurization of a ready-to-drink green tea beverage.

A recent study from Japan suggests that consuming seven cups of green tea a day over a long period can reduce the risk of dying from colorectal cancer (Suzuki et al., 2009). Epidemiological and preclinical data suggest that the consumption of green tea may prevent oral cancer. Tsao et al. (2009) carried out a randomized, placebo-controlled clinical trial in patients with high-risk oral premalignant lesions. Their results indicate that the consumption of green tea extract may suppress oral premalignant lesions, in part through reducing angiogenic stimulus. Green tea is also now known to offer protection to the heart, as evidenced by results from experimentally induced myocardial infarction in rats (Upaganlawar and Balaraman, 2009).

Tea drinking is now known to be associated with a reduction in body fat. A cross-sectional epidemiological study of 1103 Taiwanese adults reported that habitual tea drinkers who consumed green tea or oolong tea for more than 10 years had a lower percentage of body fat and waist:hip ratio compared with nonhabitual consumers (Wu et al., 2003). A large number of intervention studies that have been carried out on humans support the hypothesis that tea and tea phenols have beneficial effects on weight loss and the prevention of obesity (Kao et al., 2006; Thielecke and Boschmann, 2009; Grove and Lambert, 2010).

Thielecke et al. (2010) studied the effect of epigallocatechin-3-gallate (EGCG) on postprandial fat oxidation in overweight/obese men. Subjects were administered supplements such as low EGCG (300 mg), high EGCG (600 mg), caffeine (200 mg), EGCG/caffeine (300 mg/200 mg), or a placebo for 3 days. On the third day of supplementation, oxygen consumption and CO_2 production were measured by indirect calorimetry to assess energy expenditure and fat oxidation. The results showed that EGCG can increase fat oxidation in obese men within 2 h after the intake of a meal. Within this period, EGCG was equipotent with caffeine.

The beneficial effects of green tea extract in obese, hypertensive patients have been reported by Bogdanski et al. (2012) who administered the extract for 3 months. The supplementation had significant effects on cardiovascular risk factors such as insulin resistance, blood pressure, inflammation, and oxidative stress. Senger et al. (2012) have also reported on the anti-obesity effect of green tea. Drinking three cups of green tea daily for 60 days resulted in significant weight loss.

Ide et al. (2014) clinically evaluated the effects of green tea consumption on cognitive dysfunction. Twelve elderly, nursing home resident Japanese subjects participated in the study. The consumption of 2 g of green tea powder for 3 months improved their cognitive function, as judged by the Mini-Mental State Examination.

The performance benefits of tea were identified in several studies, with particularly consistent evidence for improved attention. Tea drinking also consistently improves alertness and arousal (Einöther and Martens, 2013). Black tea is known to improve cognitive functions. De Bruin et al. (2011) conducted a double-blind, randomized, placebo-controlled clinical trial on the effects of tea on attention. Participants made more correct responses on the intersensory subtasks and responded faster after consuming tea. Task switching improved and participants felt significantly more alert. The results of a clinical trial reported by Hodgson et al. (2013) suggest that a component of black tea solids, other than caffeine, can influence the rate of blood pressure variation during the nighttime.

6.3.1.5 Herbal tea

This is a popular product subsegment in the West. Herbal teas are formulated by blending the coarse powders of leaves, roots, bark, and flowers of different herbs. Thus, unlike conventional tea, herbal teas come in numerous varieties and they do not contain tea leaves or caffeine. Herbal tea powder or tea bags are steeped in hot water and the infusion is drunk hot. Herbal teas have been in vogue in Europe since ancient times. Some of the perennial favorites are those made from chamomile, peppermint, and rose hip (Parker, 2001).

The herbs that are used in herbal tea are generally cultivated. The plants are harvested at the appropriate ontogenetic stage, dried under shade or at low temperature, and then milled, sifted, and blended into different flavor combinations (Parker, 2001). The popularity of herbal teas can be gauged from the success of Celestial Seasonings of Boulder, Colorado. The company sells around 54 brands of herbal tea that account for the one billion cups it serves every year (Parker, 2001).

Evidence-based information on the therapeutic effects of medicinal teas is limited. However, some recent reports indicate their preventive and therapeutic value. A common ingredient of herbal tea blends sold in the West is calyx of *Hibiscus sabdariffa* L. Teas made from *H. sabdariffa* lower blood pressure in patients suffering from hypertension and type 2 diabetes (Faraji and Tarkhani, 1999; Mozaffari-Khosravi et al., 2009). The action of these teas is similar to that of common hypotensive drugs (Herrera-Arellano et al., 2004, 2007). From a randomized, double-blind, placebo-controlled clinical study of 65 human subjects, McKay et al. (2010) concluded that daily consumption of *Hibiscus* tea, in an amount readily incorporated into the diet, lowers blood pressure in pre- and mildly hypertensive adults.

Bhat et al. (2009) reported the effect of a tea fortified with *W. somnifera*, *Glycyrrhiza glabra*, *Zingiber officinale*, *Ocimum sanctum*, and *Eletettaria cardamomum* on innate immunity. In a pilot study with 32 volunteers, the consumption of this tea significantly improved the natural killer cell activity when compared with a group who consumed regular tea. This finding was validated in an independent crossover study with 110 volunteers. These results indicate that regular consumption of tea fortified with these ayurvedic herbs improves natural killer cell activity, which is an important aspect of the body's response to infections. The formulation of herbal teas with herbal extracts is, therefore, to be seriously attempted.

6.3.1.6 Coffee

A common beverage all over the world today, coffee has its origins in the highlands of Ethiopia, where it has been cultivated since 900 AD. Its berries were roasted and ground and the decoction was consumed throughout the Arab world. The Ottoman Turks were also avid drinkers of black coffee. After the capitulation of Austria following the battle of Vienna in 1683, Jerzy Franciszek Kulczycki, a Polish soldier of the Polish-Hapsburg army, opened sacks of coffee beans left behind by the fleeing Turkish army and which were mistaken for camel feed. Kulczycki opened the first coffee house in Vienna and introduced the innovation of adding milk and sugar to coffee (Belachew, 2003; Segel, 1993).

Coffee drinking appears to bestow some benefits on its users. A recent clinical trial involving 47 habitual drinkers showed that the consumption of this beverage for 3 months led to a decrease in interleukin-18 and 8-isoprostane and a significant increase in adiponectin and HDL cholesterol (Kempf et al., 2010). The results suggest that coffee may curb subclinical inflammation, which represents one important mechanism in the development of type 2 diabetes (Kolb and Mandrup-Poulsen, 2005; Kempf et al., 2008).

Postoperative ileus is a delayed return of physiologically coordinated gastrointestinal motility following intra-abdominal surgery and is caused by inflammatory reactions among others (Resnick et al., 1997). In a recent clinical trial, it was demonstrated that coffee drinking significantly shortens the time to first bowel movement during the postoperative period after an elective colectomy (Müller et al., 2012).

Caffeine has complex effects on the central nervous system, mediated by antagonists of adenosine A2a and A1 receptors, including a high turnover of several monoamine transmitters such as serotonin, dopamine, and norepinephrine (Fredholm et al., 1999; Ferre, 2008; Ferre et al., 2008). There is a possibility that a deficiency of central monoamines may be improved by caffeine, enhancing dopaminergic neurotransmission (Fredholm et al., 1999; Ferre, 2008; Ferre et al., 2008). A deficiency of central monoamines is a feature of depression (Belmaker and Agam, 2008). These studies suggest that caffeine could be a mild antidepressant and this explains the lower risk of depression among coffee drinkers in epidemiological studies (Ruusunen et al., 2010; Lucas et al., 2011). Interestingly, Lucas et al. (2013) accessed and analyzed the data of 43,599 men and 164,825 women and came to the conclusion that the consumption of coffee lowers the risk of suicide.

Epidemiological studies indicate that caffeine is a protective agent against cognitive impairment and Alzheimer's disease (AD). Less cognitive decline over a 4–10 year period is reported in elderly men drinking three cups of coffee every day (van Gelder et al., 2007) and elderly women drinking more than three cups of coffee every day (Ritchie et al., 2007). The long-term administration of caffeine in drinking water protects AD transgenic mice (experimental model for AD) from memory impairment in old age (Arendash et al., 2006) and reverses already-present memory impairment in aged AD mice (Arendash et al., 2009). Presumably, caffeine brings about these protective changes through its unique ability to suppress β- and γ-secretase enzymes that are needed for the production of the abnormal amyloid-β protein, resulting in very low aggregation and deposition of amyloid-β protein in the brain cells (Arendash et al., 2006, 2009). Cao et al. (2012a, 2012b) recently reported that caffeine/coffee may reduce the risk and/or delay the onset of dementia in individuals having mild cognitive impairment, which is the prelude to AD. In addition to caffeine, coffee contains antioxidants and anti-inflammatory agents that may potentiate the ability of caffeine to reduce the risk of AD (Higdon and Frei, 2006; Halvorsen et al., 2006; Herder et al., 2010; Pham et al., 2010; Bakuradze et al., 2011).

Recently, Borota et al. (2014) reported that caffeine could have a positive effect on long-term memory. Their observations suggest that the compound can enhance memories for up to 24 h at least after it is consumed. The team conducted a double-blind trial on participants who did not eat or drink caffeinated products regularly. The subjects received either a placebo or a 200 mg caffeine tablet 5 min after studying a series of images. The caffeine in subjects' saliva was monitored before and 1, 3, and 24 h after caffeine ingestion. The next day, both groups were tested for their ability to recognize images from the study session of the previous day. Some of the visuals were the same, some were new, and some were similar but not the same. The authors noted that more members of the caffeine group were able to correctly identify the new images as 'similar' to previously viewed images. They concluded that the higher ability to recognize the differences between two similar but not identical objects (pattern separation) in the caffeine group indicates a deeper level of memory retention.

The reactive hyperemia index is an indicator of endothelial function and predicts future coronary events such as atherosclerosis (Schoenenberger et al., 2012). Ochiai et al. (2013) conducted a randomized, acute clinical intervention study with crossover design and measured the reactive hyperemia index to assess the acute effects of a 75-g glucose load with coffee polyphenols in healthy, nondiabetic adult men. In the presence of coffee polyphenols, the reactive hyperemia index increased significantly over baseline after glucose loading. This finding suggests that a single ingestion of coffee polyphenols improves peripheral endothelial function after glucose loading in healthy subjects.

With the advent of the nutraceutical revolution, a functional coffee has also been formulated. Javafit functional coffee is said to contain ingredients such as caffeine, green tea extract, *Garcinia cambogia* extract, and niacin, which are beneficial to health and promote weight loss (Antonio and Sanchez, 2005).

A clinical study of a group of regular coffee drinkers showed that Javafit Energy Extreme coffee is more effective than a conventional coffee at increasing resting energy expenditure for up to 3 h following its consumption, devoid of any adverse hemodynamic effects such as an irregularity of the heartbeat or hypertension (Taylor et al., 2007). Another clinical study revealed that Javafit Energy Extreme coffee significantly increases maximal oxygen consumption (VO_2) at 3 min postexercise. This effect may be beneficial in enhancing fat metabolism after exercise (Roberts et al., 2007). These reports lend a scientific basis for the weight loss property attributed to this product.

According to *Nutrition Business Journal*, the sports nutrition industry is set to grow at a steady rate and there is still room for products supported by scientific research. This is indicated by the entry of Boaters Coffee and Fusion Energy Coffee, two other brands of functional coffee in the US market (Spano and Antonio, 2008).

6.3.1.7 Cocoa

Chocolate beverages and chocolate are now popular all over the world. The first chocolate beverage is believed to have been created by the Mayans of South America over 2000 years ago, and a cocoa beverage was an essential part of the Aztec culture of Mexico. The beverage became popular in Europe after its introduction from Mexico. Chocolate drinking houses were established in London in 1657 and are mentioned in Pepys's diary. The first British factory to manufacture eating chocolate was founded in Bristol by John Fry. He was followed by Cadbury and Rowntree, names that are synonymous with chocolate (Beckett, 2004).

Cocoa beans are known to have the highest content of flavan-3-ols (Arts et al., 1999; Lee et al., 2003), and epidemiological studies show an inverse relationship between the consumption of a flavonoid-rich diet and the risk of hypertension. Impaired endothelial vasodilator function is an important factor that contributes to the development of cardiovascular diseases (Brown and Hu, 2001). It is well known that obesity and hypertension are associated with impaired vasodilatation (Park et al., 2001; van Gaal et al., 2006). A short-term intake of cocoa polyphenols is reported to lower blood pressure and improve endothelium-dependent vasodilatation (Grassi et al., 2005a, 2005b; Taubert et al., 2007; Davison et al., 2008). It is believed that this beneficial effect of cocoa polyphenols is brought about by an increase in the nitric oxide synthase activity in endothelial cells, leading to an increase in nitric oxide concentration (Keen et al., 2005; Balzer et al., 2006; Heiss et al., 2003, 2005; Karim et al., 2000; Berry et al., 2010).

Martin et al. (2010) reported that the daily consumption of 40 g of dark chocolate for 14 days could reduce the urinary excretion of the stress hormone cortisol and catecholamines in 30 stressed human volunteers. The therapy also normalized stress-related differences in energy metabolism and gut microbial activities. Tomas-Barberan et al. (2007) and Cienfuegos-Jovellanos et al. (2009) have shown that flavonoids from cocoa beans can be incorporated into beverages with enhanced bioavailability in humans.

An improved cognitive performance was demonstrated by aged Wistar–Unilever rats treated with a cocoa polyphenolic extract (Bisson et al., 2008). Dietary flavanols are reported to improve endothelial function and lower blood pressure by causing vasodilation in the peripheral vasculature and in the brain (Sorond et al., 2008; Corti et al., 2009). Cocoa flavanols exhibit an antidepressant action (Smith, 2013). Recent reports suggest that cocoa powder causes neuroprotective and preventive effects in AD (Cimini et al., 2013).

The formation of protein aggregates in the form of extracellular amyloid-β (Aβ) plaques and intracellular tau neurofibrillary tangles are the two major pathological characteristics of AD. Oligomers of Aβ peptides and tau protein are toxic (Klein, 2002; Cleary et al., 2005; Santacruz et al., 2005; Berger et al., 2007; Shankar et al., 2008). Therefore, reducing these oligomer species is a potential mode of intervention. Wang et al. (2014) tested three different cocoa extracts on $A\beta_{40}$ and $A\beta_{42}$ oligomerization using the photoinduced cross-linking of unmodified proteins technique. The extracts were found to be effective in preventing the oligomerization of Aβ.

By calculating the chocolate consumption and the total number of Nobel laureates in various European countries, Messerli (2012) has come to the conclusion that there is a powerful correlation between per capita chocolate consumption and the number of Nobel laureates!

Peripheral arterial disease (PAD) is a widespread atherosclerotic condition, marked by a high incidence of coronary and cerebral events. The typical symptom of this disease is intermittent claudication, which causes impaired blood flow to the limbs during physical exercise. Loffredo et al. (2014) carried out an interventional trial on 20 PAD patients to find out the acute effect of 40 g of dark chocolate or white chocolate. Flow-mediated dilation, serum levels of isoprostanes, nitrite/nitrate, sNOX2-dp, maximal walking distance (MWD), and maximal walking time (MWT) were assessed at baseline and 2 h after chocolate ingestion. The consumption of dark chocolate significantly increased the MWD, MWT, and sNOX2-dp. However, milk chocolate did not produce any of these effects.

Botelho et al. (2014) fortified dark chocolate with phytosterols to produce a cholesterol-lowering product capable of a Food and Drug Administration (FDA) health claim on reduced cholesterol. The daily intake of one bar (30 g) provided about 2.2 g of phytosterol esters, which is higher than the amount required by the FDA (1.3 g). Additionally, the chocolate bar did not contain sugar and made use of xylitol, erythritol, maltitol, and sucralose to give sweetness and bulking. It was formulated with 50 g/100 g of cocoa, thereby making it ideal for individuals with dyslipidaemia. The dark chocolate bar also contained ascorbic acid and tocopherol in the filling and maintained its potential functionality after 5 months of storage.

6.3.2 Soups

The word *soup* is derived from the French *soupe*, which means "a broth." Soups are liquid food that is prepared by boiling vegetables and/or meat in water. Soups have formed an important part of European cuisine since ancient times. Soup as a commercial product has been popular ever since Nicolas Appert invented canning technology in 1795 (Goody, 1997). Dr. John T. Dorrance (1873–1930), a chemist working for the Campbell Soup Company, is the inventor of condensed soup, which is cooked with minimum water, forming a kind of thick stock that can be diluted with hot water and consumed (Hallet and Hallet, 1997). In 1873, Karl Heinrich Knorr pioneered the manufacture of soup powders (Brown, 2003). More recently, frozen soups have also become popular (Komarik et al., 1975).

Chinese medicine recommends the drinking of soups to prevent or treat diseases. Chinese medicinal herbs are grouped into four categories, namely, tonifying, eliminating, harmonizing, and fortifying. Many soups are prepared by combining herbs and meat. Medicinal soups are very popular in China (Au et al., 2008). The recipe for one such soup—Six Herb Regulating Soup—is given here.

Cornelian cherry fruit (3/4 oz.), alisma (1/2 oz.), treated rehmannia (3/4 oz.), *Poria cocos* sclerotium (3/4 oz.), white peony root (1/2 oz.), and wild yam (3/4 oz.) are put in a ceramic pot, brought to the boil over a high heat, and left to simmer for 30 min until the mixture is reduced to two cups. The filtered soup, when consumed twice a day, is said to improve physical and mental stamina (Chen, 2009).

Soup powder is a convenient form on account of its longer shelf life and ease of transportation. The common ingredients used are starch, spices such as ginger, garlic, onion, milk powder, salt, monosodium glutamate (MSG), ascorbic acid, sugar, dehydrated vegetables, and/or meat (Sachindra et al., 2007). Although MSG is considered to be a safe flavor additive (Hanas, 1994), it is known to have some shortcomings (Hegenbart, 1998). Recently, Abeysinghe and Illeperuma (2006) reported the development of a vegetable soup powder devoid of MSG, but having appreciable flavor on account of the colorant lycopene that is present in tomatoes.

Soup powders are usually prepared by blending dehydrated vegetables or meat with other ingredients (Chacko et al., 2005; Abeysinghe and Illeperuma, 2006). They can also be prepared by grinding fried vegetables followed by blending with other ingredients and spray drying (Singh et al., 2003).

Soups do have detectable physiological effects, as was shown by Midoh and Noguchi (2009). They reported that a 2-week consumption of chicken soup could reduce the tension-anxiety score and increase the peripheral blood flow in healthy human volunteers. Azhar et al. (2013) conducted a clinical study on the cognition-enhancing benefits of an aqueous extract of chicken. They observed that subjects who were supplemented with the extract showed significantly better performance on all cognitive tests after 6 weeks' supplementation. The antioxidant activity of soups has also been reported (Rekha et al., 2008). Thus, it is evident that soups can be transformed into fortified food by replacing conventional ingredients with those having proven health benefits.

Nagatsuka et al. (2006) reported that gelatin gel *Nikogori* soup, which is prepared from chicken wing meat, has high peroxyl and hydroxyl radical scavenging activities, as evidenced by chemiluminescence and electron spin resonance methods. The addition of soy sauce enhances the hydroxyl radical scavenging activity.

Drinking the Mediterranean soup *gazpacho*, which is prepared from tomato, cucumber, green pepper, onion, and garlic, significantly increases the plasma concentrations of vitamin C (Sanchez-Moreno et al., 2006). By a multiple-adjusted logistic regression analysis of data from 3995 individuals, Medina-Remón et al. (2012) inferred that the consumption of *gazpacho* was associated with a lower prevalence of hypertension. This effect is attributed to a synergy among several bioactive compounds in the vegetable ingredients. Similarly, the consumption of quercetin-rich onion soup inhibited some aspects of collagen-induced platelet aggregation. This observation from a double-blind, randomized, crossover, pilot study substantiates the epidemiological data, suggesting that individuals who preferentially consume large amounts of quercetin-containing foods have a reduced risk of succumbing to thrombosis and cardiovascular diseases (Hubbard et al., 2006). In the same way, *Corchorus olitorius*, a leafy vegetable popular in Asia and Africa and used in soups, has significant antioxidant activity as judged by the 2,2-diphenyl-1-picrylhydrazyl (DPPH) method (Oboh et al., 2009).

A recent study reports that consuming *Chaibu Shibuang* soup of China once a day for seven days significantly reduces tumor necrosis factor (TNF)-alpha and interleukin-6 in patients suffering from severe acute pancreatitis (Wang et al., 2009).

There is a growing trend among food manufacturers to offer natural options with functional ingredients that provide additional health benefits at lower cost. This has generated an array of soups, such as soups with low sodium, with omega-3 fatty acids, with added fiber, without fats, gluten, or trans fats, and so on. Distriborg Groupe, a French manufacturer has introduced two fortified soups (Gayelord Hauser Minceur Veloute Potiron & Noix De Muscade and Gayelord Hauser Minceur Veloute Legumes Verts & Basilic). Both products are positioned as soups for weight management and contain an extract of NeOpuntia, which is patented, dehydrated *Opuntia ficus-indica* leaves (http://www.distriborg.com). A randomized, placebo-controlled, double-blind clinical study showed that NeOpuntia could cause a pronounced reduction in LDL cholesterol (Linarès et al., 2007). There is scope for developing fortified soups and the formula for one such product is provided in **Table 6.3**.

6.3.3 Yogurt

Yogurt is a modern version of curd. It is produced by the bacterial fermentation of milk. Lactic acid, which is formed from the lactose in milk, curdles the milk proteins and the resultant is a soft, white-colored product (Vaclavik and Christian. 2008). The word *yogurt* is derived from the Turkish word *yoğurt*,

Table 6.3 Fortified Soup for Digestive Efficiency

Ingredients	Composition (%)
Water extract of *Zingiber officinale*	1.00
Water extract of *Cuminum cyminum* seeds	0.40
Water extract of *Trachyspermum ammi* seeds	0.20
Wheat powder	18.00
Corn starch	18.00
Milk powder	12.00
Edible vegetable fat	11.00
Onion powder	2.40
Pepper powder	0.40
Sugar	1.00
Table salt	6.00
Beetroot powder	3.60
Tomato powder	26.00

Preparation: Suspend 16 g of the mixture in 200 mL water. Stir well so that no lumps form and the solution is homogeneous. Heat over a slow fire and bring to the boil, stirring continuously. Simmer for 3 min.

meaning "dense" or "thick." Armenian immigrants brought yogurt to America and its commercial manufacture started in 1929.

Sodini and Tang (2006) describe at length the composition and possible ways to modify the appearance, quality, and texture of yogurt. The organic ingredients defined by the FDA (2004) for inclusion in yogurt are organic fruits, organic cane sugar juice, citric acid, vegetable colors, flavors, ascorbic acid, low methoxyl pectin, high methoxyl pectin, sodium citrate, and locust bean gum.

It is known that probiotic bacteria such as bifidobacteria and lactobacilli possess antimicrobial properties. *L. acidophilus* and *Bifidobacterium bifidum* are found to be inhibitory toward many common food-borne pathogens. Many studies indicate the preventive control of intestinal infections through the administration of milk cultured with *L. acidophilus* or *B. bifidum* (Lourens-Hattingh and Viljoen, 2001).

Several studies have reported the successful fortification of yogurt. Dave and Shah (1998) and Bhullar et al. (2002) have reported the preparation of yogurt fortified with whey powder. Attempts have also been made to achieve fortification with calcium, iron, fiber, mango, and soy (Hekmat and McMahon, 1997; Fleury et al. 1998; Singh et al., 2005; Aportela-Palacios et al., 2005; Kumar and Mishra, 2006). However, fortification of yogurt with extracts of medicinal herbs can be attempted by imaginative food technologists (see Table 1.1). Recently, Adegoke et al. (2013) attempted to flavor yogurt with the spice *Aframomum danielli*. At a relatively low concentration, *A. danielli* seeds imparted good sensory characteristics comparable with strawberry and vanilla. The appearance, texture, consistency, and palatability of the product were also good. The characteristic aroma of the product was attributed to flavor components such as 2-methyl-1-propen-1-yl, 3-methylbut-2-enyl, and 1-methyl-4-(1-methylethyl).

Arjmand (2011) attempted to develop a probiotic yogurt by incorporating pomegranate juice and pomegranate juice concentrate. The results showed that up to 20% of pomegranate juice and 6% of pomegranate juice concentrate could be incorporated post-heat treatment, without compromising the probiotic potential of the product, which had a higher color index and more than 37% total phenolic compounds.

Robert et al. (2010) encapsulated pomegranate juice and its ethanolic extracts with maltodexterin and soy protein isolates, using spray drying. The stability of the encapsulated polyphenols and anthocyanins was studied for 56 days. The polyphenols' encapsulating efficiency was better in the soy protein isolates matrix, while maltodextrin was found to be better for encapsulating anthocyanins. The microcapsules were thereafter added to yogurt samples and the stability of the polyphenols was studied for 56 days at 60°C. The rate of degradation of the polyphenols was the same for the encapsulated and nonencapsulated pomegranate juice.

There has been much interest in fortifying yogurt with soya products and phytosterols. The cholesterol-lowering effects of lactic acid fermented soy milk, in which part of the soy milk was replaced by soy yogurt, were studied in rats (Kitawaki et al., 2009). The results indicate that soy yogurt is helpful in preventing hepatic lipid accumulation in rats. Based on a study carried out on rabbits, Cavallini et al. (2009) suggest that soy yogurt could be consumed to reduce the risk of cardiovascular disease as it improves the lipid profile. Their study also shows that isoflavone supplementation enhances the anti-atherogenic effect of soy yogurt. The results of two clinical studies suggest that phytosterol-enriched yogurt has the ability to lower commonly measured lipid parameters (Nittynen et al., 2008; Ruiu et al., 2009).

In addition to their hypercholesterolemic effect, yogurts made with plant-derived lactic acid bacteria have the ability to increase defecation frequency in patients suffering from constipation (Higashikawa et al., 2010). Yogurts prepared from cow or goat's milk are also effective in offering protection against experimental acute liver injury in mice (Haro et al., 2009).

Many studies have been conducted on the physiological effects of yogurt. Kieling et al. (2002) conducted a randomized, crossover, placebo-controlled clinical trial with 29 women to evaluate the hypocholesterolemic property of yogurt supplemented with *L. acidophilus* 145 and *B. longum* 913. The results showed that HDL cholesterol was increased significantly and the ratio of LDL/HDL cholesterol decreased after treatment for 21 weeks. Ataie-Jafari et al. (2009) reported that yogurt containing the probiotic bacterial strains *L. acidophilus* and *Bifidobacterium lactis* lowered the serum cholesterol of hypercholesterolemic human volunteers.

Experimental studies in mice offer evidence for the anti-inflammatory effect of yogurt in inflammatory bowel disease. The effect is believed to be mediated by an increase in the number of immunoglobulin A (IgA)+ cells, a decrease in the CD8+ population, and an enhancement of apoptosis of infiltrative cells in the large intestine (Gobbato et al., 2008).

Recent research indicates that the consumption of yogurt could offer protection from type 2 diabetes. O'Connor et al. (2014) sampled the EPIC-Norfolk study, a cohort of more than 25,000 men and women from Norfolk, England. The authors compared the weekly food diaries of 752 people who developed new onset type 2 diabetes, over 11 years of follow-up, with 3502 randomly selected subjects. An analysis of the data showed that the consumption of four and a half standard pots of yogurt per week could reduce the risk of developing type 2 diabetes by a quarter.

Although many health benefits are attributed to yogurt, the alleged cholesterol-increasing property of milk fat is an impediment to its wider use. Therefore, several studies have been reported on the improvement of the physical, textural, and rheological properties of reduced-fat yogurts by incorporating fat replacers and manipulating the parameters of manufacture. Whey protein concentrate can be used in the manufacture of nonfat yogurt with satisfactory physical and organoleptic properties (Aziznia et al., 2008). Singh and Kim (2009) reported that a starch–lipid composite could be used as a fat replacer and a stabilizer in yogurts.

Inulin is a polysaccharide isolated from the roots of the *Chicorium intybus* and is used as a fat replacer in water-based foods. Because of its rheological properties, inulin in an aqueous medium gives a fatlike mouthfeel and texture (Zimeri and Kokini, 2003). Inulin has been successfully incorporated into yogurts (Aryana et al., 2007; Guven et al., 2005; Paseephol et al., 2008; Mazloomi et al., 2011). It is found to be an efficacious functional ingredient having a prebiotic effect (Marinangeli and Jones, 2010). In a clinical study, Perrigue et al. (2009) observed that yogurt beverages containing inulin could significantly suppress the appetite and promote satiety.

The fortification of yogurt has been attempted with a view to preventing weight gain by controlling the appetite. Lluch et al. (2010) investigated the short-term, appetite-reducing effects of yogurt fortified with protein and guar gum. From two randomized crossover studies, the authors concluded that the consumption of the fortified yogurt significantly reduced short-term appetite.

6.3.4 Sauces

These are liquid or semisolid preparations served on food. Occasionally, sauces serve as ingredients of other food as well. According to the famous chefs Carême and Soyer, sauces are to cookery what grammar is to language and melody is to music. They add flavor and visual appeal to the basic food. Sauces are said to have originated in France during the Middle Ages, as testified by a cookery book by Montardier-Gilde, published in 1394 (Senn, 1915).

The ingredients of a representative sauce are natural thickeners such as xanthan or guar gum, butter, cane sugar, glucose syrup, fruit juice, tomato concentrate, citric acid, and salt. White sauces contain milk (Sikora et al., 2004; Sworn, 2007).

Some sauces are proved to have remarkable biological effects. The best one in this regard is the Japanese-style fermented soy sauce (*shoyu*). An antihypertensive compound having the ability to inhibit angiotensin I-converting enzyme has been detected in soy sauce. The responsible compound is nicotianamine derived from soybeans. Soy sauce exhibits significant anticarcinogenic effects and acts against bacteria such as *Staphylococcus aureus, Shigella flexneri, Vibrio cholerae, Salmonella enteritidis*, nonpathogenic *Escherichia coli*, and pathogenic *E. coli*. Three tartaric isoflavones that are present in fermented soy sauce have the ability to inhibit the enzyme histidine carboxylase, which produces histamine, the mediator of inflammation, allergy, and gastric secretion. Interestingly, although soybean and wheat, the major ingredients of soy sauce, are allergenic, soy sauce does not contain any allergens. This has been confirmed by enzyme-linked immunosorbent assay (Kataoka, 2005). The addition of soy sauce slightly enhances the hydroxyl radical scavenging activity of chicken jelly soup (Nagatsuka et al., 2006).

Oke et al. (2010) evaluated the effects of the external supplementation of soy lecithin on the physicochemical properties of sauces. The levels of protein, soluble solids, and ash content were significantly increased by the addition of lecithin. The bulk viscosity and yellow color of the sauce were also enhanced.

Occasionally, sauces are used along with salads and they are called *salad dressings*. Studies show that these low-fat foods are ideal carriers of n-3 fatty acids (Mantzioris et al., 2000) and plant sterols (Judd et al., 2002). Abumweiss et al. (2008) carried out a meta-analysis of 59 eligible, randomized clinical trials published from 1992 to 2006. They observed that reductions in LDL were greater when the plant sterols were incorporated into foods such as salad dressings. These fortified dressings are well tolerated and do not show any adverse effects even at a daily intake of 9 g/day for 8 weeks (Davidson et al., 2001).

The usefulness of the *Eleutherine americana* crude extract as an antibacterial agent in salad dressings was investigated by Ifesan et al. (2009). The extract significantly reduced the *S. aureus* population, exerted a strong antioxidant action, had good retention of organoleptic properties, and good acceptability.

Mellies et al. (1985) have reported that sucrose polyester can effectively replace dietary fat in salad dressings. It can provide lubricity and organoleptic benefits devoid of the high caloric content that is characteristic of digestible fats.

6.3.5 Mayonnaise

Mayonnaise is a creamy, pale yellow, flavored sauce made of oil, water, vinegar, and egg yolk. This unique emulsion is prepared by slowly adding the oil to the egg yolk, accompanied by vigorous whisking. It is served in sandwiches and salads (Abu-Salem and Abou-Arab, 2008). At the industrial level, mayonnaise is manufactured in stainless steel Dixie mixers and Charlotte colloid mills (Duncan, 2004).

Recently, a few attempts have been made to develop low-fat mayonnaise. Mun et al. (2009) developed an acceptable, low-fat mayonnaise using 5.6 wt% of 4-alphaGTase-treated starch and 0.1 wt% of xanthan gum. β-Glucan prepared from spent brewer's yeast was used by Worrasinchai et al. (2006) as a fat replacer in mayonnaise. Sensory evaluation and stability studies showed that mayonnaise substituted with not more than 50% β-glucan was acceptable to consumers.

Su et al. (2010) developed a low-fat mayonnaise with polysaccharide gums as the functional ingredients. Xanthan gum, citrus fiber, and guar gum were used in the development of the product. Using these polysaccharides, the authors could reduce the fat content in the mayonnaise to 50% while maintaining ideal rheological properties. The innovative product had acceptable organoleptic properties as well.

The cholesterol-lowering property of a fortified mayonnaise was reported by Saito et al. (2006). Various quantities of plant sterol ester were dissolved in 15 g of a diacylglycerol-containing mayonnaise. Every day for 4 weeks, 15 g of the product was administered to human volunteers. At the conclusion of this randomized, placebo-controlled, double-blind, parallel study, total serum cholesterol was found to be decreased significantly by the ingestion of the fortified mayonnaise containing at least 0.4 g/day of the phytosterols.

6.3.6 Pickles

Pickles are strongly flavored adjuvants that are consumed to enhance the flavor and acceptability of the basic food. European pickles contain cooked or uncooked vegetables such as onions, cauliflower, gherkins (cucumber), walnuts, olives, beetroot, red cabbage, carrot, and so on. Uncooked vegetables are usually packed in a clear vinegar-salt-spice liquor, and cooked vegetables are packed in a thick sauce. Many pickles contain acetic acid, salt, natural fruit acids, stabilizers, pectin, and extracts of fruits and vegetables. The commonly used preservatives are sulfur dioxide (up to 100 mg/kg), benzoic acid (250 mg/kg), or sorbic acid (1 g/kg) (Anderson, 1990; Fleming et al., 2001). Pickles have a high nutritive value. It is said that during the construction of the Great Wall of China, a mixture of pickled vegetables such as radish, cabbage, turnip, and beetroot was paid to the workmen as part of their wages (Joshi and Bhat, 2000).

Indian pickles are invariably pungent and spicy. They are made of vegetables and sour fruits such as lemon, lime, unripe mango, ginger, Indian gooseberry, and bitter gourd. There are several ways of making pickles. One method is to cook the major ingredient in a mixture of vegetable oil, salt, chilli powder, turmeric, and spices and bring to an optimum consistency or moisture content. The oiliness of pickles is an ideal property that helps in the delivery of extracts of nutritional value.

6.3.7 Chutneys

Chutneys are spicy preparations that also form an accompaniment for a main dish. Fresh herbs, grated coconut, chillies, turmeric, spices, and salt are ground to a fine or coarse paste. They are usually prepared fresh. Nevertheless, chutneys with a shelf life of 2–3 years can be produced with the addition of preservatives (Marwaha and Marwaha, 2000). Date fruits and flaxseed are also major ingredients of some chutneys and oil-based pickles. Date fruits at all stages of maturity are known to contain numerous beneficial phytochemicals (Sidhu and Al-Hooti, 2005). The flaxseed chutney consumed in southern India is known to have significant hypolipidemic and hepatoprotective properties (Shakir and Madhusudhan, 2007).

6.3.8 Jams, jellies, and marmalades

Jams, jellies, and marmalades are sweet preparations made by concentrating fruits to nearly 70% total suspended solids by adding sugar and heating the mixture. The high content of sugar and the moisture reduce the chances of microbial spoilage. Jams are pectin gels and jelly is jam-like, but transparent. Marmalades are jellies prepared from citrus fruits such as oranges and lemons. However, the shredded peels of the fruits are suspended in them (Bhardwaj, 2000).

In recent years, the fortification of jams and jellies has been attempted. Qasi et al. (2003) claimed in their patent application, the invention of a nutraceutical jam with proven memory enhancement, antidepressant, adaptogenic, and immunomodulatory properties. Similarly, Toves (2004) reported the fortification of jams with soluble dietary fibers.

A novel way to fortify apple fruits was attempted by Lavelli et al. (2010). They fortified apple puree with green tea extract equivalent to that which is present in a cup of green tea and freeze-dried the material. After 1 month of storage, the green tea–fortified product retained almost all its monomeric flavan-3-ols and total procyanidins. Green tea fortification was found to increase the antioxidant potential of the fruit puree three- to sixfold. The authors remark that the novel fortification strategy adopted by them would be advantageous for two reasons. From an economic point of view, it would facilitate the appearance of more dehydrated apple products available in the market. Secondly, on nutritional grounds, the novel product could offer consumers a simple opportunity to regularly consume green tea as a health-enhancing dietary aid (Lavelli et al., 2010).

The consumption of fruit juice powders has obvious health benefits. Jin et al. (2010) administered encapsulated fruit powders to 117 human volunteers for 2 months in a randomized, double-blind, placebo-controlled clinical study. The fruit powders were derived from acerola, cherry, apple, cranberry, orange, peach, papaya, pineapple, bilberry, blackberry, black currant, blueberry, Concord grape, elderberry, raspberry, and red currant. An analysis of blood samples showed that markers of inflammation, such as monocyte chemotactic protein-1, macrophage inflammatory protein 1-β, and regulated on activation, normal T cell expressed and secreted (RANTES) were significantly reduced, and superoxide dismutase and micronutrients (β-carotene, vitamin C, tocopherol) were increased. Fruit juice concentrate powders can be used in food fortification, as they reduce the inflammatory load in healthy humans.

6.3.9 Cheese

Cheese is a protein and fat-rich product made from the milk of cows, buffalo, goats, and sheep. Milk is coagulated with the addition of the enzyme rennet. The curdled portion is collected and pressed into desired forms (Snyder, 2008). The various steps in the production of cheese are described by Fox et al. (2000). Cheese is a good medium for carrying oil-soluble natural colorants and phytochemicals. Carotenoids and bixin are extensively used in this regard.

Salt is an important ingredient in the cheese-making process. It contributes to a desirable flavor and texture profile and to microbial safety. As cheese is rich in salt and fats, it is generally not recommended in heart-healthy diets. However, the traditional Norwegian cheese Gamalost is rich in angiotensin-converting, enzyme-inhibiting peptides. According to a recent study, the consumption of Gamalost may reduce blood pressure. Nilsen et al. (2014) questioned 168 male and female participants from one Norwegian town about their dietary habits and other health-related factors. Based on their responses and measurements of their blood pressure, the authors observed that the intake of Gamalost was inversely associated with SBP and an increase in a frequency unit of Gamalost corresponded to a reduction in SBP of 0.72 mmHg.

Lutein is known to be a dietary factor that can prevent the appearance of age-related macular degeneration (Seddon et al., 1994; Snodderly, 1995; Yao et al., 2013). Considering this aspect, Jones et al. (2005) incorporated various amounts of lutein isolated from corn into cheddar cheese and assessed its stability during storage. A significant quantity of lutein was recovered and no lutein disintegration products like lutein 5,6 epoxide were detected. While there was a difference in color on account of the color of lutein, the pH remained unchanged and products were free from pathogenic microbes. The authors suggest that cheese is a good medium for the delivery of lutein.

Martini et al. (2009) fortified 50% reduced-fat cheddar cheese with the omega 3-fatty acids: docosahexanoic acid and eicosapentanoic acid. Although a fishy smell was perceivable in the products, the off-flavor disappeared after 3 months. This study shows that 50% reduced-fat cheddar cheese aged for 3 months can be fortified with omega 3-fatty acids, without generating off-flavors.

An attempt at developing a functional cheese product containing polyphenolic compounds was reported by Han et al. (2011). Phenolic compounds such as catechin, EGCG, tannic acid, homovanillic acid, hesperetin, flavones, grape extract, green tea extract, and dehydrated cranberry powder were added to the prepared cheese as functional ingredients. These ingredients showed different levels of retention in the cheese curd. Cheese curds containing the polyphenolic compounds at a concentration of 0.5 mg/mL showed significant free radical scavenging activity, suggesting that this technology can be extended to the fortification of dairy products such as yogurts and milk shakes.

6.3.10 Margarine

Margarine resembles butter in appearance, consistency, and composition. It is used as a substitute for butter. Margarine was invented in 1869 by the French chemist Hippolyte Mège-Mouriès, shortly before the Franco-Prussian War. Meges-Mouries was awarded a prize of 50,000 francs for his innovation, which was made of processed bovine fat mixed with a paste of cow's udders. Milk was added to this mixture and churned to give a butter-like appearance (Morris and Lichtenwalter, 1984).

Today, margarine is made of a blend of vegetable oils and other ingredients. The final product is used as a spread or for baking purposes. According to the US Standards of Identity, margarine should have 80% fat. The second component is water; cow's milk was used in the original formula. The third ingredient is an emulsifier such as lecithin or mono/diglycerides to prevent phase separation. Preservatives such as sodium benzoate, benzoic acid, or potassium sorbate are added to act as antimicrobials. A butter-like flavor is imparted by diacetyl and rancidity is prevented by antioxidants (Morris and Lichtenwalter, 1984).

Corn oil is used as the major oil, as it is a polyunsaturated fat and thereby beneficial for health. However, due to growing concerns about the harmful effects of the trans-fatty acids in margarines and spreads, other vegetable oils having medicinal value are also being used in the manufacture of margarines and spreads. Important among them are soybean oil, sunflower oil, safflower oil, canola oil, and rice bran oil (Moreau, 2005; Chrysan, 2005). Recently, El-Haddad et al. (2011) formulated a functional chocolate spread replacing butter fat with red palm olein. This substitution was found to significantly increase the content of tocopherols, tocotrienol, and carotenes. Herbal extracts can take the place of emulsifiers, antioxidants, antimicrobials, and colorants. Such innovations can popularize the consumption of analogue foods such as fortified spreads.

6.3.11 Sausages

Sausages are food products made from ground meat. A casing, which is traditionally made from sheep intestines, is filled with cooked or uncooked meat. The word *sausage* is derived from the Latin root *salsus*, which literally means *salted* or *preserved*. Sausages have been in vogue in Europe since pre-Christian times and are always known by the place where they originated (Predika, 1983). The frankfurter is a kind of spiced and smoked sausage packed in a thin casing. It was first introduced in 1852 by the butchers' guild of Frankfurt, Germany.

The addition of functional ingredients is now advocated as an approach to the development of healthier meat products (Jiménez-Colmenero et al., 2001). Moreover, meat products are viewed as an ideal matrix to incorporate oil-soluble ingredients such as lutein, as they contain zinc, selenium, and fat, which enhance the absorption of lutein (Csapo et al., 2006). Granado-Lorencio et al. (2010) fortified frankfurters with lutein and studied their physicochemical characteristics and the *in vitro* bioaccessibility of lutein. Lutein imparted a red color to the frankfurters. On *in vitro* digestion, there was a greater than 84% recovery of lutein, indicating the suitability of this type of meat product as a carrier of lutein.

Andrés et al. (2009) formulated chicken sausages with pre-emulsified squid oil and studied their physicochemical properties after vacuum storage. The sausages had 40 g/100 g of polyunsaturated fatty acids, of which docasahexanoic acid was the predominant acid. Low lipid oxidation was also observed in them. This study shows that sausages can be fortified with oils rich in polyunsaturated fatty acids.

Sausages have been formulated with many nonmeat proteins from vegetable sources. Examples are soy proteins (Gujral et al., 2002; Pietrasik and Duda, 2000; Porcella et al., 2001), buckwheat protein (Bejosano and Corke, 1998), samh flour (El-Gasim and Al-Wesali, 2000), common bean flour (Dzudie et al., 2002), Bengal gram, green gram, black gram (Modi et al., 2003; Bhat and Pathak, 2009), and corn flour (Serdaroglu and Degirmencioglu, 2004).

Various plant products have been studied for their ability to replace fats. Rye bran was used as a fat substitute in the production of meatballs (Yilmaz, 2004). Garcia et al. (2002) reported that oat bran and oat fiber provide fat-like flavor, texture, and mouthfeel to ground beef and pork sausages. Inulin was added to pork meatballs and Chinese-style sausages (Flaczyk et al., 2009; Huang et al., 2011). In addition to functioning as fat replacers, these ingredients increase the fiber content in the products (Bhat and Bhat, 2011). Chevance et al. (2000) reported that tapioca starch slows down the release of some Maillard reaction products, thereby improving the aroma and quality of beefburgers, salamis, and frankfurters.

Citrus by-products such as lemon albedo pectin and orange fiber were added in different concentrations to cooked and dry-cured sausages. These ingredients have positive health effects on account of their ability to lower the levels of residual nitrite and delay oxidation (Aleson-Carbonell et al., 2003, 2004; Fernandez-Gines et al., 2003, 2004).

In a turkey frankfurter system, Gadang et al. (2008) evaluated the effectiveness of whey protein isolate coatings containing GSE, nisin, malic acid, and ethylenediamine tetraacetic acid in inhibiting the growth of *Listeria monocytogenes*, *E. coli* O157:H7, and *Salmonella typhimurium*. Their observations suggest that an edible film coating containing nisin, organic acids, and herbal extracts can control the growth and recontamination of these pathogens in ready-to-eat meat products such as frankfurters.

Natural carotenoids such as zeaxanthin and norbixin seem to act as natural preservatives, improving the shelf life of refrigerated sausages. Mercadante et al. (2010) prepared six sausage formulations with sodium erythorbate, norbixin, lycopene, zeaxanthin, β-carotene, and dextrose at 0.05 g/100 g. Following refrigerated storage for 45 days, the oxidative stability of the products was evaluated by measuring the levels of malondialdehyde. Zeaxanthin and norbixin were found to be the most effective in improving shelf life and were superior to sodium erythorbate, a synthetic compound used in sausage formulations. Similar results were reported by Barros et al. (2011), who was able to prevent lipid peroxidation in beefburger patties by incorporating extracts of the edible mushroom *Boletus edulis*.

Mladenoska et al. (2012) added monolaurin to sausages as a food emulsifier. The pasteurized sausages were inoculated with cultures from *Aspergillus niger* and *Saccharomyces cerevisiae*. The infected sausages were stored at 28°C for 72 h. At the end of the experiment, it was observed that monolaurin inhibited the growth of the fungi. The authors conclude that monolaurin can be used as an emulsifier and antifungal agent, preventing the spoilage of sausages from mold and prolonging their shelf life. However, it is to be used in low concentrations and in combination with some other food additives and spices that can mask its soapy smell and taste.

The beef patty is a Caribbean delicacy that originated as a product of colonialism and the migration of various ethnic groups into the region. It is a *samosa* (turnover)-like pastry filled with seasoned ground beef. The patties are prepared by cooking at 191°C. A class of compounds called *heterocyclic amines* (HCA) are known to be generated during the cooking of muscle foods at such high temperatures. Puangsombat and Smith (2010) incorporated five extracts of rosemary into beef patties, which were cooked at 191°C for 6 min each side and 204°C for 5 min each side. All the extracts significantly reduced the levels of HCAs. However, 10% and 20% ethanol extracts showed the highest inhibition. These extracts contain a mixture of rosmarinic acid, carnosol, and carnosic acid. These compounds may be exerting the beneficial effect on account of their antioxidant activity.

Li et al. (2010) demonstrated the beneficial effect of adding antioxidant-rich spice mixtures to hamburgers. Eleven healthy human subjects consumed two types of hamburgers. One type was seasoned with a spice blend (**Table 6.4**) while the other one was not. The production of malondialdehye in the hamburgers and the concentration of malondialdehyde in the subjects' plasma and urine after consumption were measured by HPLC. It was observed that after the addition of the spice mixture, there was a significant reduction in the malondialdehyde concentration in the hamburgers as well as in the subjects' plasma and urine after ingestion of the cooked product. Forty percent of the rosmarinic acid, attributed to Mediterranean oregano and rosemary, remained in the hamburgers after cooking. The presence of rosmarinic acid in the spice mix and the ability of polyphenols to inhibit the formation and absorption

Table 6.4 Composition of the Spice Mixture

Spice	Percentage	Weight (g/hamburger)
Ground cloves	4.34	0.5
Ground cinnamon	4.34	0.5
Ground Mediterranean oregano	26.17	3.0
Ground rosemary	4.34	0.5
Ground ginger	10.86	1.2
Ground black pepper	6.51	0.7
Ground paprika	30.44	3.4
Powdered garlic	12.99	1.5
Total	100.0	11.3

Source: Data from Li, Z., et al., 2010, *American Journal of Clinical Nutrition* 91, 1180–1184.

of lipid peroxidation products such as malondialdehyde seem to be reasons for the reported effect (Kanner and Lapidot, 2001; Gorelik et al., 2005, 2008). Malondialdehyde is implicated in the oxidation of lipids and the consequent appearance of atherosclerosis (Fogelman et al., 1980; Haberland et al., 1988; Palinski et al., 1989; Yla-Herttuala et al., 1989). The findings of Li et al. (2010) are significant in this perspective.

6.3.12 Bread

Nowadays, many different kinds of breads are available with different shapes, sizes, textures, crusts, colors, elasticity, eating qualities, and flavors. Bread making has a long history. Neolithic communities (New Stone Age) prepared the first bread, which was a kind of flat bread. It was the people of Sumeria, in southern Mesopotamia, who baked the first leavened bread. Around 3000 BC, the Sumerians passed on their style of bread making to the Egyptians. The Egyptians refined the process and started adding yeast to the flour (Jacob, 2007; Arzani, 2011). The industrial production of bread was pioneered by Otto Frederick Rohwedder (1880–1960). In 1912, he invented a machine that sliced and wrapped the bread (Tannahill, 1973).

Bread dough is a versatile matrix; therefore, attempts have been made to fortify flour with herb powders. Considerable success has been achieved in the fortification with ingredients such as amaranth (Singh and Singh, 2011), sorghum flour (Taylor and Anyango, 2011), buckwheat (Takahama et al., 2011), okra seed (*Abelmoschus esculentus*) (Adelakun and Oyelade, 2011), potato (Ezekiel and Singh, 2011), chempedak (*Artocarpus integer*), seed flour (Aziz and Zabidi, 2011), chickpea flour (Miñarroa et al., 2012), and ground onion skin (Gawlik-Dzikia et al., 2013).

Pepper (*Capsicum annum*) is a vegetable rich in carotenoids such as β-carotene, capsorubin, violaxanthin, capsanthin, zeaxanthin, antheroxanthin, β-cryptoxanthin, and cucurbitaxanthin (Ravishankar et al., 2003). Therefore, Danza et al. (2014) prepared a fortified bread by blending yellow pepper flour to durum wheat semolina. By studying the textural characteristics, carotenoid content, and glycemic response, the authors concluded that hydrated yellow pepper flour at a concentration of 25% can be successfully incorporated into bread to enhance the content of functional components with antioxidant properties.

Šimurina et al. (2008) tried to fortify bread with 45% propylene glycol extracts as well as the powders of Frangulae cortex, *Mentha piperita folium*, *Carvi fructus*, and *Petroselini fructus*. While the bread fortified with extracts showed poor sensory qualities, the bread prepared with herb powder had a pleasant aroma. The elasticity and compressibility of the product were excellent, the crumb was of high quality, and there was an increase in the loaf volume.

In recent years, the fortification of bread with phytochemicals has been attempted and the topic has been reviewed by Hayta and Őzuğur (2011).

Green tea extract was incorporated into bread (50, 100, and 150 mg/100 g of flour). There was no detectable loss of tea catechins in the bread during storage at room temperature for 4 days (Wang and Zhou, 2004).

Peng et al. (2010) fortified bread with GSE, which contains catechins and proanthocyanidins having a strong antioxidant and free radical scavenging activity. The GSE-fortified bread had a stronger antioxidant activity than the blank bread. The antioxidant property of the bread increased with the increasing level of GSE. Nevertheless, thermal processing caused a 30%–40% decline in the antioxidant activity.

Lemon flavonoids also seem to be useful for fortifying bread (Kanae et al., 2008). Lemon flavonoid extract prepared from lemon peel contains 30% eriocitrin, which is a potent antioxidant. Sixty-five percent of eriocitrin was retained in the bread and 78% of antioxidant activity remained after the baking process. Sensory tests showed that up to 0.50% of lemon flavonoid extract could be added and that larger amounts impart a bitter taste to the bread.

Abd El-Megeid et al. (2009) investigated the protective effect of 2% and 4% green tea–fortified bread on renal failure induced by excessive dietary arginine. Eating green tea–fortified bread could ameliorate the pathological effects on kidney enzymes, uric acid, urea nitrogen, and creatinine.

Several clinical studies have been reported on the effect of bread fortification. Clifton et al. (2004) studied the effect of phytosterol-enriched bread in mildly hypercholesterolemic men and women. On consumption of this bread, the plasma sitosterol content increased by 23% and campesterol by 52%. There was a significant reduction in the serum LDL cholesterol level.

Söderholm et al. (2012) studied the effect of high-fiber rye bread enriched with nonesterified plant sterols on the major serum lipids and apolipoproteins (apo) in normocholesterolemic subjects. Sixty-eight subjects were randomized to receive a rye bread enriched with plant sterols. Compliance was monitored with 3-day food diaries and alkylresorcinol metabolite concentrations. Consumption of the fortified bread for 2 weeks significantly reduced serum total and LDL cholesterol, apoB/apoA1, and total cholesterol:HDL cholesterol ratios by 5.1%, 8.1%, 8.3%, and 7.2%, respectively, when compared with controls. The doubling of the dose in the following 2 weeks resulted in a 6.5%, 10.4%, 5.5%, and 3.7% difference.

Hobbs et al. (2013) conducted a randomized clinical trial to study the effect of acute ingestion of beetroot bread on endothelium-independent vasodilation and diastolic blood pressure. Twenty-three healthy men received 200 g of bread containing 100 g of beetroot and postprandial microvascular vasodilation was measured by laser Doppler iontophoresis. Arterial stiffness was measured by pulse wave analysis and blood pressure was measured by a noninvasive ambulatory blood pressure monitor. The authors observed a significantly greater microvascular vasodilation response and a lower diastolic blood pressure after acute consumption of beetroot-fortified bread.

Extracts of green tea and grape seed offer a unique flavor and a powerful antioxidant action. However, there are some basic issues that need to be addressed. The most important are the stability of the active compounds and their interaction with other components of the food matrix (Sharma et al., 2011).

The roller milling of wheat produces a fraction known as *wheat bran*, which contains the outer layers (pericarp) in addition to the hyaline and aleuronic layers of the grain. Bran has a higher content of fiber, vitamins, and minerals than the endosperm. These qualities attribute interesting nutritional properties to it (Verma et al., 2008; Kumar et al., 2011). In addition to this, bran has the ability to prevent the appearance of diseases (McIntyre et al., 1991; Anderson et al., 2000; Haripriya and Premakumari, 2010). On account of these interesting features, bran is also a good candidate for the fortification of breads (Lai et al., 1989; Shenoy and Prakash, 2002; Karaoglu, 2006; Gómez et al., 2011).

Martinex-Cerbera et al. (2014) carried out some studies on the development of sugar-free muffins. They observed that maltitol, isomalt, and sorbitol are suitable as sucrose replacers in muffins. These polyols have similar batter thermosetting properties to those of sucrose and the prepared muffins showed good instrumental texture properties and a sensory acceptability score. The use of polyols as sucrose replacers also has health advantages. They do not promote the development of dental caries, as dental bacteria are not able to ferment these sugar alcohols. Additionally, they produce a lower glycemic response than sucrose and are ideal for diabetics. Thirdly, they are lower in calories when compared with sucrose.

6.3.13 Biscuits

Biscuits are small, crisp, and flat baked products made from wheat flour, starch, powdered sugar, glucose syrup, emulsifiers, food color, and flavors. Originally spelled *bisket*, the word *biscuit* is derived from the Old French *bescuit* and Medieval Latin *biscoctum*, both meaning “baked twice” (Manley, 2003a), alluding to the now-discontinued process of baking first at high temperature, followed by drying in a cooler oven. In conventional biscuits, the ingredients are mixed and the dough is cut into pieces of desired size. The dough pieces are baked carefully at 180°C for about 8 min (Manley, 2005). On account of their high sugar and fat content, biscuits are sometimes viewed as an unhealthy food. However, biscuits are an ideal medium for carrying many nutritionally beneficial ingredients. Manley (2003b) suggests several ways in which biscuits can be fortified.

Several studies have been carried out on the substitution of wheat flour with other flours. Rababah et al. (2006) attempted the fortification of biscuits using chickpea flour, broad bean flour, and soy protein. Their study shows that high-protein biscuits can be produced using these ingredients. Studies conducted in India show that 30% soy flour can be used in the baking of biscuits, enhancing their nutritive value without any loss of their physical characteristics (Singh and Sabapathy, 2006). Salama et al. (1997) incorporated the rootlets of malt sprouts into biscuits. Biscuits that contained 15% roots were acceptable. However, their sensory qualities were affected.

The flour of the African plant *Afzelia africana* was incorporated into dough at various levels of substitution. Sensory scores showed high acceptability for the biscuits and the flour of this protein-rich legume can be substituted at the 10%–30% level (Onweluzo and Morakinyo, 1997).

The fortification of biscuits with raw, soaked, and germinated fenugreek (Hooda and Jood, 2003), sesame seed flour (Alobo, 2001), malted wheat, Bengal gram, and colocasia leaf powder (Goyle and Gujral, 1992) has also been reported. Uchoa et al. (2009) incorporated cashew apple and guava residues from the fruit juice industry into biscuit dough. The fruit powder improved the protein and fiber content of the cookies.

The usefulness of fiber-incorporated biscuits in the treatment of constipation was reported by Odes et al. (1986), who used spent grain, which was obtained by decanting the fermented distillate of barley. The consumption of cookies fortified with fresh corn pericarp improved the fecal parameters in individuals suffering from constipation (Sevá-Pereira et al., 1991).

The addition of red palm oil as an alternative fortificant for addressing vitamin A deficiency has been proposed by van Stuijvenberg et al. (2001), on the basis of a randomized, controlled clinical study.

Tyagi et al. (2006) studied the nutritional, sensory, and textural characteristics of defatted mustard flour–fortified biscuits. The wheat flour in the biscuit recipe was replaced by defatted mustard flour at four incorporation levels. The protein content of the fortified biscuits increased by 2.5 times and there was a reduction in the fat content and an increase in the fiber content. The study showed that the incorporation of 15% defatted mustard flour was ideal for the production of fortified biscuits with desirable nutritional, sensory, and textural attributes.

Hussain et al. (2006) reported that flaxseed flour can be blended with wheat flour in the production of cookies. Cookies containing 20% and lower amounts of full-fat flaxseed flour were found to be acceptable to consumers. Zucco et al. (2011) reported that cookies with acceptable physical characteristics and an improved nutritional profile could be produced with partial or complete replacement of the wheat flour with a blend of navy bean, pinto bean, green lentil, and yellow pea.

Akkinawande et al. (2008) explored the fortification of biscuits with cassava and soy flour blends. The addition of soy flour increases the nutritive value of the biscuits, and ginger powder added to the dough effectively masks the beany flavor of soybean flour. Sharma and Zhou (2011) attempted the fortification of biscuits with green tea extract. They observed that tea catechins were relatively stable in dough.

Boobier et al. (2006) converted a traditional biscuit to a functional food by adding B_{12}, folic acid, vitamin C, and prebiotic fiber. The consumption of these fortified biscuits by human volunteers reduced their blood levels of homocysteine and glucose, suggesting these biscuits may be able to reduce risk factors associated with heart diseases such as myocardial infarction, stroke, and venous thrombosis (Boobier et al., 2007; Hortin, 2006).

The replacement of sugar and fat in biscuits has also been a subject of investigation (Barndt and Antenucci, 1996). Gallagher et al. (2003) reported that sugar in biscuits can be reduced by 20%–30% with the incorporation of the oligofructose Raftilose. Sucralose is a sugar substitute that is currently being used in bakery products in Europe. It is heat stable and no loss occurs during the baking processes (see Barndt and Antenucci, 1996). Savitha et al. (2008) studied the replacement of 30% sugar with 0.5% sucralose and of different levels of maltodextrin on dough rheology and the quality of biscuits. The results show that sucralose and maltodextrin can replace sugar in biscuits.

Sharif et al. (2005) observed that superior-quality cookies can be produced by replacing the normal shortening in the recipe with rice bran oil. Rice bran oil is a good source of polyunsaturated fatty acids and lowers blood cholesterol better than sunflower, corn, or safflower oils (Suzuki and Oshima, 1962).

The fortification of biscuits with guar gum was attempted several years ago on account of its beneficial effect on diabetes mellitus. Tredger and Ransley (1978) incorporated guar gum into cheese biscuits and observed that they were acceptable to diabetic patients. Another type of guar gum–fortified biscuit was developed by Smith et al. (1982). It reduced fasting blood glucose levels and decreased 24-h urinary glucose excretion in human volunteers. The usefulness of guar gum–fortified biscuits in the management of diabetes was further confirmed by Ellis et al. (1988), who reported on the basis of a linear regression model that a reduction of 209 mU/min/L in integrated insulin can be achieved for every 1 g of guar gum incorporated into the biscuit. The prebiotic potential of partially hydrolyzed guar gum (PHGG) and fructo-oligosaccharides in biscuits was assessed in human volunteers by Tuohy et al. (2001). Bifidobacteria significantly increased in number on the consumption of the biscuits and returned to pretreatment levels after cessation of intake.

There is a growing demand for biscuits for diabetics. This can be met by replacing sugar with polyols such as mannitol or polydextrose and including in the recipe, dietary fibers that have a proven hypolipidemic or antidiabetic property. Jenkins et al. (2008) extracted glucomannan, a glucose-mannose polysaccharide from the tuberous roots of the *Amorphophallus konjac*, and incorporated it into the dough of biscuits. The consumption of these biscuits significantly reduced the glycemic index by 74% in healthy human volunteers and by 63% in participants with diabetes mellitus. This study suggests that biscuits fortified with other viscous and starchy foods can replace high glycemic index snack foods.

Many varieties of starches are used as ingredients in food products. Starches can escape digestion if they are embedded in a matrix that renders them inaccessible to digestive enzymes. Some others such as the starch from potato, green bananas, and maize also resist digestion on account of their structural properties. Foods formulated with such resistant starches are effective in lowering the glycemic index of the food, despite their formulation with high glycemic index ingredients. This was proven in a recent clinical study (Haub et al., 2010). As resistant starch is known to reduce food intake, it can also be considered in food fortification aimed at weight management (Bodinham et al., 2010).

Replacing fat with equal quantities of maltodextrin and polydextrose reduces the consistency and hardness of dough. However, a significant improvement in biscuit texture can be achieved by using maltodextrin along with glycerol monostearate or guar gum (Sudha et al., 2007).

6.4 Stability of food

Except for wine and cheese, all food and beverages undergo degradation on storage. All manufacturers strive to present consumers with food and beverages that have an acceptable shelf life. According to the definition offered by the Institute of Food Science and Technology (1993), shelf life is defined as the period during which the food will remain safe, be certain to retain its desired sensory, chemical, physical, and microbiological characteristics, and comply with declared nutritional data, when stored under the recommended conditions (Kilcast and Subramaniam, 2000).

Many factors influence shelf life and they can be intrinsic or extrinsic (IFST, 1993). Intrinsic factors are the properties of the final product and include water activity (available water) (a_w), pH value, total acidity, redox potential (E_h), available oxygen, nutrients, surviving microbiological counts, enzymes, chemical reactants, and preservatives (Valero et al., 2012).

Extrinsic factors are those factors that are encountered by the final product as it moves through the food chain. They include the time–temperature profile during processing; temperature control during storage and distribution; relative humidity during processing, storage, and distribution; exposure to ultraviolet and infrared light during processing, storage, and distribution; composition of atmosphere within packaging; subsequent heat treatment such as reheating or cooking; and consumer handling (Valero et al., 2012). The interaction of these intrinsic and extrinsic factors either inhibits or stimulates many processes that limit shelf life. These processes can be microbiological, chemical, physical, and temperature related.

6.4.1 Microbiological changes

The growth of specific organisms depends on several factors. The most important among them are the initial microbial load at the start of storage; the physicochemical properties of the food such as moisture, pH, and preservatives; the processing method used; the external environment of the food (gas composition); and the storage temperature. The growth of food poisoning bacteria such as *Salmonella* species and *L. monocytogenes* need not be accompanied by changes in appearance, odor, flavor, or texture detectable by human senses. The changes caused by the growth of spoilage organisms include visual mold growth, the generation of off-odors, and changes in texture (Kilcast and Subramaniam, 2000).

6.4.2 Chemical changes

Important deteriorative changes may occur from reactions within the food or from the interaction of food components with external entities such as oxygen. Fat-containing foods can develop rancidity. Chemical hydrolysis may occur in food containing strong sweeteners and Maillard reactions may cause nonenzymatic browning. Exposure to light can cause loss of color in natural food colors and milk products may develop off-flavors (Kilcast and Subramaniam, 2000).

6.4.3 Physical changes

Deteriorative physical changes happen in stored food through moisture migration. This is evident in fresh produce through moisture loss and dry food such as breakfast cereals may lose their crispness through moisture uptake. The chemical components in packaging material may migrate to food and cause taints (Kilcast and Subramaniam, 2000).

6.4.4 Temperature-dependent changes

Both high and low temperatures can cause deterioration. An increase in temperature increases the rate of chemical reactions, resulting in deterioration. At higher temperatures, food containing sugar syrups may change their crystallization characteristics and fluctuating temperatures can cause the destabilization of emulsion systems (Kilcast and Subramaniam, 2000).

These factors may result in a range of deteriorative changes, depending on the type of food. For example, starch retrogradation and moisture migration occur in bread, resulting in mold growth and a stale texture and flavor. Syneresis and oxidation may occur in yogurt, causing serum separation and rancidity. Fruit juices may undergo oxidation and enzymatic reactions. Consequently, cloud instability will occur and the product may lose flavor and nutrients (Kilcast and Subramaniam, 2000).

6.4.5 Shelf-life measurement

6.4.5.1 Sensory panels

Changes in the eating quality of stored food can be measured with sensory techniques, for which trained panels and naïve consumers are used. An appropriate set of ethical procedures is a *sine qua non* for sensory testing. To remove doubts regarding safety, microbiological testing should be carried out in advance (Kilcast and Subramaniam, 2000).

6.4.5.2 Instrumental methods

Instrumental methods are used to measure sensory quality factors. But they are reliable only when the measured parameter has been validated against sensory measurements. Computerized texture analyzers, rheometers, and volatile detectors (electronic noses) are valuable only when the measured parameters are correlated with sensory attributes.

The deterioration of food can be detected by the "marker" approach, according to which a physical or chemical property is identified and linked to the process of deterioration. Thereafter, a sensor is designed to measure some aspect of this property. The evolution of cadaverine and putrescine is an indicator of the spoilage of meat. Biosensors are used to measure meat freshness based on levels of glucose. Similarly, mechanical resonance probes measure the viscosity increase of oil that is occurring during oxidation and polymerization in frying-induced deterioration (Kilcast and Subramaniam, 2000).

Several indicators of spoilage have been developed based on volatile compounds generated during the deteriorative processes. A myoglobin-based indicator has been developed for detecting the presence of hydrogen sulfide formed during the spoilage of poultry meat (Smolander et al., 2002). Pacquit et al. (2006, 2007) developed a colorimetric dye–based sensor that can detect the presence of total volatile basic nitrogen produced during the deterioration of fish.

Carbon dioxide is known to form during the growth of bacteria and molds in food. Therefore, the detection of carbon dioxide levels is an ideal way to detect microbial spoilage (Puligundla et al., 2012). Mattila and Ahvenainen (1989) and Mattila et al. (1990) made a correlation between carbon dioxide concentration and the growth of microorganisms in pea or tomato soup packed aseptically either in air or in a mixture of 5% oxygen and nitrogen. The pH-sensitive dye bromothymol blue was used as an indicator to detect the formation of carbon dioxide.

6.4.5.3 Physical measurements

Measuring the changes in the texture of products is a commonly used physical test. Several instruments are available for texture measurement and measuring hardness or crispness (Kilcast and Subramaniam, 2000).

6.4.5.4 Chemical measurements

Chemical analysis is used to measure the deterioration changes that are happening in stored food. Thus, the peroxide value and the free fatty acid content are markers for the level of rancidity in food (Kilcast and Subramaniam, 2000).

6.4.5.5 Microbiological measurements

The microbiological stability of a product can be determined on the basis of the microbial growth leading to spoilage and the growth of microbial pathogens. The extent of the microorganisms that are likely to grow in a food product depends on the water activity, storage temperature, duration of storage, and pH. The "time of spoilage" is determined by storing the food product at the appropriate temperature and measuring the microbial load at staged intervals. The time taken to reach a predetermined level of microbial load is considered the end point. Generally, 70% of the time to spoilage is taken as the storage life (Kilcast and Subramaniam, 2000).

6.4.6 Prediction of shelf life

Food manufacturers introduce new and attractive products with minimum delay. Knowledge of the storage characteristics over the intended shelf-life period is essential to achieve success. Therefore, as in the pharmaceutical industry, accelerated shelf-life studies are carried out. Such studies are designed on the assumption that by changing a storage condition, the chemical or physical process that leads to deterioration is accelerated. However, the validity of any accelerated condition is to be tested against known deterioration characteristics under ambient storage conditions. If it is not possible to identify the ambient storage characteristics of a product due to time pressure, then a comparison is to be made between the test product and an equivalent product with a similar structure and for which a shelf life has already been established. Mizrahi (2000) describes in detail the study of the accelerated stability of food products.

6.5 Official methods of analysis

All food products need to be analyzed through the phases of product development, commercial production, and after introduction to the market. The functional characteristics, nutritive value, and acceptability by consumers are determined by the chemical composition and physical properties of the food. The analytical methods to be adopted in this regard should have speed, precision, accuracy, and ruggedness (Nielsen, 2003a). The compendia published by many professional associations describe the various methods used in the analysis of food substances and their ingredients. Prominent among such compendia are the following:

6.5.1 The Association of Analytical Communities International (AOAC International)

The AOAC International was founded in 1884 and was known as the Association of Official Agricultural Chemists. With the patronage of the US Department of Agriculture (USDA), the AOAC evolved analytical methods for government regulatory and research organizations. Analytical methods that are recognized by the AOAC International and data on their validation are available in the *Journal of the AOAC International.* Such methods are validated in interlaboratory studies before their inclusion in the AOAC's *Official Methods of Analysis* (Vasconcellos, 2005; Latimer, 2012).

6.5.2 The American Association of Cereal Chemists (AACC)

Established in 1915 for the standardization of analytical methods for cereal laboratories, the AACC publishes approved methods for the analysis of cereals and cereal products, which are periodically reviewed and updated (Anonymous, 2000).

6.5.3 The American Oil Chemists' Society (AOCS)

The AOCS publishes official methods and recommended practices containing around 400 methods to analyze fats, oils, oilseed proteins, and oil-derived products (Anonymous. 1996a).

6.5.4 The American Public Health Association (APHA)

The APHA, which is concerned with many public health–related issues, publishes several methods for the analysis of food and water (Downes and Ito, 1998; Marshall, 1993; Clesceri et al., 1998).

6.5.5 The American Spice Trade Association (ASTA)

The ASTA was founded in 1907 and now represents nearly 34 spice-producing nations. The analytical methods published by the ASTA form the official methods for the analysis of spices and products derived from them (Anonymous, 1997).

6.5.6 Food Chemicals Codex

Established half a century ago, *Food Chemicals Codex* provides standards for the purity of food chemicals. This voluminous work lays out specifications for the chemicals and additives that are used in the food industry (Anonymous, 1996b). In addition to these, the various aspects of food analysis are treated exhaustively by Wildman (2001), Gunstone (2003), Nielsen (2003b), Nollet (2004), and Hurst (2007).

The Pharmacopoeial Laboratory for Indian Medicine, Ghaziabad, Uttar Pradesh, has published a protocol for testing Ayurveda, Siddha, and Unani medicines. This excellent document details the methods for determining the various parameters for evaluating the quality and safety of single and compound formulations (Lohar, 2010).

6.6 Management of product quality

The quality of a food product should be excellent if it is to be accepted by consumers. A total quality program is implemented in professionally managed companies to achieve this target. In a company managed on the concept of total quality management (TQM), workers assume the responsibilities, risks, and rewards associated with the decisions that they make themselves (Vasconcellos, 2005). As workers are the only personnel who can make detailed observations on a daily basis, the empowerment of workers is essential if TQM is to be achieved. They should be trained to take decisions that will improve the quality of their work and products, thereby enhancing the commercial success of the company. Good managers should investigate genuine complaints of lack of facilities, maintenance, and mismanagement, rather than viewing those workers as troublemakers (Vasconcellos, 2005). Vasconcellos (2005) lists numerous common causes of manpower wastage in production plants, leading to a decline in the quality of the products and a dwindling in the fortunes of the company (**Table 6.5**). Good management practices and building teams of empowered employees are essential for the continuous success of a company.

6.7 Concluding remarks

Food and beverages with enhanced nutritional and functional values can be formulated using ingredients with a long history of use as a food. An example is the pea (*Pisum sativum*), which has been cultivated by man since 6000 BC. Both immature and dry seeds are consumed. However, the consumption of this nutritious bean declined after the end of World War II, more as it acquired the image of a "poor man's food." With newfound interest in healthy, natural foods, there has been a revival in the consumption of peas in various forms (Sandberg, 2008).

The acceptability of pea protein as a food ingredient was achieved by the selection of cultivars with a high content of protein and amino acids; the removal of trypsin inhibitors, lectins, phytic acids, and saponins to improve digestibility and palatability and reduce antigenicity; the removal of oligosaccharides by wet processing to reduce flatulence arising from microbial fermentation; and the enhancement of organoleptic characteristics, especially by reducing the strong pea flavor. Pea protein can be added to flour (in bread or pasta), meat (in patties, hamburgers), in textured products (desserts), soups, snacks, beverages, and milk (Sandberg, 2008).

Table 6.5 Common Causes of Wastage of Manpower

Failure to get the best from competent employees
Scanty/incomplete job instructions for new employees
Failure to explain in detail the nature of the work
Failure to explain what is expected from the employee
Failure to select qualified and experienced employees
Impatience with new recruits who learn slowly
Failure to integrate new employees into the working environment
Failure to get other workers to show a friendly attitude toward new employees
Failure to establish sound relationship with new employees
Poorly maintained or calibrated equipment and instruments
Failure to train an understudy
Failure to recognize or commend exceptional performance
Failure to promote an employee when it is possible and appropriate
Lack of interest in the progress and affairs of employees
Not providing employees with the required support and help
Lack of attention to employee's ability and temperament
Keeping an employee in a job for which he or she is not mentally or physically suitable
Failure to view employees as individuals and the resultant demotivation
Lack of due consideration to problems resulting from uncomfortable working conditions such as humidity, noise, confusion, temperature (heat or cold), poor ventilation, poor light, and dirt
Failure to admit mistakes
Failure to control turnover of efficient employees
Failure to interpret correctly to the employees the company's real aims and policies
Not inducting new employees appropriately and not educating them on the policies, procedures, pay, terms, and conditions of employment and the facilities of the company
Making promises that cannot be fulfilled
Not keeping promises that could have been fulfilled
Not appreciating the direct and indirect costs of employee turnover
Dismissing employees without sufficient cause
Improper use of the dismissal procedures as a penalty
Too much "bossing"
Keeping an employee in a job which he or she dislikes
Too strict or too lax enforcement of discipline
Criticizing a worker in front of others
Favoritism or treating one person better than another
Taking sides in disputes between employees

Source: Adapted from Vasconcellos, J.A., *Quality Assurance for the Food Industry—A Practical Approach.* CRC Press, Boca Raton, FL, 2005. With permission.

Functional beverages are a growing segment, as they can provide the necessary nutrients to prevent nutrition-related diseases. Beverages are considered to be an excellent medium for delivering nutraceutical components, such as soluble fibers and herbal extracts. Many innovative ideas have already appeared. For example, Thongsombat et al. (2007) fortified guava juice with a soluble dietary fiber such as pectin. Based on the hypoglycemic property of bitter gourd (*Momordica charantia*), Din et al. (2011) formulated a palatable beverage with 15% of the fruit extract. The extract of bamboo leaves (*Bambusa vulgaris*) has been incorporated into orange and pineapple juices with enhanced antimicrobial activity (Owokotomo and Owoeye, 2011).

The functional beverage market includes several categories such as sports/energy drinks, herbal teas, fruit juices, and vegetable juices. However, product formulators always bear in mind the fact that the consumer is unwilling to sacrifice taste for health benefits. Important in this context is the all-natural sweetener, erythritol, which has taste- and mouthfeel-enhancing properties. It is a natural ingredient of many fruits and fermented foods. It is suitable for fortified beverages as it is noncaloric, noncarcinogenic, and nonglycemic, and it has high digestive tolerance and antioxidant properties (De Cock and Bechert, 2002).

A few interesting functional foods have also made their appearance recently. The stem bark of the magnolia has been used in the treatment of coughs, diarrhea, and allergic rhinitis in the traditional medicine of China, Japan, and Korea. Recent research indicates that magnolol and honokiol in the bark have strong anti-inflammatory and antimicrobial activities. They are active against an array of organisms such as *E. coli*, *Pseudomonas aeruginosa*, *Trichophyton mentagrophytes*, *Porphyromonas gingivalis*, *Epidermophyton floccosum*, *A. niger*, *Cryptococcus neoformans*, and *Candida albicans* (Lee et al., 2009). Considering these activities, the gum giant Wrigley has been granted novel food approval in the European Union for *Magnolia officinalis* extract to be used in chewing gums and mints with perceived breath-freshening benefits (Anonymous, 2011c). Chewing gum is also intended to be a dosage form to deliver the peptide YY (PYY) hormone that is reported to promote a feeling of fullness and thus act as a weight loss agent (Fazen et al., 2011).

A prominent manufacturer of chewing gum, Fertin Pharma A/S, Denmark, has come up with an immunity-enhancing chewing gum containing *Echinacea purpurea* extract. A clinical trial was conducted in 246 subjects with the common cold. It was observed that treating them with chewing gum containing 13.56 mg of *E. purpurea* extract, three times daily for 7 days caused cold symptoms to disappear more than when treated with a placebo. The gum is sweetened with xylitol, sorbitol, aspartame, and acesulfame-K. Fertin Pharma has also developed a chewing gum for stress relief and weight management. It contains saffron (*Crocus sativus*) extract (Anonymous, 2013a). The company offers chewing gums as "a dosage form that delivers active ingredients in an effective and consumer-friendly way" (Anonymous, 2013b).

Over the years, green tea has migrated from hot drinks to other foods and has made its appearance in breakfast cereal. In October 2013, the South African company Herbex Health introduced their product Slimmers Cereal containing extracts of green tea and *G. cambogia* (Anonymous, 2014a). *G. cambogia* helps the body to burn fat and prevents the formation of new fat (Heymsfield et al., 1998). The use of these two ingredients in combination with a breakfast cereal is an innovation that is set to benefit from current global interest in achieving weight loss with natural ingredients.

A similar trend was initiated earlier in China by Quaker Oats, who started offering functional benefits to consumers through ingredients rooted in traditional health concepts. Based on their R&D efforts since 2005, Quaker Oats started incorporating ingredients from Chinese traditional medicine. Their products now contain ingredients such as red dates, wolfberry, and white fungus. These ingredients are now incorporated into China's traditional rice-based porridge (congee). Quaker Oats has started marketing a brand of oats containing lotus starch (Anonymous, 2014b).

Since 1982, electronic noses have been used in the quality control of food and beverages (Persaud and Dodd, 1982). These devices identify and classify aroma mixtures. Numerous different prototypes of artificial nose have been developed to discriminate complex vapor mixtures containing many different types of volatile organic compounds (Wilson and Baietto, 2009). Recent research suggests that robots with a sense of taste could one day supplant human food tasters (Del Valle, 2010, 2012). Cetó et al. (2013) recently reported the classification of beer using an instrument that could distinguish between different types of beers with 82% accuracy. This device could be used to improve product quality and detect fraud. According to the authors, "supervised learning" and linear discriminant analysis allowed the electronic tongue to distinguish between Schwarzbier, lager, double malt, Pilsen, Alsatian, and low-alcohol beer. In the future, it is hoped that the electronic tongue will become a powerful tool in the quality control of food and beverages.

The components of functional foods can enhance vital functions in the body and these beneficial effects can be assessed by the identification, characterization, measurement, and validation of relevant markers. Consumers in the developed world are slowly migrating to healthier foods and beverages for controlling diseases, rather than taking several pills per day. Functional food science is predicted to contribute to human health in the coming decades, by reducing the risk of diseases (Roberfroid, 2008).

References

Abd El-Megeid, A.A., AbdAllah, I.Z.A., Elsadek, M.F., Abd El-Moneim, Y.F. 2009. The protective effect of the fortified bread with green tea against chronic renal failure induced by excessive dietary arginine in male albino rats. *World Journal of Dairy & Food Sciences* 4: 107–117.

Abeysinghe, C.P., Illeperuma, C.K. 2006. Formulation on an MSG (monosodium glutamate) free instant vegetable mix. *Journal of National Science Foundation of Sri Lanka* 34: 91–95.

Abumweiss, S.S., Barake, R., Jones, P.J. 2008. Plant sterols/stanols as cholesterol lowering agents—A meta-analysis of randomized controlled trials. *Food and Nutrition Research* 52: 1–17.

Abu-Salem, F.M., Abou-Arab, A.A. 2008. Chemical, microbiological and sensory evaluation of mayonnaise prepared from ostrich eggs. *Grasas y Aceites* 59: 352–360.

Adegoke, G.O., Musenbi, S., Wachira, S., et al. 2013. Production of probiotic yoghurt flavored with the spice, *Aframomum danielli*, strawberry and vanilla. *Food and Public Health* 3: 92–96.

Adelakun, O.E., Oyelade, O.J. 2011. Potential use of okra seed (*Abelmoschus esculentus* Moench) flour for food fortification and effects of processing. In *Flour and Breads and Their Fortification in Health and Disease Prevention*, ed. V.R. Preedy, R.R. Watson, and V.B. Patel, 205–212. San Diego, CA: Academic Press.

Ade-Omowaye, B.I.O., Olaniyan, S.A., Adeyemi, I.A., Isola, O.O. 2006. Development and quality evaluation of non-alcoholic beverages from maize-based products. *Nutrition and Food Science* 36: 183–190.

Akingbala, J.O., Adeyami, I.A., Sangodoyin, S.O., Oke, L. 1994. Evaluation of amaranth grain for *ogi* manufacture. *Plant Foods for Human Nutrition* 46: 19–26.

Akkinawande, B.A., Ade-Omowaye, B.I.O., Olaniyan, S.A., Akintaro, O.O. 2008. Quality evaluation of ginger-flavoured soy-cassava biscuit. *Nutrition and Food Science* 38: 473–481.

Aleson-Carbonell, L., Fernandez-Lopez, J., Sayas-Barbera, E., Sendra, E., Perez-Alvarez, J.A. 2003. Utilization of lemon albedo in dry-cured sausages. *Journal of Food Science* 68: 1826–1830.

Aleson-Carbonell, L., Fernandez-Lopez, J., Sendra, E., Sayas-Barbera, E., Perez-Alvarez, J.A. 2004. Quality characteristics of a non-fermented dry-cured sausage formulated with lemon albedo. *Journal of the Science of Food and Agriculture* 84: 2077–2084.

Alobo, A.P. 2001. Effect of sesame seed flour on millet biscuit characteristics. *Plant Foods for Human Nutrition* 56: 195–202.

Anderson, J.W., Hanna, T.J., Peng, X., Kryscio, R. 2000. Whole grain foods and heart disease risk. *Journal of the American College of Nutrition* 19 (Suppl. 3): 291S–299S.

Anderson, K.G. 1990. Pickles, sauces and dips. In *Snack Food*, ed. R.G. Booth, 139–158. New York: Van Nostrand Reinhold.

Andrés, S.C., Zaritzky, N.E., Califano, A.N. 2009. Innovations in the development of healthier, chicken sausages formulated with different lipid sources. *Poultry Science* 88: 1755–1764.

Anonymous. 1992. Camellia. *The Wealth of India*, Vol. 3, 96–177. New Delhi: C.S.I.R.

Anonymous. 1996a. *Official Methods and Recommended Practices*, 4th edn. Champaign, IL: American Oil Chemists' Society.

Anonymous. 1996b. *Food Chemicals Codex*, 4th edn. Washington, DC: Food and Nutrition Board.

Anonymous. 1997. *ASTA Analytical Methods*, 4th edn. Englewood Cliffs, NJ: American Spice Trade Association.

Anonymous. 2000. *Approved Methods of Analysis*, 10th edn. St. Paul, MN: American Association of Cereal Chemists.

Anonymous. 2011a. Archaeological oncology project uncovers cancer-fighting compounds in ancient herbal beverages. *Herbalgram* 89: 15–16.

Anonymous. 2011b. Nexcite: Refreshing beverage. www.nexcite.se. Accessed July 25, 2011.

Anonymous. 2011c. Wrigley gains novel food approval for magnolia bark extract in EU. http://www.confectionerynews.com/Regulation-Safety/Wrigley-gains-novel-food-approval-for-magnolia-bark-extract-in-EU. Accessed November 26, 2011.

Anonymous. 2012. Aqua Libra: Delicious sparkling temperance drinks. http://www.delicioussparklingtemperancedrinks.net/AquaLibra.html. Accessed January 19, 2012.

Anonymous. 2013a. Chewing gum developed to battle common cold. http://www.confectionerynews.com/R-D/Chewing-gum-developed-to-battle-common-cold. Accessed May18, 2013.

Anonymous. 2013b. Medicated chewing gum development. http://www.fertin.com/Products/Nutraceutical/Pages/default.aspx. Accessed December 25, 2013.

Anonymous. 2014a. Slimmers Cereal (Vanilla & Cinnamon). http://herbexhealth.com/shop/slimmers-cereal/. Accessed February 19, 2014.

Anonymous. 2014b. Quaker thrives in China with traditional medicine cereal fortification. http://www.nutraingredients.com/Industry/Quaker-thrives-in-China-with-traditional-medicine-cereal-fortification/?utm_source=newsletter_daily&utm_medium=email&utm_campaign=Newsletter%2BDaily&c=i8ceY2CuOjuxrIbwpCujxQ%3D%3D. Accessed February 22, 2014.

Antonio, J., Sanchez, C. 2005. *Javalution—Fitness and Weight Loss Through Functional Coffee*. Laguna Beach, CA: Basic Health Publications.

Aportela-Palacios, A., Sosa-Morales, M.E., Velez-Ruiz, J.F. 2005. Rheological and physicochemical behaviour of fortified yogurt with fiber and calcium. *Journal of Texture Studies* 36: 333–349.

Arendash, G.W., Mori, T., Cao, C., et al. 2009. Caffeine reverses cognitive impairment and decreases brain A levels in aged Alzheimer's mice. *Journal of Alzheimer's Disease* 17: 661–680.

Arendash, G.W., Schleif, W., Rezai-Zadeh, K., et al. 2006. Caffeine protects Alzheimer's mice against cognitive impairment and reduces brain beta-amyloid production. *Neuroscience* 142: 941–952.

Arjmand, A. 2011. Antioxidant activity of pomegranate (*Punica granatum* L.) polyphenols and their stability in probiotic yoghurt, 1–132. Thesis submitted to School of Applied Sciences, RMIT University, Australia.

Arts, I.C., Hollman, P.C., Kromhout, D. 1999. Chocolate as a source of tea flavonoids. *Lancet* 354: 488.

Arvanitoyannis, I.S., van Houwelingen-Koukaliaroglou, M. 2005. Functional foods—A survey of health claims, pros and cons and current legislation. *Critical Reviews in Food Science and Nutrition* 45: 385–404.

Aryana, K.J., Plauche, S., Rao, R.M., McGrew, P., Shah, N.P. 2007. Fat-free plain yogurt manufactured with inulins of various chain lengths and *Lactobacillus acidophilus*. *Journal of Food Science* 72: M79–M84.

Arzani, A. 2011. Emmer (*Triticum turgidum* spp. *dicoccum*) flour and breads. In *Flour and Breads and Their Fortification in Health and Disease Prevention*, ed. V.R. Preedy, R.R. Watson, and V.B. Patel, 69–78. San Diego, CA: Academic Press.

Ataie-Jafari, A., Larijani, B., Alavi Majd, H., Tahbaz, F. 2009. Cholesterol-lowering effect of probiotic yogurt in comparison with ordinary yogurt in mildly to moderately hypercholesterolemic subjects. *Annals of Nutrition and Metabolism* 54: 22–27.

Au, D.T., Wu, J., Jiang, Z., Chen, H., Lu, G., Zhao, Z. 2008. Ethnobotanical study of medicinal plants used by Hakka in Guangdong, China. *Journal of Ethnopharmacology* 117: 41–50.

Azhar, Z.M., Zubaidah, J.O., Norjan, K.O., Zhuang, C.Y., Tsang, F. 2013. A pilot placebo-controlled, double-blind, and randomized study on the cognition-enhancing benefits of a proprietary chicken meat ingredient in healthy subjects. *Nutrition Journal* 12: 121.

Aziz, N.A.A., Zabidi, M.A. 2011. Partial substitution of wheat flour with chempadek (*Artocarpus integer*) seed flour in bread. In *Flour and Breads and Their Fortification in Health and Disease Prevention*, ed. V.R. Preedy, R.R. Watson, and V.B. Patel, 365–374. San Diego, CA: Academic Press.

Aziznia, S., Khosrowshahi, A., Madadlou, A., Rahimi, J. 2008. Whey protein concentrate and gum tragacanth as fat replacers in nonfat yogurt—Chemical, physical and microstructural properties. *Journal of Dairy Science* 91: 2545–2552.

Bakuradze, T., Boehm, N., Janzowski, C., et al. 2011. Antioxidant-rich coffee reduces DNA damage, elevates glutathione status and contributes to weight control—Results from an intervention study. *Molecular Nutrition & Food Research* 55: 793–797.

Balzer, J., Heiss, C., Schroeter, H. 2006. Flavanols and cardiovascular health—Efforts on the circulating NO pool in humans. *Journal of Cardiovascular Pharmacology* 47 (Suppl. 2): S122–S127.

Banihani, S.A., Makahleh, S.M., El-Akawi, Z., et al. 2014. Fresh pomegranate juice ameliorates insulin resistance, enhances β-cell function, and decreases fasting serum glucose in type 2 diabetic patients. *Nutrition Research* 34: 862–867.

Barndt, R.L., Antenucci, R.N. 1996. Fat and calorie-modified baked products. In *Low-Calorie Foods and Food Ingredients*, ed. R. Khan, 106–132. Glasgow: Blackie Academic & Professional.

Barros, L., Barreira, J.C.M., Grangeia, C., Batista, C., Cadavez, V.A. P., Ferreira, I.C.F.R. 2011. Beef burger patties incorporated with *Boletus edulis* extracts—Lipid peroxidation inhibition effects. *European Journal of Lipid Science and Technology* 113: 737–743.

Beckett, S.T. 2004. The history of chocolate. In *The Science of Chocolate*, 1–3. Cambridge: The Royal Society of Chemistry.

Bejosano, F.P., Corke, H. 1998. Amaranthus and buckwheat protein concentrate effects on an emulsion-type meat product. *Meat Science* 50: 343–353.

Belachew, M. 2003. Coffee. In *Encyclopaedia Aethiopica*, 763. Wiesbaden: Harrassowitz Verlag.

Belmaker, R.H., Agam, G. 2008. Major depressive disorder. *New England Journal of Medicine* 358: 55–68.

Berger, Z., Roder, H., Hanna, A., et al. 2007. Accumulation of pathological tau species and memory loss in a conditional model of tauopathy. *The Journal of Neuroscience* 27: 3650–3662.

Berry, N.M., Davison, K., Coates, A.M., Buckley, J.D., Howe, P.R.C. 2010. Impact of cocoa flavanol consumption on blood pressure responsiveness to exercise. *British Journal of Nutrition* 103: 1480–1484.

Bhardwaj, J.C. 2000. Jams, jellies and marmalades. In *Postharvest Technology of Fruits and Vegetables*, Vol. 2, ed. L.R. Verma and V.K. Joshi, 597–636. New Delhi: Indus.

Bhat, J., Damle, A., Vaishnav, P.P., Albers, R., Joshi, M., Banerjee, G. 2009. *In vivo* enhancement of natural killer cell activity through tea fortified with ayurvedic herbs. *Phytotherapy Research* 24: 129–135.

Bhat, Z.F., Bhat, H. 2011. Functional meat products—A review. *International Journal of Meat Science* 1: 1–14.

Bhat, Z.F., Pathak, V. 2009. Effect of mung bean (*Vigna radiata*) on quality characteristics of oven roasted chicken *seekh kababs*. *Fleischwirtschaft International* 6: 58–60.

Bhullar, Y.S., Uddin, M.A., Shah, N.P. 2002. Effects of ingredients supplementation on textural characteristics and microstructure of yoghurt. *Milchwissenschaft* 57: 328–332.

Bisson, J.F., Nejdi, A., Rozan, P., Hidalgo, S., Lalonde, R., Messaoudi, M. 2008. Effects of long-term administration of a cocoa polyphenolic extract (Acticoa powder) on cognitive performances in aged rats. *British Journal of Nutrition* 100: 94–101.

Bodinham, C.L., Frost, G.S., Robertson, M.D. 2010. Acute ingestion of resistant starch reduces food intake in healthy adults. *British Journal of Nutrition* 103: 917–922.

Bogdanski, P., Suliburska, J., Szulinska, M., Stepien, M., Pupek-Musialik, D., Jablecka, A. 2012. Green tea extract reduces blood pressure, inflammatory biomarkers, and oxidative stress and improves parameters associated with insulin resistance in obese, hypertensive patients. *Nutrition Research* 32: 421–427.

Boobier, W.J., Baker, J.S., Davies, B. 2006. Development of a healthy biscuit—An alternative approach to biscuit manufacture. *Nutrition Journal* 5: 7.

Boobier, W.J., Baker, J.S., Hullen, D., Graham, M.R., Davies, B. 2007. Functional biscuits and coronary heart disease risk factors. *British Food Journal* 109: 260–267.

Borota, D., Murray, E., Keceli, G., et al. 2014. Post-study caffeine administration enhances memory consolidation in humans. *Nature Neuroscience* 17: 201–203.

Botelho, P.B., Galasso, M., Dias, V., et al. 2014. Oxidative stability of functional phytosterol-enriched dark chocolate. *LWT: Food Science and Technology* 55: 444–451.

Bouckley, B. 2013. World's first stevia-sweetened Coke hits Argentina. http://www.beveragedaily.com/Markets/World-s-first-stevia-sweetened-Coke-hits-Argentina?utm_source=newsletter_breaking_news&utm_medium=email&utm_campaign=Newsletter%2BBreaking%2BNews. Accessed June 28, 2013.

Brown, A.A., Hu, F.B. 2001. Dietary modulation of endothelial function—Implications for cardiovascular disease. *American Journal of Clinical Nutrition* 73: 673–686.

Brown, E. 2003. Mixes unmasked. In *The Complete Idiot's Guide to Cooking with Mixes*, 5. New York: Alpha Books.

Cao, C., Loewenstein, D.A., Lin, X. 2012a. High blood caffeine levels in MCI linked to lack of progression to dementia. *Journal of Alzheimer's Disease* 30: 559–572.

Cao, C., Loewenstein, D.A., Lin, X., et al. 2012b. High blood caffeine levels in MCI linked to lack of progression to dementia—Supplementary data. *Journal of Alzheimer's Disease* 30: 1–2.

Carducci, M.A., Paller, C.J., Wozniak, P., et al. 2011. A phase II study of pomegranate extract for men with rising prostate-specific antigen following primary therapy. *Journal of Clinical Oncology* 29 (Suppl. 7): Abstr 11.

Carillon, J., Notin, C., Schmitt, K., Simoneau, G., Lacan, D. 2014. Dietary supplementation with a superoxide dismutase-melon concentrate reduces stress, physical and mental fatigue in healthy people—A randomised, double-blind, placebo-controlled trial. *Nutrients* 6: 2348–2359.

Cavallini, D.C., Abdalla, D.S., Vendramini, R.C., et al. 2009. Effects of isoflavone-supplemented soy yogurt on lipid parameters and atherosclerosis development in hypercholesterolemic rabbits—A randomized double- blind study. *Lipids in Health and Disease* 8: 40.

Cetó, X., Gutiérrez-Capitán, M., Calvo, D., del Valle, M. 2013. Beer classification by means of a potentiometric electronic tongue. *Food Chemistry* 141: 2533–2540.

Chacko, D., Renitta, R.E., Patterson, J. 2005. Development of soup powder from squid *Sepioteuthis lessoniana* and shelf-life assessment during storage in laminated packaging material. *Journal of Food Technology* 3: 449–452.

Chen, T.M. 2009. Medicinal soups. In *A Tradition of Soup—Flavors from China's Pearl River Delta*, 259–272. Berkeley, CA: North Atlantic Books.

Chevance, F.F., Farmer, L.J., Desmond, E.M., Novelli, E., Troy, D.J., Chizzolini, R. 2000. Effect of some fat replacers on the release of volatile aroma compounds from low-fat meat products. *Journal of Agricultural and Food Chemistry* 48: 3476–3484.

Chrysan, M. 2005. Margarines and spreads. In *Bailey's Industrial Oil and Fat Products*, Vol. 4, ed. F. Shahidi, 33–82. Hoboken, NJ: Wiley.

Chudnovskiy, R., Thompson, A., Tharp, K., Hellerstein, M., Napoli, J.L., Stahl, A. 2014. Consumption of clarified grapefruit juice ameliorates high-fat diet induced insulin resistance and weight gain in mice. *PLoS ONE* 9(10): e108408.

Cienfuegos-Jovellanos, E., Quinones, M.M., Muguerza, B., Moulay, L., Miguel, M., Aleixandre, A. 2009. Antihypertensive effect of a polyphenol-rich cocoa powder industrially processed to preserve the original flavonoids of the cocoa beans. *Journal of Agricultural and Food Chemistry* 57: 6156–6162.

Cimini, A., Gentile, R., D'Angelo, B., et al. 2013. Cocoa powder triggers neuroprotective and preventive effects in a human Alzheimer's disease model by modulating BDNF signaling pathway. *Journal of Cellular Biochemistry* 114: 2209–2220.

Cleary, J.P., Walsh, D.M., Hofmeister, J.J., et al. 2005. Natural oligomers of the amyloid-β protein specifically disrupt cognitive function. *Nature Neuroscience* 8: 79–84.

Clesceri, L.S., Greenberg, A.E., Eaton, A.D. 1998. *Standard Methods for the Examination of Water and Wastewater*, 20th edn. Washington, DC: American Public Health Association.

Clifton, P.M., Noakes, M., Sullivan, D., et al. 2004. Cholesterol-lowering effects of plant sterol esters differ in milk, yoghurt, bread and cereal. *European Journal of Clinical Nutrition* 58: 503–509.

Collins, J.K., Wu, G., Perkins-Veazie, P., et al. 2007. Watermelon consumption increases plasma arginine concentrations in adults. *Nutrition* 23: 261–266.

Corti, R., Flammer, A.J., Hollenberg, N.K., Lüscher, T.F. 2009. Cocoa and cardiovascular health. *Circulation* 119: 1433–1441.

Csapo, I., Incze, K., Kovacks, A., Zelenak, L., Zsigo, J. 2006. Development of meat products with lutein for eye health. In *52nd International Congress of Meat Science and Technology*, ed. D. Troy, R. Pearce, B. Byrne, and J. Kerry, 687–688. Wageningen: Wageningen Academic.

Dahl, W.J., Whiting, S.J., Isaac, T.M., Weeks, S.J., Arnold, C.J. 2005. Effects of thickened beverages fortified with inulin on beverage acceptance, gastrointestinal function, and bone resorption in institutionalized adults. *Nutrition* 21: 308–311.

Danza, A., Mastromatteo, M., Cozzolino, F., et al. 2014. Processing and characterization of durum wheat bread enriched with antioxidant from yellow pepper flour. *LWT: Food Science and Technology* 59: 479–485.

Dartsch, P.C., Kler, A., Kriesl, E., Bonnländer, B. 2010. Evaluation of the health benefits of different functional drink concepts. *NutraCos* 9: 2–5.

Dauchet, L., Ferrieres, J., Arveiler, D., et al. 2004. Frequency of fruit and vegetable consumption and coronary heart disease in France and Northern Ireland—The PRIME study. *British Journal of Nutrition* 92: 963–972.

Dave, R.I., Shah, N.P. 1998. The effect of ingredient supplementation on the textural characteristics of yogurt. *Australian Journal of Dairy Technology* 53: 180–184.

Davidson, M.H., Maki, K.C., Umporowicz, D.M., et al. 2001. Safety and tolerability of esterified phytosterols administered in reduced-fat spread with salad dressing to healthy adult men and women. *Journal of the American College of Nutrition* 4: 307–319.

Davison, K., Coates, A.M., Buckley, J.D., Howe, P.R.C. 2008. Effect of cocoa flavanols and exercise on cardiometabolic risk factors in overweight and obese subjects. *International Journal of Obesity* 32: 1289–1296.

De Bruin, E.A., Rowson, M.J., Van Buren, L., Rycroft, J.A., Owen, G.N. 2011. Black tea improves attention and self-reported alertness. *Appetite* 56: 235–240.

De Cock, P., Bechert, C.L. 2002. Erythritol. Functionality in noncaloric functional beverages. *Pure and Applied Chemistry* 74: 1281–1289.

Del Valle, M. 2010. Electronic tongues employing electrochemical sensors. *Electroanalysis* 22: 1539–1555.

Del Valle, M. 2012. Sensor arrays and electronic tongue systems. *International Journal of Electrochemistry* 2012: 986025, doi:10.1155/2012/986025

Din, A., Bukhari, S.A.H., Salam, A., Ishfaq, B. 2011. Development of functional and dietetic beverage from bitter gourd. *Internet Journal of Food Safety* 13: 355–360.

Diplock, A.T., Aggett, P.J., Ashwel, M., Bornet, F., Fern, E.B., Roberfroid, M.B. 1999. Scientific concepts of functional foods in Europe—Consensus document. *British Journal of Nutrition* 81 (Suppl. 1): S1–S27.

Downes, F.P., Ito, K. 1998. *Compendium of Methods for the Microbiological Examination of Foods*, 4th edn. Washington, DC: American Public Health Association.

Duncan, S.E. 2004. Fats—Mayonnaise. In *Food Processing—Principles and Applications*, ed. J.S. Smith and Y.H. Hui, 329–342. Ames, IA: Blackwell.

Dzudie, T., Scher, J., Hardy, J. 2002. Common bean flour as an extender in beef sausages. *Journal of Food Engineering* 52: 143–147.

Einöther, S.J., Martens, V.E. 2013. Acute effects of tea consumption on attention and mood. *The American Journal of Clinical Nutrition* 98 (Suppl. 6): 1700S–1708S.

El-Gasim, E.A., Al-Wesali, M.S. 2000. Water activity and Hunter colour values of beef patties extended with samh (*Mesembryanthemum forsskalei* Hochst) flour. *Food Chemistry* 69: 181–185.

El-Haddad, N.N.M., Youssef, M.M., Abd El-Aal, M.H., Abou-Gharbia, H.H. 2011. Utilisation of red palm olein in formulating functional chocolate spread. *Food Chemistry* 124: 285–290.

Ellis, P.R., Kamalanathan, T., Dawoud, F.M., Strange, R.N., Coultate, T.P. 1988. Evaluation of guar gum biscuits for use in the management of diabetes: Tests of physiological effects and palatability in non-diabetic volunteers. *European Journal of Clinical Nutrition* 42: 425–435.

Evans, M., Wilson, D., Guthrie, N. 2014a. A randomized, double-blind, placebo-controlled, pilot study to evaluate the effect of whole grape extract on antioxidant status and lipid profile. *Journal of Functional Foods* 7: 680–691.

Evans, S.F., Meister, M., Mahmood, M., et al. 2014b. Mango supplementation improves blood glucose in obese individuals. *Nutrition and Metabolic Insights* 7: 77–84.

Ezekiel, R., Singh, N. 2011. Use of potato flour in bread and flat bread. In *Flour and Breads and Their Fortification in Health and Disease Prevention*, ed. V.R. Preedy, R.R. Watson, and V.B. Patel, 247–262. San Diego, CA: Academic Press.

Faraji, H.M., Tarkhani, A.H. 1999. The effect of sour tea (*Hibiscus sabdariffa*) on essential hypertension. *Journal of Ethnopharmacology* 65: 231–236.

Fazen, C.H., Valentin, D., Fairchild, T.J., Doyle, R.P. 2011. Oral delivery of the appetite suppressing peptide hPYY(3–36) through the Vitamin B_{12} uptake pathway. *Journal of Medicinal Chemistry* 54: 8707–8711.

FDA. 2004. *Code of Federal Regulations*. Washington, DC: US Department of Health and Human Services.

Fernandez-Gines, J.M., Fernandez-Lopez, J.M., Sayas-Barbera, E., Sendra, E., Perez-Alvarez, J.A. 2003. Effect of storage conditions on quality characteristics of bologna sausages made with citrus fiber. *Journal of Food Science* 68: 710–714.

Fernandez-Gines, J.M., Fernandez-Lopez, J.M., Sayas-Barbera, E., Sendra, E., Perez-Alvarez, J.A. 2004. Lemon albedo as a new source of dietary fiber—Application to bologna sausages. *Meat Science* 67: 7–13.

Ferre, S. 2008. An update on the mechanisms of the psychostimulant effects of caffeine. *Journal of Neurochemistry* 105: 1067–1079.

Ferre, S., Ciruela, F., Borycz, J., et al. 2008. Adenosine A1–A2A receptor heteromers—New targets for caffeine in the brain. *Frontiers in Bioscience* 13: 2391–2399.

Figueroa, A., Sanchez-Gonzalez, M.A., Perkins-Veazie, P.M., Arjmandi, B.H. 2011. Effects of watermelon supplementation on aortic blood pressure and wave reflection in individuals with prehypertension—A pilot study. *American Journal of Hypertension* 24: 40–44.

Figueroa, A., Sanchez-Gonzalez, M.A., Wong, A., Arjmandi, B.H. 2012. Watermelon extract supplementation reduces ankle blood pressure and carotid augmentation index in obese adults with prehypertension or hypertension. *American Journal of Hypertension* 25: 640–643.

Flaczyk, E., Górecka, D., Kobus, J., Szymandera-Buszka, K. 2009. The influence of inulin addition as fat substitute on reducing energy value and consumer acceptance of model pork meatballs. *ZYWNOŚ Nauka Technologia Jakość* 4: 41–46.

Fleming, H.P., McFeeters, R.F., Breidt, F. 2001. Fermented and acidified vegetables. In *Compendium of Methods for the Microbiological Examination of Foods*, ed. F.P. Downes and K. Ito, 520–532. Washington, DC: American Public Health Association.

Fleury, A.R., Funk, D.F., Patel, M.T., Vala, W.D. 1998. Calcium fortified yogurt and methods of preparation. US Patent No. 5820903.

Fogelman, A.M., Schechter, I., Seager, J., Hokom, M., Child, J.S., Edwards, P.A. 1980. Malondialdehyde alteration of low density lipoproteins leads to cholesteryl ester accumulation in human monocyte-macrophages. *Proceedings of the National Academy of Sciences (USA)* 77: 2214–2218.

Fox, P.F., Guinee, T.P., Cogan, T.M., McSweeney, P.L.H. 2000. Overview of cheese manufacture. In *Fundamentals of Cheese Science*, 10–18. Gaithersburg, MD: Aspen.

Fredholm, B.B., Battig, K., Holmen, J., Nehlig, A., Zvartau, E.E. 1999. Actions of caffeine in the brain with special reference to factors that contribute to its widespread use. *Pharmacological Reviews* 51: 83–133.

Frontela, C., Ros, G., Martínez, C., Sánchez-Siles, L.M., Canali, R., Virgilia, F. 2010. Stability of Pycnogenol® as an ingredient in fruit juices subjected to *in vitro* gastrointestinal digestion. *Journal of the Science of Food and Agriculture* 91: 286–292.

Gadang, V.P., Hettiarachchy, N.S., Johnson, M.G., Owens, C. 2008. Evaluation of antibacterial activity of whey protein isolate coating incorporated with nisin, grape seed extract, malic acid and EDTA on a turkey frankfurter system. *Food Science* 73: M389–394.

Gallagher, E., O'Brien, C.M., Scannell, A.G.M., Arendt, E.K. 2003. Evaluation of sugar replacers in short dough biscuit production. *Journal of Food Engineering* 56: 261–263.

Garcia, M.L., Dominguez, R., Galvez, M.D., Casas, C., Selgas, M.D. 2002. Utilization of cereal and fruit fibres in low fat, dry fermented sausages. *Meat Science* 60: 227–236.

Gawlik-Dzikia, U., Świecaa, M., Dzikib, D., Baraniaka, B., Tomiłob, J., Czyżc, J. 2013. Quality and antioxidant properties of breads enriched with dry onion (*Allium cepa* L.) skin. *Food Chemistry* 138: 1621–1628.

Gobbato, N., Rachid, M., Perdigon, G. 2008. Anti-inflammatory effect of yoghurt in an experimental inflammatory bowel disease in mouse. *Journal of Dairy Science* 75: 497–504.

Goldberg, I. 1999. Introduction. In *Functional Foods: Designer Foods, Pharmafoods, Nutraceuticals.* ed. I. Goldberg, 3–16. Gaithersburg, MD: Aspen.

Gómez, M., Jiménez, S., Ruiz, E., Oliete, B. 2011. Effect of extruded wheat bran on dough rheology and bread quality. *LWT: Food Science and Technology* 44: 2231–2237.

Gonzalez-Molina, E., Moreno, D.A., Garcia-Viguera, C. 2009. A new drink rich in healthy bioactives combining lemon and pomegranate juices. *Journal of Food Chemistry* 115: 1364–1372.

Goody, J. 1997. Industrial food. In *Food and Culture—A Reader,* 2nd edn, ed. C. Counihan and P. van Esterik, 340–343. New York: Routledge.

Gorelik, S., Lapidot, T., Shaham, I., et al. 2005. Lipid peroxidation and coupled vitamin oxidation is stimulated and human gastric fluid inhibited by dietary polyphenols—Health implications. *Journal of Agricultural and Food Chemistry* 53: 3397–3402.

Gorelik, S., Ligumsky, M., Kohen, R., Kanner, J. 2008. The stomach as a "bioreactor"—When red meat meets red wine. *Journal of Agricultural and Food Chemistry* 56: 5002–5007.

Goyle, A., Gujral, S. 1992. Sensory evaluation of and acceptability trials on biscuits prepared from raw and malted wheat (*Triticum aestivum*), Bengal gram (*Cicer arietinum*) mixes with or without a green leafy vegetable. *Plant Foods for Human Nutrition* 42: 291–296.

Graham, H.N. 1992. Green tea composition, consumption, and polyphenol chemistry. *Preventive Medicine* 21: 334–350.

Granado-Lorencio, F., Lopez-Lopez, I., Herrero-Barbudo, C., et al. 2010. Lutein-enriched frankfurter-type products—Physicochemical characteristics and lutein *in vitro* bioaccessibility. *Food Chemistry* 120: 741–748.

Grassi, D., Lippi, C., Necozione, S., Desideri, G., Ferri, C. 2005a. Short-term administration of dark chocolate is followed by a significant increase in insulin sensitivity and a decrease in blood pressure in healthy persons. *American Journal of Clinical Nutrition* 81: 611–614.

Grassi, D., Necozione, S., Lippi, C., et al. 2005b. Cocoa reduces blood pressure and insulin resistance and improves endothelium-dependent vasodilation in hypertensives. *Hypertension* 46: 398–405.

Grivetti, L.E., Wilson, T. 2004. A brief history of human beverage consumption. In *Beverages in Nutrition and Health*, ed. T. Wilson and N.J. Temple, 1–18. Totowa, MA: Humana Press.

Grove, K.A., Lambert, J.D. 2010. Laboratory, epidemiological, and human intervention studies show that tea (*Camellia sinensis*) may be useful in the prevention of obesity. *The Journal of Nutrition* 140: 446–453.

Gujral, H.S., Kaur, A., Singh, N., Sodhi, S.N. 2002. Effect of liquid whole egg, fat and textured soy protein on the textural and cooking properties of raw and baked patties from goat meat. *Journal of Food Engineering* 53: 377–385.

Gunstone, F.D. 2003. *Lipids for Functional Foods and Nutraceuticals*. Bridgewater: The Oily Press.

Guven, M., Yasar, K., Karaca, O.B., Hayaloglu, A.A. 2005. The effect of inulin as a fat replacer on the quality of set-type low-fat yogurt manufacture. *International Journal of Dairy Technology* 58: 180–184.

Haberland, M.E., Fong, D., Cheng, L. 1988. Malondialdehyde-altered protein occurs in atheroma of Watanabe heritable hyperlipdemic rabbits. *Science* 241: 215–218.

Hallet, A., Hallet, D. 1997. *Entrepreneur Magazine Encyclopedia of Entrepreneurs*, 169–172. Hoboken, NJ: Wiley.

Halvorsen, B., Carlsen, M., Phillips, K., et al. 2006. Content of redox-active compounds (ie. antioxidants) in foods consumed in the United States. *American Journal of Clinical Nutrition* 84: 95–135.

Han, J., Britten, M., St-Gelais, D., et al. 2011. Polyphenolic compounds as functional ingredients in cheese. *Food Chemistry* 124: 1589–1594.

Hanas, O.P. 1994. Seasoning ingredients. In *Handbook of Industrial Seasonings*, ed. E.W. Underreiner and I.R. Hume, 20–41. Glasgow: Chapman & Hall.

Hannig, C., Sorg, J., Spitzmüller, B., Hannig, M., Al-Ahmad, A. 2009. Polyphenolic beverages reduce initial bacterial adherence to enamel *in situ*. *Journal of Dentistry* 37: 560–566.

Harbowy, M.E., Balentine, D.A. 1997. Tea chemistry. *Critical Reviews in Plant Sciences* 16: 415–480.

Haripriya, S., Premakumari, S. 2010. Effect of wheat bran on diabetic subjects. *Indian Journal of Science and Technology* 3: 284–286.

Harms-Ringdahl, M., Jenssen, D., Haghdoost, S. 2012. Tomato juice intake suppressed serum concentration of 8-oxodG after extensive physical activity. *Nutrition Journal* 11: 29.

Haro, C., Lazarte, S., Zelaya, H., Alvarez, S., Aguero, G. 2009. Yogurt—Effect on leukocytes and blood coagulation in an acute liver injury model. *Journal of Medicinal Food* 12: 796–802.

Haub, M.D., Hubach, K.L., Al-Tamimi, E.K., Ornelas, S., Seib, P.A. 2010. Different types of resistant starch elicit different glucose responses in humans. *Journal of Nutrition and Metabolism* 2010: 230501. doi:10.1155/2010/230501

Hayta, M., Őzuğur, G. 2011. Phytochemical fortification of flour and bread. In *Flour and Breads and Their Fortification in Health and Disease Prevention*, ed. V.R. Preedy, R.R. Watson, and V.B. Patel, 293–300. San Diego, CA: Academic Press.

Hegenbart, S.L. 1998. Alternative enhancers. *Food Product Design* 7: 60–71.

Heiss, C., Dejam, A., Kleinbongard, P., Schewe, T., Sies, H., Kelm, M. 2003. Vascular effects of cocoa rich in flavan-3-ols. *Journal of the American Medical Association* 290: 1030–1031.

Heiss, C., Kleinbongard, P., Dejam, A., et al. 2005. Acute consumption of flavanol-rich cocoa and the reversal of endothelial dysfunction in smokers. *Journal of the American College of Cardiology* 46: 1276–1283.

Hekmat, S., McMahon, D.J. 1997. Manufacture and quality of iron-fortified yogurt. *Journal of Dairy Science* 80: 3114–3122.

Herder, K., Erlund, I., Kolb, H., et al. 2010. Effects of coffee consumption on subclinical inflammation and other risk factors for type 2 diabetes—A clinical trial. *American Journal of Clinical Nutrition* 91: 950–957.

Herrera-Arellano, A., Flores-Romero, S., Chavez-Soto, M.A., Tortoriello, J. 2004. Effectiveness and tolerability of a standardized extract from *Hibiscus sabdariffa* in patients with mild to moderate hypertension—A controlled and randomized clinical trial. *Phytomedicine* 11: 375–382.

Herrera-Arellano, A., Miranda-Sanchez, J., Avila-Castro P., et al. 2007. Clinical effects produced by a standardized herbal medicinal product of *Hibiscus sabdariffa* on patients with hypertension. A randomized, double-blind, lisinopril-controlled clinical trial. *Planta Medica* 73: 6–12.

Heymsfield, S.B., Allison, D.B., Vasselli, J.R., Pietrobelli, A., Greenfield, D., Nunez, C. 1998. *Garcinia cambogia* (hydroxycitric acid) as a potential antiobesity agent—A randomized controlled trial. *Journal of American Medical Association* 280: 1596–1600.

Hietanen, H., Pääkkönen, R., Salomaa, V. 2008. Ankle blood pressure as a predictor of total and cardiovascular mortality. *BMC Cardiovascular Disorders* 8: 3.

Higashikawa, F., Noda, M., Awaya, T., Nomura, K., Oku, H., Sugiyama, M. 2010. Improvement of constipation and liver function by plant-derived lactic acid bacteria—A double-blind, randomized trial. *Nutrition* 26: 367–374.

Higdon, J., Frei, B. 2006. Coffee and health—A review of recent human research. *Critical Reviews in Food Science and Nutrition* 46: 101–123.

Hobbs, D.A., Goulding, M.G., Nguyen, A., et al. 2013. Acute ingestion of beetroot bread increases endothelium-independent vasodilation and lowers diastolic blood pressure in healthy men—A randomized controlled trial. *The Journal of Nutrition* 43: 1399–1405.

Hodgson, J.M., Croft, K.D., Woodman, R.J., et al. 2013. Black tea lowers the rate of blood pressure variation—A randomized controlled trial. *American Journal of Clinical Nutrition* 97: 943–950.

Hooda, S., Jood, S. 2003. Physicochemical, rheological and organoleptic characteristics of wheat-fenugreek supplemented blends. *Nahrung* 47: 265–268.

Hortin, G.L. 2006. Homocysteine—Clinical significance and laboratory measurement. *Laboratory Medicine* 37: 551–553.

Huang, S.C., Tsai, Y.F., Chen, C.M. 2011. Effects of wheat fiber, oat fiber and inulin on sensory and physicochemical properties of Chinese-style sausages. *Asian–Australian Journal of Animal Sciences* 24: 875–880.

Hubbard, G.P., Wofram, S., de Vos, R., Bovy, A., Gibbins, J.M., Lovegrove, J.A. 2006. Ingestion of onion soup high in quercetin inhibits platelet aggregation and essential components of the collagen-stimulated platelet activation pathway in man—A pilot study. *British Journal of Nutrition* 96: 482–488.

Hurst, W.J. 2007. *Methods of Analysis for Functional Foods and Nutraceuticals*, 2nd edn. Boca Raton, FL: CRC Press.

Hussain, S., Anjum, F.M., Butt, M.S., Khan, M.I., Asghar, A. 2006. Physical and sensoric attributes of flaxseed flour supplemented cookies. *Turkish Journal of Biology* 30: 87–92.

Ide, K., Yamada, H., Takuma, N., et al. 2014. Green tea consumption affects cognitive dysfunction in the elderly—A pilot study. *Nutrients* 6: 4032–4042.

Ifesan, B.O., Siripongvutikorn, S., Voravuthikunchal, S.P. 2009. Application of *Eleutherine americana* crude extract in homemade salad dressing. *Journal of Food Protection* 72: 650–655.

IFST. 1993. *Shelf Life of Foods–Guidelines for Its Determination and Prediction*, 1–84. London: Institute of Food Science and Technology.

Jacob, H.E. 2007. *Six Thousand Years of Bread*, 1–109. New York: Skyhorse.

Jacobson, M.F. 2004. Liquid candy—How soft drinks harm the health of Americans. In *Beverages in Nutrition and Health*, ed. T. Wilson and N.J. Temple, 351–366. Totowa, MA: Humana Press.

Jenkins, A.L., Jenkins, D.J.A., Wolever, T.M.S. 2008. Comparable postprandial glucose reductions with viscous fiber blend enriched biscuits in healthy subjects and patients with diabetes mellitus—Acute randomized controlled clinical trial. *Croatian Medical Journal* 49: 772–782.

Jensen, G.S., Ager, D.M., Redman, K.A., Mitzner, M.A., Benson, K.F., Schauss, A.G. 2011. Pain reduction and improvement in range of motion after daily consumption of an açai (*Euterpe oleracea* Mart.) pulp-fortified polyphenolic-rich fruit and berry juice blend. *Journal of Medicinal Food* 14: 702–711.

Jensen, G.S., Wu, X., Patterson, K.M., et al. 2008. In vitro and in vivo antioxidant and anti-inflammatory capacities of an antioxidant-rich fruit and berry juice blend. Results of a pilot and randomized, double-blinded, placebo-controlled, crossover study. *Journal of Agricultural and Food Chemistry* 56: 8326–8333.

Jiménez-Colmenero, F., Carballo, J., Cofrades, S. 2001. Healthier meat and meat products—Their role as functional foods. *Meat Science* 59: 5–13.

Jin, Y., Cui, X., Singh, U.P., et al. 2010. Systemic inflammatory load in humans is suppressed by consumption of two formulations of dried, encapsulated juice concentrate. *Molecular Nutrition & Food Research* 54: 1–9.

Johnsen, S.P., Overvad, K., Stripp, C., Tjonneland, A., Husted, S.E., Sorensen, H.T. 2003. Intake of fruit and vegetables and the risk of ischemic stroke in a cohort of Danish men and women. *American Journal of Clinical Nutrition* 78: 57–64.

Jones, S.T., Aryana, K.J., Losso, J.N. 2005. Storage stability of lutein during ripening of cheddar cheese. *Journal of Dairy Science* 88: 1661–1670.

Joshi, V.K., Bhat, A. 2000. Pickles–Technology of preparation. In *Postharvest Technology of Fruits and Vegetables*, Vol. 2, ed. L.R. Verma and V.K. Joshi, 777–820. New Delhi: Indus Publishing.

Judd, J.T., Baer, D.J., Chen, S.C. 2002. Plant sterol esters lower plasma lipids and most carotenoids in mildly hypercholesterolemic adults. *Lipids* 37: 33–42.

Kanae, O., Tomoko, K., Yoshiaki, M. 2008. Effect of lemon flavonoid on the properties of bread. *Japan Society of Cookery Science* 41: 297–303.

Kanner, J., Lapidot, T. 2001. The stomach as a bioreactor: Dietary lipid peroxidation in the gastric fluid and the effects of plant-derived antioxidants. *Free Radical Biology & Medicine* 31: 1388–1395.

Kao, Y.H., Chang, H.H., Lee, M.J., Chen, C.L. 2006. Tea, obesity and diabetes. *Molecular Nutrition & Food Research* 50: 188–210.

Karaoglu, M.M. 2006. Effect of baking procedure and storage on the pasting properties and staling of part-baked and rebaked wheat bran bread. *International Journal of Food Science and Technology* 41: 77–82.

Karim, M., McCormick, K., Kappagoda, C.T. 2000. Effects of cocoa extracts in endothelium-dependent relaxation. *Journal of Nutrition* 130: 2105S–2108S.

Kataoka, S. 2005. Functional effects of Japanese style fermented soy sauce (shoyu) and its compounds. *Journal of Bioscience and Bioengineering* 100: 227–234.

Keen, C.L., Holt, R.R., Oteiza, P.I., Fraga, C.G., Schmitz, H.H. 2005. Cocoa antioxidants and cardiovascular health. *American Journal of Clinical Nutrition* 81: 298S–303S.

Kempf, K., Herder, C., Erlund, I., et al. 2010. Effects of coffee consumption on subclinical inflammation and other risk factors for type 2 diabetes—A clinical trial. *American Journal of Clinical Nutrition* 91: 950–957.

Kempf, K., Rathmann, W., Herder, C. 2008. Impaired glucose regulation and type 2 diabetes in children and adolescents. *Diabetes Metabolism Research and Reviews* 24: 427–437.

Khaleghi, M., Kullo, I.J. 2007. Aortic augmentation index is associated with the ankle-brachial index—A community-based study. *Atherosclerosis* 195: 248–253.

Kieling, G., Schneider, J., Jahreis, G. 2002. Long-term consumption of fermented dairy products over 6 months increases HDL cholesterol. *European Journal of Clinical Nutrition* 56: 843–849.

Kilcast, D., Subramaniam, P. 2000. Introduction. In *The Stability and Shelf-Life of Food*, ed. D. Kilcast and P. Subramaniam, 1–22. Boca Raton, FL: CRC Press.

Kim, E.S., Liang, Y.R., Jin, J., et al. 2006. Impact of heating on chemical compositions of green tea liquor. *Food Chemistry* 103: 1263–1267.

Kitawaki, R., Nishimura, Y., Takagi, N., Iwasaki, M., Tsuzuki, K., Fukuda, M. 2009. Effects of lactobacillus fermented soymilk and soy yogurt on hepatic lipid accumulation in rats fed a cholesterol-free diet. *Bioscience, Biotechnology and Biochemistry* 73: 1484–1488.

Klein, W.L. 2002. Aβ toxicity in Alzheimer's disease—Globular oligomers (ADDLs) as new vaccine and drug targets. *Neurochemistry International* 41: 345–352.

Kolb, H., Mandrup-Poulsen, T. 2005. An immune origin of type 2 diabetes? *Diabetologia* 48: 1038–1050.

Komarik, S.L., Luh, B.S., Sutton, A.K. 1975. Vegetables in soups. In *Commercial Vegetable Processing*, ed. B.S. Luh and J.G. Woodroof, 729–745. Westport, CT: AVI.

Koyama, S., Cobb, L.J., Metha, H.H., et al. 2010. Pomegranate extract induces apoptosis in human prostate cancer cells by modulation of the IGF–IGFBP axis. *Growth Hormone & IGF Research* 20: 55–62.

Kriangkraiphiphat, K., Cheowtirakul, C. 2000. Ready-to-drink- green tea with ginger and honey beverage. In *Proceedings of the 4th International Conference on Food Science and Technology*, 377–396, Wuxi, China.

Kubomara, K. 1998. Japan redefines functional foods. *Prepared Foods* 167: 129–132.

Kumar, P., Mishra, H.N. 2006. Moisture sorption characteristics of mango-soy-fortified yogurt powder. *International Journal of Dairy Technology* 59: 22–28.

Kumar, P., Yadava, R.K., Gollen, B., Kumar, S., Verma, R.K., Yadav, S. 2011. Nutritional contents and medicinal properties of wheat—A review. *Life Sciences and Medicine Research* 22: 1–10.

Ladas, S.D., Kamberoglou, D., Karamanolis, G., Vlachogiannakos, J., Zouboulis-Vafiadis, I. 2013. Systematic review—Coca-Cola can effectively dissolve gastric phytobezoars as a first-line treatment. *Alimentary Pharmacology and Therapeutics* 37: 169–173.

Lai, C.S., Hoseney, R.C., Davis, A.B. 1989. Effects of wheat bran in breadmaking. *Cereal Chemistry* 66: 217–219.

Lansley, K.L., Winyard, P.G., Fulford, J., et al. 2011. Dietary nitrate supplementation reduces the O_2 cost of walking and running—A placebo-controlled study. *Journal of Applied Physiology* 110: 591–600.

Latimer, G.W. 2012. *Official Methods of Analysis of AOAC International*, 19th edn, Vols. 1 & 2. Gaithersburg, MD: AOAC International.

Lavelli, V., Vantaggi, C., Corey, M., Kerr, W. 2010. Formulation of a dry green tea-apple product—Study on antioxidant and color stability. *Journal of Food Science* 75: C184–190.

Lee, J., Jung, E., Hur, S., Park, D. 2009. Antibacterial and anti-inflammatory effects of a magnolia extract. *Cosmetics & Toiletries* 124: 53–60.

Lee, K.W., Kim, Y.J., Lee, H.J., Lee, C.Y. 2003. Cocoa has more phenolic phytochemicals and higher antioxidant capacity than teas and red wines. *Journal of Agricultural and Food Chemistry* 51: 7292–7295.

Li, Z., Henning, S.M., Zhang, Y., et al. 2010. Antioxidant-rich spice added to hamburger meat during cooking results in reduced meat, plasma and urine malondialdehyde concentrations. *American Journal of Clinical Nutrition* 91: 1180–1184.

Linarès, E., Thimonier, C., Degre, M. 2007. The effect of NeOpuntia on blood lipid parameters—Risk factors for the metabolic syndrome (syndrome X). *Advances in Therapy* 24: 1115–1125.

Llanos, A.A., Peng, J., Pennell, M.L., et al. 2014. Effects of tomato and soy on serum adipokine concentrations in postmenopausal women at increased breast cancer risk—A cross-over dietary intervention trial. *Journal of Clinical Endocrinology & Metabolism* 99: 625–632.

Lluch, A., Hanet-Geisen, N., Salah, S., Salas-Salvado, J., L'Heureux-Bouron, D., Halford, J.C.G. 2010. Short-term appetite-reducing effects of a low-fat dairy product enriched with protein and fibre. *Food Quality and Preference* 21: 402–409.

Loffredo, L., Perri, L., Catasca, E., et al. 2014. Dark chocolate acutely improves walking autonomy in patients with peripheral artery disease. *Journal of the American Heart Association* 3: e001072. doi: 10.1161/JAHA.114.001072.

Lohar, D.R. 2010. *Protocol for Testing Ayurveda, Siddha & Unani Medicines*. Ghaziabad: Pharmacopoeial Laboratory for Indian Medicines.

Lourens-Hattingh, A., Viljoen, B.C. 2001. Yogurt a probiotic carrier food. *International Dairy Journal* 11: 1–17.

Lucas, M., Mirzaei, F., Pan, A. 2011. Coffee, caffeine, and risk of depression among women. *Archives of Internal Medicine* 171: 1571–1578.

Lucas, M., O'Reilly, E.J., Pan, A., et al. 2013. Coffee, caffeine, and risk of completed suicide—Results from three prospective cohorts of American adults. *The World Journal of Biological Psychiatry* 15: 377–386.

Manley, D. 2003a. Setting the scene: History and position of biscuits. In *Technology of Biscuits, Crackers and Cookies*, 3rd edn, 1–8. Boca Raton, FL: CRC Press.

Manley, D. 2003b. Position of biscuits in nutrition. In *Technology of Biscuits, Crackers and Cookies*, 3rd edn, 307–313. Boca Raton, FL: CRC Press.

Manley, D. 2005. Recipes for dietetic biscuits. In *Biscuit, Cracker and Cookie Recipes for the Food Industry*, 168–173. Boca Raton, FL: CRC Press.

Mantzioris, E., Cleland, L.G., Gibson, R.A., Neumann, M.A., Demasi, M., James, M.J. 2000. Biochemical effects of a diet containing foods enriched with n-3 fatty acids. *American Journal of Clinical Nutrition* 72: 42–48.

Marinangeli, C.P.F., Jones, P.J.H. 2010. The use of functional plant ingredients for the development of efficacious functional foods. In *Functional Food Product Development*, ed. J. Smith and E. Charter, 110–134. Chichester: Wiley-Blackwell.

Markman, G., Livney, Y.D. 2012. Maillard-conjugate based core–shell co-assemblies for nanoencapsulation of hydrophobic nutraceuticals in clear beverages. *Food & Function* 3: 262–270.

Marshall, R.T. 1993. *Standard Methods for the Examination of Dairy Products*. Washington, DC: APHA.

Martin, F.J., Rezzi, S., Trepat, E.P., et al. 2010. Metabolic effects of dark chocolate consumption on energy, gut microbiota and stress-related metabolism in free-living subjects. *Journal of Proteome Research* 8: 5568–5579.

Martinex-Cerbera, S., Salvador, A., Sanz, T. 2014. Comparison of different polyols as total sucrose replacers in muffins—Thermal, rheological, texture and acceptability properties. *Food Hydrocolloids* 35: 1–8.

Martini, S., Thurgood, J.E., Brothersen, C., Ward, R., McMahon, D.J. 2009. Fortification of reduced-fat Cheddar cheese with n-3 fatty acids—Effect on off-flavor generation. *Journal of Dairy Science* 92: 1876–1884.

Marwaha, U., Marwaha, S.S. 2000. Production of chutneys and sauces. In *Postharvest Technology of Fruits and Vegetables*, Vol. 2, ed. L.R. Verma and V.K. Joshi, 742–776. New Delhi: Indus Publishing.

Mathews, A.C. 1999. Beverage flavorings and their applications. In *Food Flavorings*, 3rd edn, ed. P.R. Ashurst, 199–228. Gaithersburg, MD: Aspen.

Mattila, T., Ahvenainen, R. 1989. Preincubation time and the use of oxygen indicators in determining the microbiological quality of aseptically packed pea and tomato soup. *International Journal of Food Microbiology* 9: 205–214.

Mattila, T., Tawast, J., Ahvenainen, R. 1990. New possibilities for quality control of aseptic packages: Microbiological spoilage and seal defect detection using head-space indicators. *Lebensmittel-Wissenschaft & Technologie* 23: 246–251.

Mazloomi, S.M., Shekarforoush, S.S., Ebrahimnejad, H., Sajedianfard, J. 2011. Effect of adding inulin on microbial and physicochemical properties of low fat probiotic yogurt. *Iranian Journal of Veterinary Research* 12: 93–98.

McGovern, P.E., Christofidou-Solomidou, M., Wang, W., Dukes, F., Davidson, T., El-Deiry, W.S. 2010. Anticancer activity of botanical compounds in ancient fermented beverages (Review). *International Journal of Oncology* 37: 5–14.

McIntyre, A., Young, G.P., Taranto, T., Gibson, P.R., Ward, P.B. 1991. Different fibers have regional effects on luminal contents of rat colon. *Gastroenterology* 101: 1274–1281.

McKay, D.L., Chen, C.Y., Saltzman, E., Blumberg, J.B. 2010. *Hibiscus sabdariffa* L. tea (Tisane) lowers blood pressure in prehypertensive and mildly hypertensive adults. *The Journal of Nutrition* 140: 298–303.

Medina-Remón, A., Vallverdú-Queralt, A., Arranz, S., et al. 2012. Gazpacho consumption is associated with lower blood pressure and reduced hypertension in a high cardiovascular risk cohort. Cross-sectional study of the PREDIMED trial. *Nutrition, Metabolism & Cardiovascular Diseases* 23: 944–952.

Mellies, M.J., Vitale, C., Jandacek, R.J., Lamkin, G.E., Glueck, C.J. 1985. The substitution of sucrose polyester for dietary fat in obese, hypocholesterolemic outpatients. *American Journal of Clinical Nutrition* 41: 1–12.

Mercadante, A.Z., Capitani, C.D., Decker, E.A., Castro, I.A. 2010. Effect of natural pigments on the oxidative stability of sausages stored under refrigeration. *Meat Science* 84: 718–726.

Messerli, F.H. 2012. Chocolate consumption, cognitive function, and Nobel laureates. *The New England Journal of Medicine* 367: 1562–1564.

Midoh, N., Noguchi, T. 2009. Effect of chicken soup intake on mood states and peripheral blood flow in humans. *Journal of Health Science* 55: 56–61.

Miñarroa, B., Albanellb, E., Aguilara, N., Guamisa, B., Capellas, M. 2012. Effect of legume flours on baking characteristics of gluten-free bread. *Journal of Cereal Science* 56: 476–481.

Mizrahi, S. 2000. Accelerated shelf-life tests. In *The Stability and Shelf-Life of Food*, ed. D. Kilcast and P. Subramaniam, 107–128. Boca Raton, FL: CRC Press.

Mladenoska, I., Nikolovska, V., Puzderliska, L. 2012. Model meat pasteurized sausages enriched with monolaurin as nutraceuticals with pronounced antimicrobial properties. *Food & Feed Research* 39: 69–71.

Modi, V.K., Mahendrakar, N.S., Rao, N.D., Sachindra, N.M. 2003. Quality of buffalo meat burger containing legume flours as binders. *Meat Science* 66: 143–149.

Mohagheghi, M., Rezaei, K., Labbafi, M., Mousavi, S.M.E. 2011. Pomegranate seed oil as a functional ingredient in beverages. *European Journal of Lipid Science and Technology* 113: 730–736.

Morand, C., Dubray, C., Milenkovic, D., et al. 2010. Hesperidin contributes to the vascular protective effects of orange juice—A randomized crossover study in healthy volunteers. *American Journal of Clinical Nutrition* 93: 73–80.

Moreau, R.A. 2005. Corn oil. In *Bailey's Industrial Oil and Fat Products*, Vol. 2, ed. F. Shahidi, 149–172. Hoboken, NJ: Wiley.

Morris, C., Lichtenwalter, G. 1984. Fats and fatty acids. In *Encyclopedia of Chemical Processing and Design*, No. 21, ed. J.M. McKetta, 193–257. New York: Marcel Dekker.

Mousavi, Z.E., Mousavi, S.M., Razavi, S.H., Emam-Djomeh, Z., Kiani, H. 2011. Fermentation of pomegranate juice by probiotic lactic acid bacteria. *World Journal of Microbiology and Biotechnology* 27: 123–128.

Mozaffari-Khosravi, H., Jalali-Khanabadi, B.A., Afkhami-Ardekani, M., Fatehi, F., Noori-Shadkam, M. 2009. The effects of sour tea (*Hibiscus sabdariffa*) on hypertension in patients with type II diabetes. *Journal of Human Hypertension* 23: 48–54.

Mullen, W., Gonzalez, J., Siwy, J., et al. 2011. A pilot study on the effect of short-term consumption of a polyphenol rich drink on biomarkers of coronary artery disease defined by urinary proteomics. *Journal of Agricultural and Food Chemistry* 59: 12850–12857.

Müller, S.A., Rahbari, N.N., Schneider, F., et al. 2012. Randomized clinical trial on the effect of coffee on postoperative ileus following elective colectomy. *British Journal of Surgery* 99: 1530–1538.

Mun, S., Kim, Y.L., Kang, C.G., Park, K.H., Shim, J.Y., Kim, Y.R. 2009. Development of reduced-fat mayonnaise using 4αGTase-modified rice starch and xanthan gum. *International Journal of Biological Macromolecules* 44: 400–407.

Nagatsuka, N., Harada, K., Ando, M., Nagao, K. 2006. Measurement of the radical scavenging activity of chicken jelly soup, a part of the medicated diet Yazuken made from gelatin gel food Nikogori, using chemiluminiscence and electron spin resonance methods. *International Journal of Molecular Medicine* 18: 107–111.

Nemzer, B.V., Rodriguez, L.C., Hammond, L., DiSilvestro, R., Hunter, J.M., Pietrzkowski, Z. 2011. Acute reduction of serum 8-iso-PGF2-alpha and advanced oxidation protein products *in vivo* by a polyphenol-rich beverage—A pilot clinical study with phytochemical and *in vitro* antioxidant characterization. *Nutrition Journal* 10: 67.

Nielsen, S.S. 2003a. Introduction to food analysis. In *Food Analysis*, 3rd edn, 3–13. New York: Plenum.

Nielsen, S.S. 2003b. *Food Analysis*, 3rd edn, 1–557. New York: Plenum.

Nilsen, R., Pripp, A.H., Høstmark, A.T., Haug, A., Skeie, S. 2014. Short communication—Is consumption of a cheese rich in angiotensin-converting enzyme-inhibiting peptides, such as the Norwegian cheese Gamalost, associated with reduced blood pressure? *Journal of Dairy Science* 97: 2662–2668.

Nittynen, L.H., Jauhiainen, T.A., Poussa, T.A., Korpela, R. 2008. Effects of yoghurt enriched with free plant sterols on the levels of serum lipids and plant sterols in moderately hypercholesterolaemic subjects on a high-fat diet. *International Journal of Food Sciences & Nutrition* 59: 357–367.

Nollet, L.M.L. 2004. *Handbook of Food Analysis*, Vols. 1–3. Boca Raton, FL: CRC Press.

Nowicka, G., Naruszewicz, M. 2004. Assessing health claims for functional foods. In *Functional Foods, Cardiovascular Disease and Diabetes*, ed. A. Arnoldi, 10–18. Boca Raton, FL: Woodhead/CRC Press.

O'Connor, L.M., Lentjes, M.A.H., Luben, R.N., Khaw, K.T., Wareham, N.J., Forouhi, N.G. 2014. Dietary dairy product intake and incident type 2 diabetes—A prospective study using dietary data from a 7-day food diary. *Diabetologia* 57: 909–917.

Oboh, G., Raddatz, H., Henle, T. 2009. Characterization of the antioxidant properties of hydrophilic and lipophilic extracts of jute (*Corchorus olitorius*) leaf. *International Journal of Food Sciences & Nutrition* 60 (Supplement 2): 124–134.

Ochiai, M., Hayashi, T., Morita, M., et al. 2010. Short-term effects of l-citrulline supplementation on arterial stiffness in middle-aged men. *International Journal of Cardiology* 155: 257–261.

Ochiai, R., Sugiura, Y., Shioya, Y., Otsuka, K., Katsuragi, Y., Hashiguchi, T. 2013. Coffee polyphenols improve peripheral endothelial function after glucose loading in healthy male adults. *Nutrition Research* 34: 155–159.

Odes, H.S., Madar, Z., Trop, M., Namir, S., Gross, J., Cohen, T. 1986. Pilot study of the efficacy of spent grain dietary fiber in the treatment of constipation. *Israeli Journal of Medical Sciences* 22: 12–15.

Oke, M., Jacob, J.K., Paliyath, G. 2010. Effect of soy lecithin in enhancing fruit juice/sauce quality. *Food Research International* 43: 232–240.

Onweluzo, J.C., Morakinyo, A.O. 1997. Effect of pre-dehulling treatments on the composition of seeds of the legume *Afzelia africana* and its potential use in pastries. *Plant Foods for Human Nutrition* 50: 203–210.

Orea-Tejeda, A., Orozco-Gutiérrez, J.J., Castillo-Martínez, L., et al. 2010. The effect of L-arginine and citrulline on endothelial function in patients in heart failure with preserved ejection fraction. *Cardiology Journal* 17: 464–470.

Owokotomo, I.A., Owoeye, G. 2011. Proximate analysis and antimicrobial activities of *Bambusa vulgaris* L. leaves' beverage. *African Journal of Agricultural Research* 6: 5030–5032.

Pacquit, A., Frisby, J., Diamond, D., Lau, K.T., Farrell, A., Quilty, B. 2007. Development of a smart packaging for the monitoring of fish spoilage. *Food Chemistry* 102: 466–470.

Pacquit, A., Lau, K. T., McLaughlin, H., Frisby, J., Quilty, B., Diamond, D. 2006. Development of a volatile amine sensor for the monitoring of fish spoilage. *Talanta* 69: 515–520.

Palinski, W., Rosenfeld, M.E., Yla-Herttuala, S., et al. 1989. Low density lipoprotein undergoes oxidative modification *in vivo*. *Proceedings of the National Academy of Sciences (USA)* 86: 1372–1376.

Pantuck, A.J., Leppert, J.T., Zomorodian, N., et al. 2006. Phase II study of pomegranate juice for men with rising prostate-specific antigen following surgery or radiation for prostate cancer. *Clinical Cancer Research* 12: 4018–4026.

Park, J.B., Charbonneau, F., Schiffrin, E.L. 2001. Correlation of endothelial function in large and small arteries in human essential hypertension. *Journal of Hypertension* 19: 415–420.

Parker, R. 2001. Beverages. *Introduction to Food Science*, 431–433. New York: Delmar Thomson Learning.

Paseephol, T., Small, D.M., Sherkat, F. 2008. Rheology and texture of set yogurt as affected by inulin addition. *Journal of Texture Studies* 39: 617–634.

Peng, X., Ma, J., Cheng, K.W., Jiang, Y., Chen, F., Wang, M. 2010. The effects of grape seed extract fortification on the antioxidant activity and quality attributes of bread. *Food Chemistry* 119: 49–53.

Perrigue, M.M., Monsivais, P., Drewnowski, A. 2009. Added soluble fiber enhances the satiating power of low-energy-density liquid yogurts. *Journal of American Dietetic Association* 109: 1862–1868.

Persaud, K., Dodd, G. 1982. Analysis of discrimination mechanisms in the mammalian olfactory system using a model nose. *Nature* 299: 352–355.

Peters, C.M., Green, R.J., Janle, E.M., Ferruzzi, M.G. 2010. Formulation with ascorbic acid and sucrose modulates catechin bioavailability from green tea. *Food Research International* 43: 95–102.

Pham, M., Yosida, D., Yin, G., Ohnaka, K., Takayanagi, R., Kono, S. 2010. The relationship of coffee and green tea consumption with high-sensitivity C-reactive protein in Japanese men and women. *Clinical Chemistry and Laboratory Medicine* 48: 849–854.

Pietrasik, Z., Duda, Z. 2000. Effect of fat content soy protein carragenan mix on the quality characteristics of comminuted scalded sausages. *Meat Science* 56: 181–188.

Pinna, M., Roberto, S., Milia, R., et al. 2014. Effect of beetroot juice supplementation on aerobic response during swimming. *Nutrients* 6: 605–615.

Porcella, M.I., Sanchez, G., Vaudagna, S.R., et al. 2001. Soy protein isolate added to vacuum packaged chorizos—Effect on drip loss, quality characteristics and stability during refrigerated storage. *Meat Science* 57: 437–443.

Predika, J. 1983. The history of sausage. *The Sausage-Making Cookbook*, 1–3. Harrisburg, PA: Stackpole Books.

Pszczola, D.E. 1998. The ABC's of nutraceutical ingredients. *Food Technology* 52: 30–37.

Puangsombat, K., Smith, J.S. 2010. Inhibition of heterocyclic amine formation in beef patties by ethanolic extracts of rosemary. *Journal of Food Science* 75: T40–T47.

Puligundla, P., Jung, J., Ko, S. 2012. Carbon dioxide sensors for intelligent food packaging applications. *Food Control* 25: 328–333.

Qasi, G.N., Tickoo, C.L., Gupta, A.K., et al. 2003. Herbal nutraceuticals and a process for preparing the same. US Patent Application No. 20030185911A1.

Rababah, T.M., Al-Mahasneh, M.A., Ereifej, K.I. 2006. Effect of chickpea, broad bean or isolated soy protein additions on the physicochemical and sensory properties of biscuits. *Journal of Food Science* 71: S438–S442.

Ravishankar, G.A., Suresh, B., Giridhar, P., Rao, S.R., Johnson, T.S. 2003. Biotechnological studies on *Capsicum* for metabolite production and plant improvement. In *Capsicum—The genus Capsicum*, ed. A.K. De, 129–138. Boca Raton, FL: CRC Press.

Rekha, M.N., Yadav, A.R., Dharmesh, S., Chauhan, A.S., Ramteke, R.S. 2008. Evaluation of antioxidant properties of dry soup mix extracts containing dill (*Anethum sowa* L.) leaf. *Food and Bioprocess Technology* 3: 441–449.

Resnick, H.E., Lindsay, R.S., McDermott, M.M., et al. 2004. Relationship of high and low ankle brachial index to all-cause and cardiovascular disease mortality—The Strong Heart Study. *Circulation* 109: 733–739.

Resnick, J., Greenwald, D.A., Brandt, L.J. 1997. Delayed gastric emptying and postoperative ileus after nongastric abdominal surgery–Part I. *American Journal of Gastroenterology* 92: 751–762.

Riemersma, R.A. 1996. A fat little earner. *Lancet* 347: 775–776.

Ritchie, K., Carriere, I., de Mendonca, A., et al. 2007. The neuroprotective effects of caffeine—A prospective population study (the Three City Study). *Neurology* 69: 536–545.

Roberfroid, M.B. 2008. Defining functional foods. In *Functional Foods—Concept to Product*, ed. G.R. Gibson and C.M. Williams, 9–27. Boca Raton, FL: CRC Press.

Robert, P., Gorena, T., Romero, N., Sepulveda, E., Chavez, J., Saenz, C. 2010. Encapsulation of polyphenols and anthocyanins from pomegranate (*Punica granatum*) by spray drying. *International Journal of Food Science and Technology* 45: 1386–1394.

Roberts, M.D., Taylor, L.W., Wismann, J.A., Wilborn, C.D., Kreider, R.B., Willoughby, D.S. 2007. Effects of ingesting JavaFit Energy Extreme functional coffee on aerobic and anaerobic fitness markers in recreationally-active coffee consumers. *Journal of the International Society of Sports Nutrition* 4: 25.

Ruiu, G., Pinach, S., Veglia, F., et al. 2009. Phytosterol-enriched yogurt increases LDL affinity and reduces CD36 expression in polygenic hypercholesterolemia. *Lipids* 44: 153–160.

Ruusunen, A., Lehto, S.M., Tolmunen, T., Mursu, J., Kaplan, G.A., Voutilainen, S. 2010. Coffee, tea and caffeine intake and the risk of severe depression in middle-aged Finnish men—The Kuopio Ischaemic Heart Disease Risk Factor Study. *Public Health Nutrition* 13: 1215–1220.

Sachindra, N.M.B., Sakhare, P.Z., Puttarajappa, P., Rao, D.N. 2007. Chicken soup mix composition and a process for preparing the same. US Patent No. 7255889.

Saito, S., Takeshita, M., Tomonobu, K., et al. 2006. Dose-dependent cholesterol—Lowering effect of a mayonnaise-type product with a main component of diacylglycerol-containing plant sterol esters. *Nutrition* 22: 174–178.

Salama, A.A., El-Sahin, M.A., Mesellam, A.S., Shehata A.M.E. 1997. Evaluation of the quality of bread, biscuit and butcher's sausage supplemented with rootlets of malt sprouts. *Nahrung* 41: 228–231.

Sanchez-Moreno, C., Cano, M.P., de Ancos, B., et al. 2006. Mediterranean vegetable soup consumption increases plasma vitamin C and decreases F2-isoprostanes, prostaglandin E2 and monocyte chemotactic protein-1 in healthy humans. *Journal of Nutritional Biochemistry* 17: 183–189.

Sandberg, A.S. 2008. Developing functional ingredients—A case study. In *Functional Foods—Concept to Product*, ed. G.R. Gibson and C.M. Williams, 209–232. Boca Raton, FL: CRC Press.

Santacruz, K., Lewis, J., Spires, T., et al. 2005. Tau suppression in a neurodegenerative mouse model improves memory function. *Science* 309: 476–481.

Savitha, Y.S., Indrani, D., Prakash, J. 2008. Effect of replacement of sugar with sucralose and maltodextrin on rheological characteristics of wheat flour dough and quality of soft dough biscuits. *Journal of Texture Studies* 39: 605–616.

Schauss, A.G., Clewell, A., Balogh, L., et al. 2010. Safety evaluation of an açai-fortified fruit and berry functional juice beverage (MonaVie Active®). *Toxicology* 278: 46–54.

Schauss, A.G., Wu, X., Jensen, G.S. 2009. Increased antioxidant capacity and inhibition of lipid peroxidation in healthy adults consuming an açai (*Euterpe oleracea*) fruit-based juice. In *Proceedings of the 2nd International Symposium on Human Health Effects of Fruits and Vegetables*, ed. B. Patil, 97–100. *Acta Horticulturae* 841.

Schauss, A.G., Wu, X., Prior, R.L., et al. 2006. Phytochemical and nutrient composition of the freeze-dried Amazonian palm berry, *Euterpe oleracea* Mart. (açai). *Journal of Agricultural and Food Chemistry* 54: 8598–8603.

Schoenenberger, A.W., Urbanek, N., Bergner, M., Toggweiler, S., Resink, T.J., Erne, P. 2012. Associations of reactive hyperemia index and intravascular ultrasound-assessed coronary plaque morphology in patients with coronary artery disease. *American Journal of Cardiology* 109: 1711–1716.

Schubert, S.Y., Lansky, E.P., Neeman, I. 1999. Antioxidant and eicosanoid enzyme inhibition properties of pomegranate seed oil and fermented juice flavonoids. *Journal of Ethnopharmacology* 66: 11–17.

Seddon, J.M., Ajani, U.A., Sperduto, R.D., et al. 1994. Dietary carotenoids, vitamins A, C, and E, and advanced age-related macular degeneration. Eye disease case control study group. *Journal of the American Medical Association* 272: 1413–1420.

Seeram, N.P., Aronson, W.J., Zhang, Y., et al. 2007. Pomegranate ellagitannin-derived metabolites inhibit prostate cancer growth and localize to the mouse prostate gland. *Journal of Agricultural and Food Chemistry* 55: 7732–7737.

Segel, H.B. 1993. Introduction—The Vienna coffeehouse in societies and cultures. In *The Vienna Coffeehouse Wits 1890–1938*, ed. H.B. Segel, 6–11. West Lafayette, IN: Purdue Research Foundation.

Senger, A.E.V., Schwanke, C.H.A., Gomes, I., Gottlieb, M.G.V. 2012. Effect of green tea (*Camellia sinensis*) consumption on the components of metabolic syndrome ion elderly. *The Journal of Nutrition, Health and Aging* 16: 738–742.

Senn, C.H. 1915. The history of sauce making. *The Book of Sauces*, 1–16. Chicago: The Hotel Monthly Press.

Serdaroglu, M., Degirmencioglu, O. 2004. Effects of fat level (5%, 10%, 20%) and corn flour (0%, 2%, 4%) on some properties of Turkish type meatballs (*koefte*). *Meat Science* 68: 291–296.

Sethi, M. 2008. Characteristics of food. In *Institutional Food Management*, 265–307. New Delhi: New Age International.

Sevá-Pereira, A., de Moraes, G.R., de Oliveira, S.P., Ryes, F.G. 1991. Use of fiber enriched biscuit in the treatment of chronic intestinal constipation. *Revista da Associafio Paulista de Medicina* 109: 265–268.

Shakir, K.A.F., Madhusudhan, B. 2007. Hypocholesterolemic and hepatoprotective effects of flaxseed chutney—Evidences from animal studies. *Indian Journal of Clinical Biochemistry* 22: 117–121.

Shankar, G.M., Li, S., Mehta, T.H., et al. 2008. Amyloid-β protein dimers isolated directly from Alzheimer's brains impair synaptic plasticity and memory. *Nature Medicine* 14: 837–842.

Sharif, K., Butt, M.S., Anjum, F.M., Nasir, M. 2005. Improved quality of baked products by rice bran oil. *Internet Journal of Food Safety* 5: 1–8.

Sharma, A., Wang, R., Zhou, W. 2011. Functional foods from green tea. In *Functional Foods of the East*, ed. J. Shi, C.T. Ho, and F. Shahidi, 173–196. Boca Raton, FL: CRC Press.

Sharma, A., Zhou, W. 2011. A stability study of green tea catechins during the biscuit making process. *Food Chemistry* 126: 568–573.

Shenoy, A., Prakash, J. 2002. Wheat bran (*Triticum aestivum*)—Composition, functionality and incorporation in unleavened bread. *Journal of Food Quality* 25: 197–211.

Sidhu, J.S., Al-Hooti, S.N. 2005. Functional foods from date fruits. In *Asian Functional Foods*, ed. J. Shi, C.T. Ho, and F. Shahidi, 491–524. Boca Raton, FL: CRC Press.

Sikora, M., Sady, M., Krawontka, J., Plaszek, P., Kowalski, S. 2004. Combinations of potato starch-xanthan gum and modified starches-xanthan gum as thickeners of sweet and sour sauces. Part I. Thickening of sauces without additives. In *Starch—From Starch Containing Sauces to Isolation of Starches and Their Applications*, ed. V.P. Yuryev, P. Tomasik, and H. Ruck, 125–142. New York: Nova Science.

Šimurina, O., Filipčev, B., Psodorov, Đ., et al. 2008. Bread supplemented with herbal blend Vitalplant®. *Food Processing, Quality and Safety* 35: 113–117.

Singh, M., Kim, S. 2009. Yogurt fermentation in the presence of starch-lipid composite. *Journal of Food Science* 74: C85–C89.

Singh, R.K.P., Sabapathy, K.K. 2006. Opportunities for processing and utilization of soybean to increase nutritional security in North-East India. In *Agriculture, Food Security, Nutrition and Health in North-East India*, ed. D. Basu, B.F. Kulirani, and B.D. Ray, 295–308. New Delhi: Mittal.

Singh, N., Singh, P. 2011. Amaranth: Potential source for flour enrichment. In *Flour and Breads and Their Fortification in Health and Disease Prevention*, ed. V.R. Preedy, R.R. Watson, and V.B. Patel, 101–112. San Diego, CA: Academic Press.

Singh, S., Ghosh, S., Patil, G.R. 2003. Development of a mushroom-whey soup powder. *International Journal of Food Science & Technology* 38: 217–224.

Singh, G., Biswas, A., Muthukumarappan, K. 2005. Development of technology for the calcium fortified mango yogurt. Paper No. 056163 presented at Annual Meeting of American Society of Agricultural and Biological Engineers, Michigan.

Smith, D.F. 2013. Benefits of flavanol-rich cocoa-derived products for mental well-being—A review. *Journal of Functional Foods* 5: 10–15.

Smith, C.J., Rosman, M.S., Levitt, N.S., Jackson, W.P. 1982. Guar biscuits in the diabetic diet. *South African Medical Journal* 61: 196–198.

Smolander, M., Hurme, E., Latva-Kala, K., Luoma, T., Alakomi, H.-L., Ahvenainen, R. 2002. Myoglobin-based indicators for the evaluation of freshness of unmarinated broiler cuts. *Innovative Food Science and Emerging Technologies* 3: 279–288.

Snodderly, D.M. 1995. Evidence for protection against age-related macular degeneration by carotenoids and antioxidant vitamins. *American Journal of Clinical Nutrition* 62: 1448S–1461S.

Snyder, H. 2008. The chemistry of cheese-making. In *The Chemistry of Dairying*, 58–77. Charleston, SC: Bibliobazaar, LLC.

Söderholm, P.P., Alfthan, G., Koskela, A.H., Adlercreutz, H., Tikkanen, M.J. 2012. The effect of high-fiber rye bread enriched with nonesterified plant sterols on major serum lipids and apolipoproteins in normocholesterolemic individuals. *Nutrition, Metabolism & Cardiovascular Diseases* 22: 575–582.

Sodini, I., Tang, P.S. 2006. Milk and milk-based diary ingredients. In *Manufacturing Yogurt and Fermented Milks*, ed. R.C. Chandan, C.H. White, A. Kilara, and Y.H. Hui, 167–178. Ames, IA: Blackwell.

Sorond, F.A., Lipsitz, L.A., Hollenberg, N.K., Fisher, N.D. 2008. Cerebral blood flow response to flavanol-rich cocoa in healthy elderly humans. *Journal of Neuropsychiatric Disease and Treatment* 4: 433–440.

Spano, M., Antonio, J. 2008. Future trends—Nutritional supplements in sports. In *Nutritional Supplements in Sports and Exercise*, ed. M. Greenwood, D.S. Kalman, and J. Antonio, 491–508. Totowa, NJ: Humana Press.

Stephen, A.M. 1998. Regulatory aspects of functional products. In *Functional Foods—Biochemical and Processing Aspects*, ed. G. Mazza, 403–438. Lancaster, PA: Technomic.

Su, H.P., Lien, C.P., Lee, T.A., Ho, J.H. 2010. Development of low-fat mayonnaise containing polysaccharide gums as functional ingredients. *Journal of the Science of Food and Agriculture* 90: 806–812.

Sudha, N.L., Srivastava, A.K., Vetrimani, R., Leelavathi, K. 2007. Fat replacement in soft dough biscuits—Its implications on dough rheology and biscuit quality. *Journal of Food Engineering* 80: 922–930.

Suzuki, E., Yorifuji, T., Takao, S., et al. 2009. Green tea consumption and mortality among Japanese elderly people—The prospective Shizuoka elderly cohort. *Annals of Epidemiology* 19: 732–739.

Suzuki, S., Oshima, S. 1962. Influence of blending oils on human serum cholesterol—Rice bran oil, safflower and sunflower oil. *Journal of Nutrition* 28: 194–198.

Sworn, G. 2007. Natural thickeners. In *Handbook of Industrial Water Soluble Polymers*, ed. P.A. Williams, 10–31. Oxford: Blackwell.

Takahama, U., Tanaka, M., Hirota, S. 2011. Buckwheat flour and bread. In *Flour and Breads and Their Fortification in Health and Disease Prevention*, ed. V.R. Preedy, R.R. Watson, and V.B. Patel, 141–152. San Diego, CA: Academic Press.

Tannahill, R. 1973. *Food in History*, 37, 61, 69. New York: Stein and Day.

Tarazona-Díaz, M.P., Alacid, F., Carrasco, M., Martínez, I., Aguayo, E. 2013. Watermelon juice: Potential functional drink for sore muscle relief in athletes. *Journal of Agricultural and Food Chemistry* 61: 7522–7528.

Taubert, D., Roesen, R., Lehmann, C., Jung, N., Schömig, E. 2007. Effects of low habitual cocoa intake on blood pressure and bioactive nitric oxide. *Journal of the American Medical Association* 298: 49–60.

Taylor, J.R.N., Anyango, J.O. 2011. Sorghum flour and flour products—Production, nutritional quality and fortification. In *Flour and Breads and Their Fortification in Health and Disease Prevention*, ed. V.R. Preedy, R.R. Watson, and V.B. Patel, 127–140. San Diego, CA: Academic Press.

Taylor, L.W., Wilborn, C.D., Harvey, T., Wismann, J., Willoughby, D.S. 2007. Acute effects of ingesting JavaFit™ Energy Extreme functional coffee on resting energy expenditure and hemodynamic responses in male and female coffee drinkers. *Journal of the International Society of Sports Nutrition* 4: 10.

Temelli, F., Bansema, C., Stobbe, K. 2004. Development of an orange-flavored barley β-glucan beverage. *Cereal Chemistry* 81: 499–503.

Thielecke, F., Boschmann, M. 2009. The potential role of green tea catechins in the prevention of the metabolic syndrome—A review. *Phytochemistry* 70: 11–24.

Thielecke, F., Rahn, G., Böhnke, J., et al. 2010. Epigallocatechin-3-gallate and postprandial fat oxidation in overweight/obese male volunteers—A pilot study. *European Journal of Clinical Nutrition* 64: 704–713.

Thongsombat, W., Sirichote, A., Chanthachum, S. 2007. The production of guava juice fortified with dietary fiber. *Songklanakarin Journal of Science and Technology* 29 (Suppl. 1): 187–196.

Tipvarakarnkoon, T., Einhorn-Stoll, U., Senge, B. 2010. Effect of modified Acacia gum (Super Gum) on the stabilization of coconut o/w emulsions. *Food Hydrocolloids* 24: 595–601.

Tomas-Barberan, F.A., Cienfuegos-Jovellanos, E., Marin, A., et al. 2007. A new process to develop a cocoa powder with higher flavonoid monomer content and enhanced bioavailability in healthy humans. *Journal of Agricultural and Food Chemistry* 55: 3926–3935.

Toniolo, G.E.C. 2012. Usefulness of pomegranate in prostate cancer. In *Proceedings of the II International Symposium on the Pomegranate*, ed. P. Melgarejo and D. Valero, 311–320. Paris: CIHEAM.

Tóthova, L., Hodosy, J., Mettenburg, K., et al. 2013. No harmful effect of different Coca-cola beverages after 6 months of intake on rat testes. *Food and Chemical Toxicology* 62: 343–348.

Toves, F.A. 2004. Nutrient-fortified, reduced calorie fruit and/or vegetable food product and processes for making same. US Patent Application No. 20040161522.

Tredger, J., Ransley, J. 1978. Guar-gum-its acceptability to diabetic patients when incorporated into baked food products. *Journal of Human Nutrition* 32: 427–432.

Tsao, A.S., Liu, D., Martin, J., et al. 2009. Phase II randomized, placebo-controlled trial of green tea extract in patients with high-risk oral premalignant lesions. *Cancer Prevention Research* 2: 931–941.

Tuohy, K.N., Kolida, S., Lustenberger, A.M., Gibson, G.R. 2001. The prebiotic effects of biscuits containing partially hydrolysed guar gum and fructo-oligosaccharides—A human volunteer study. *British Journal of Nutrition* 86: 341–348.

Tyagi, S.K., Manikantan, M.R., Oberoi, H.S., Kaur, G. 2006. Effect of mustard flour incorporation on nutritional, textural and organoleptic characteristics of biscuits. *Journal of Food Engineering* 80: 1043–1050.

Uchoa A.M., Correia da Costa, J.M., Maia, G.A., Meira, T.R., Sousa, P.H., Montenegro Brasil, I. 2009. Formulation and physicochemical and sensorial evaluation of biscuit-type cookies supplemented with fruit powders. *Plant Foods for Human Nutrition* 64: 153–159.

Upaganlawar, A., Balaraman, R. 2009. Combined effect of green tea extract and Vitamin E on serum and heart tissue lipids, lipid metabolizing enzymes and histopathological alteration in isoproterenol-induced myocardial infarction in rats. *Scientia Pharmaceutica* 77: 791–803.

Vaclavik, V.A., Christian, E.W. 2008. Milk and milk products. In *Essentials of Food Science*, 3rd edn, 237–270. Heidelberg: Springer.

Valero, A., Carrasco, E., Garcia-Gimenco, R.M. 2012. Principles and methodologies for the determination of shelf-life in foods. In *Trends in Vital Food and Control Engineering*, ed. A.H.A. Eissa, 3–42. Croatia: InTech.

van Gaal, L.F., Mertens, I.L., de Block, C.E. 2006. Mechanisms linking obesity with cardiovascular disease. *Nature* 444: 875–880.

van Gelder, B., Buijsse, B., Tijhuis, M., et al. 2007. Coffee consumption is inversely associated with cognitive decline in elderly European men—The FINE Study. *European Journal of Clinical Nutrition* 61: 226–232.

van Stuijvenberg, M.E., Dhansay, M.A., Lombard, C.J., Faber, M., Benadé, A.J. 2001. The effect of a biscuit with red palm oil as a source of beta-carotene on the vitamin A status of primary school children—A comparison with beta-carotene from a synthetic source in a randomized controlled trial. *European Journal of Clinical Nutrition* 55: 657–662.

Vasconcellos, J.A. 2005. Introduction: Concepts. In *Quality Assurance for the Food Industry—A Practical Approach*, 1–18. Boca Raton, FL: CRC Press.

Vassalle, C., Andreassi, M.G. 2004. 8-iso-prostaglandin F2α as a risk marker in patients with coronary heart disease. *Circulation* 110: e49–e50.

Venugopal, V. 2009a. Functional foods—An overview. In *Marine Products for Healthcare—Functional and Bioactive Nutraceutical Compounds from the Ocean*, 1–22. Boca Raton, FL: CRC Press.

Venugopal, V. 2009b. Marine nutraceuticals for food fortification and enrichment. In *Marine Products for Healthcare—Functional and Bioactive Nutraceutical Compounds from the Ocean*, 405–425. Boca Raton, FL: CRC Press.

Verma, B., Hucl, P., Chibbar, R.N. 2008. Phenolic content and antioxidant properties of bran in 51 wheat cultivars. *Cereal Chemistry* 85: 544–549.

Wang, R., Zhou, W. 2004. Stability of tea catechins in the bread making process. *Journal of Agricultural and Food Chemistry* 52: 8224–8229.

Wang, H., Zheng, H., Zhao, Z., Chen, P. 2009. Effect of Chaihu Shihuang soup on blood serum level TNF, IL-6 and IL-10 of severe acute pancreatitis. *Zhongguo Zhong Yao Za Zhi* 34: 1582–1584.

Wang, J., Varghese, M., Ono, K., et al. 2014. Cocoa extracts reduce oligomerization of amyloid—Implications for cognitive improvement in Alzheimer's disease. *Journal of Alzheimer's Disease* 41: 643–650.

Weinberg, B.A., Bealer, B.K. 2001. Tea—Asian origins. In *The World of Caffeine: The Science and Culture of the World's Most Popular Drug*, 27–40. London: Routledge.

Whitehead, J. 2005. Functional drinks containing herbal extracts. In *Chemistry and Technology of Soft Drinks and Fruit Juices*, ed. P.R. Ashurst, 300–335. Oxford: Blackwell.

Wildman, R.E.C. 2001. *Handbook of Nutraceuticals and Functional Foods*. Boca Raton, FL: CRC Press.

Wilson, A.D., Baietto M. 2009. Applications and advances in electronic-nose technologies. *Sensors* 9: 5099–5148.

Worrasinchai, S., Suphantharika, M., Pinjai, S., Jamnong, P. 2006. β-Glucan prepared from spent brewer's yeast as a fat replacer in mayonnaise. *Food Hydrocolloids* 20: 68–78.

Wu, C.H., Lu, F.H., Chang, C.S., Chang, T.C., Wang, R.H., Chang, C.J. 2003. Relationship among habitual tea consumption, percent body fat and body fat distribution. *Obesity Research* 11: 1088–1095.

Yao, Y., Qiu, Q.H., Wu, X.W., Cai, Z.Y., Xu, S., Liang, X.Q. 2013. Lutein supplementation improves visual performance in Chinese drivers—1-year randomized, double-blind, placebo-controlled study. *Nutrition* 29: 958–964.

Yilmaz, I. 2004. Effects of rye bran addition on fatty acid composition and quality characteristics of low-fat meatballs. *Meat Science* 67: 245–249.

Yla-Herttuala, S., Palinski, W., Rosenfeld, M.E., et al. 1989. Evidence for the presence of oxidatively modified low density lipoprotein in atherosclerotic lesions of rabbit and man. *Journal of Clinical Investigation* 84: 1086–1095.

Zhu, N., Sang, S., Huang, T.C., Bai, N., Yang, C.S., Ho, C.T. 2000. Antioxidant chemistry of green tea catechins—Oxidation products of (–)-epigallocatechin gallate and (–)-epigallocatechin with peroxidase. *Journal of Food Lipids* 7: 275–282.

Zimeri, J.E., Kokini, J.L. 2003. Rheological properties of inulin-waxy maize starch systems. *Carbohydrate Polymers* 52: 67–85.

Zucco, F., Borsuk, Y., Arntfield, S.S. 2011. Physical and nutritional evaluation of wheat cookies supplemented with pulse flours of different particle sizes. *LWT: Food Science and Technology* 44: 2070–2076.

Index

A

D

E

F

G

H

I

N

O

P

Q

R

S

T

U

V

W

X

Y

Z